Civil Engineering: Building Materials Properties and Testing

Civil Engineering: Building Materials Properties and Testing

Editor: Cayson Carter

MURPHY & MOORE
www.murphy-moorepublishing.com

www.murphy-moorepublishing.com

MURPHY & MOORE

Cataloging-in-Publication Data

Civil engineering : building materials, properties and testing / edited by Cayson Carter.
p. cm.
Includes bibliographical references and index.
ISBN 978-1-63987-761-4
1. Civil engineering. 2. Building materials. 3. Buildings--Thermal properties.
4. Testing. I. Carter, Cayson.
TA153 .C583 2023
624--dc23

Murphy & Moore Publishing
1 Rockefeller Plaza,
New York City,
NY 10020, USA

ISBN 978-1-63987-761-4

Contents

Preface

Over the recent decade, advancements and applications have progressed exponentially. This has led to the increased interest in this field and projects are being conducted to enhance knowledge. The main objective of this book is to present some of the critical challenges and provide insights into possible solutions. This book will answer the varied questions that arise in the field and also provide an increased scope for furthering studies.

Civil engineering refers to a professional engineering discipline that is primarily concerned with the design, construction and maintenance of the physical and naturally built environment. The types of structures formed in civil engineering include canals, pipelines, roads, structural components of buildings, bridges, airports, railways, dams and sewage systems. There are a wide range of building materials utilized in construction such as wood, concrete, masonry and steel. These materials have unique properties like durability, weight, cost and strength. These properties are appropriate for specific applications in construction. The testing of building materials aids in construction by revealing any flaws that could lead to future deformities. Testing can be done through two different ways which include laboratory testing and field testing. Some of the types of tests used in this field are non-destructive testing, mechanical testing and fire testing. This book contains some path-breaking studies in the field of civil engineering. It elucidates various building materials and testing methods used in this field. For all those who are interested in civil engineering, this book can prove to be an essential guide.

I hope that this book, with its visionary approach, will be a valuable addition and will promote interest among readers. Each of the authors has provided their extraordinary competence in their specific fields by providing different perspectives as they come from diverse nations and regions. I thank them for their contributions.

Editor

1

Low-Cost Sensors for Determining the Variation in Interior Moisture Content in Gypsum Composite Materials

Daniel Ferrández [1], Carlos Morón [2,*], Pablo Saiz [3], Evangelina Atanes-Sánchez [4] and Engerst Yedra [2]

1 Departamento de Ingeniería de Organización, Administración de Empresas y Estadística, Universidad Politécnica de Madrid, 28660 Madrid, Spain; daniel.fvega@upm.es
2 Departamento de Tecnología de la Edificación, Universidad Politécnica de Madrid, 28040 Madrid, Spain; e.yedra@alumnos.upm.es
3 Departamento de Economía Financiera, Contabilidad e Idioma Moderno, Universidad Rey Juan Carlos, 28032 Madrid, Spain; pablo.saiz@urjc.es
4 Departamento de Ingeniería Mecánica, Química y Diseño Industrial, Universidad Politécnica de Madrid, 28012 Madrid, Spain; evangelina.atanes@upm.es
* Correspondence: carlos.moron@upm.es

Abstract: Non-destructive testing can be used to determine some of the most relevant physical properties of building materials. In this work, two low-cost measuring devices were developed capable of determining the variation in real-time of the percentage of humidity that is produced in the construction of gypsum and plaster during the hardening process. For this, an Arduino resistive sensor and a capacitive sensor of our design were used. The results show how it is possible to determine the variations in mixing water content during the seven days of curing established by the UNE-EN 13279-2 standard as well relate to the mechanical resistance of the test specimens with the same percentage of humidity. Additionally, the study was completed with the determination of the formation of the dihydrate compound linked to this setting process in the test specimens by conducting X-ray diffraction and thermogravimetric analysis tests at different ages of the samples.

Keywords: low-cost sensors; Arduino; non-destructive evaluation; building materials; gypsum and plaster

1. Introduction

Gypsum is one of the most widely used construction materials in the building sector, being the oldest artificial conglomerate known to mankind [1]. It is a material that has been present in all periods of history [2], from its traditional use as an interior cladding due to the fact of its qualities as a hygrothermal regulator [3] to its versatile current construction applications for the manufacture of interior partitions, decorative plaster moldings or lightened plates for ceilings among others [4]. However, its relative abundance, ease of working on site, and low cost compared to other conglomerates continue to favor its direct use today [5]. In addition, this traditional application generally requires increasingly scarce qualified labor [6]; for this reason, current research lines tend to reduce the limitations of gypsum as a construction material, which implies increasing its mechanical resistance, promoting industrialization through new construction systems, reducing the weight of precast to reduce building loads, and advancing the understanding of setting times [7,8].

This work revolved around the understanding of the hardening processes of gypsum composite materials and the study of the variation that occurs in the water content from mixing during this process. Several studies have focused on the determination of an adequate water/conglomerate ratio to obtain, in optimal conditions of workability of the material in a fresh state, the best performance in terms of achieving mechanical resistance [9,10]. There are studies, such as the one carried out by Ochoa et al. [11], in which by using natural additives extracted from *Agave* plants, the application times of the gypsum materials were increased without detriment to their mechanical properties, although it is true that most of the commercial additives to accelerate or delay the setting of this type of binding materials are usually composed of chemical compounds designed for this purpose [12,13]. However, this typology of studies on the hardening and setting of gypsum materials have generally focused more on the achievement of new dosages to improve certain properties than on monitoring and continuous monitoring of the setting process. This has led to the fact that most of the tests were carried out on days set by the reference regulations, evidencing the continuous variations in humidity experienced by the samples, their relationship with the variations in the physical properties of the material, and the performance in most studies of only some complementary chemical tests to study the evolution of the dihydrate and the hemihydrate [14,15].

In this sense, the measurement techniques used to determine the moisture content in construction materials have undergone notable progress in recent years [16]. In this way, it has evolved from the traditional gravimetric techniques based on the determination of moisture content through the weight variation that occurs in the material during its hardening and drying [17] to modern georadar techniques capable of recording the variations in the speed and amplitude of the waves when they propagate through a wet material [18]. Better known than microwave reflection [19] or nuclear magnetic resonance [20] is the use of hygrometers to determine the moisture content in porous soil and materials [21]. Among the commercial hygrometers, we found two well-differentiated types:

(1) resistive ones that base their operation on the fact that, in contact to water, material increases its electrical conductivity [22]
(2) the capacities that obtain their response as a function of the variation in the dielectric properties of the material [23].

Regarding the second, their greatest advantage is that they do not physically penetrate the material, which makes them ideal for rehabilitation work [24]. In historical buildings where interventions are difficult, infrared thermography has been used to detect humidity with a high degree of precision [25].

Special mention requires the development of new measurement systems driven by Arduino technology for testing in the last decades [26,27]. In the field of moisture determination and its application in the building sector, we currently find a wide variety of studies. Some authors have used resistive sensors to better understand the setting processes of cement mortars made with recycled aggregates [28], although the use of this type of sensor to determine soil moisture and build automatic irrigation systems also stands out [29]. There are also studies using capacitive sensors in Arduino, such as the one carried out by Calva et al. [30], to determine the moisture content in lands of the Northwest of Brazil with high precision.

The objective of this work was to design, develop, and evaluate the correct operation of two low-cost pieces of measuring equipment that allowed for obtaining, in real-time, the moisture content of gypsum and plaster materials. For this, on the one hand, monitoring tests were carried out on studying the variation in the mixing water content of the samples, since they were poured in the fresh state in the mold using a resistive sensor in Arduino developed for this purpose and performing complementary thermogravimetric analysis. Moreover, with the help of a capacitive sensor of our design, the evolution in the interior moisture content of the specimens was determined from the demolding to the performance of the standardized tests after seven days, relating these measurements to the evolution of the properties' mechanics of materials.

2. Materials and Methods

2.1. Materials

To carry out this work, two differentiated gypsum materials were used for their fineness of grinding: fine white gypsum (YF) for applications in buildings, and plaster (E35) with smaller size particles. The raw material used to obtain the gypsum materials is aljez, a sedimentary rock of chemical precipitation constituted by calcium sulfate with two water molecules ($CaSO_4 \cdot 2H_2O$) called gypsum dihydrate (DH). The basis for firing gypsum dihydrate for use as a construction material is carried out with the following basic reaction scheme:

$$CaSO_4 \cdot 2H_2O \rightarrow CaSO_4 \cdot \frac{1}{2}H_2O\ (\alpha, \beta) + \frac{3}{2}H_2O \quad (1)$$

$$CaSO_4 \cdot \frac{1}{2}H_2O \rightarrow CaSO_4 + \frac{1}{2}H_2O \quad (2)$$

In such a way that the dihydrate gypsum is dehydrated according to an endothermic process at temperatures between 105–110 °C to obtain mainly calcium sulfate with 0.5 molecules of water ($CaSO_4 \cdot 0.5H_2O$) called gypsum hemihydrate (HH), in its two forms α (more compact and with greater resistance) or β (more soluble and less stable), and which is the majority component of commercial gypsum for building applications. In the implementation of commercial gypsum, mixing with water leads to the rehydration of the HH, giving DH again in the setting process.

For its part, the hemihydrate undergoes endothermic dehydration to obtain anhydrous calcium sulfate ($CaSO_4$) called soluble anhydrite or anhydrite III at a temperature below 200 °C [31,32]. The phase transformation of soluble anhydrite to insoluble anhydrite (or anhydrite II), exothermic, and that does not entail associated weight loss, occurs at a temperature, around 200 °C, in the case of hemihydrate and between 300–400 °C for the hemihydrate [32]. The exact temperature at which these dehydration reactions occur depends on experimental conditions, and the separation of events that occur at similar temperatures improves when the rate of heating is low and controlled [33].

The analysis of the raw materials of plaster E35 and fine gypsum YF, supplied by the company Saint Gobain PLACO IBERICA S.A. (Madrid, Spain), was carried out on the powder samples. First, the X-ray diffraction test was carried out using Siemens Krystalloflex D5000 equipment (Madrid, Spain) with a Cu-Kα graphite monochromator. In this way, a diffractogram has been obtained in a range between $5° \leq 2\theta \leq 100°$ every 0.04° and one second of passage as can be seen in Figure 1.

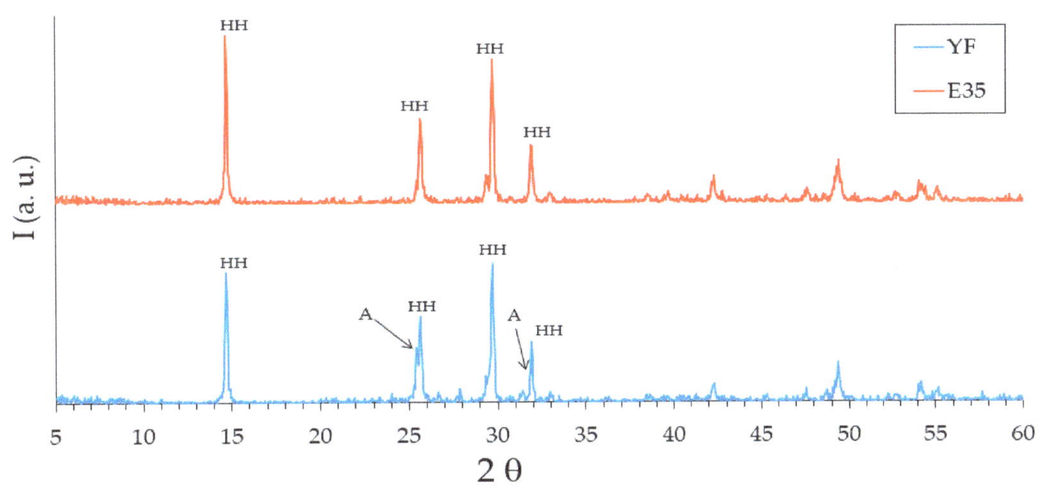

Figure 1. Analysis by X-ray diffraction (XRD) of the raw materials. HH: hemihydrate. A: anhydrite.

In Figure 1, it is observed that both the fine gypsum and the plaster used in this work show four main peaks at angles of 14.7°, 25.7°, 29.7°, and 31.9° corresponding to the hemihydrate gypsum [32].

The presence of anhydrite can be seen in the fine gypsum sample with diffraction peaks located at angles of 25.4° and 31.4°.

Besides, a thermogravimetric analysis of both raw materials was also carried out using TA Instruments SDT Q600 equipment (Madrid, Spain). Heating was carried out from room temperature to 400 °C at a rate of 5 °C/min and air atmosphere (100 mL/min), analyzing an approximate mass of 40 mg in each test. The results obtained for the two gypsum materials used are shown in Figures 2 and 3. They show the evolution of the mass of the samples versus temperature (green line), the first derivative of the mass versus temperature (blue line), and the energy per unit mass versus temperature (brown line) that monitors associated thermal events. The figures show the percentage of mass (%) lost concerning the initial mass and the mass (mg) lost in each event, as well as the temperature at which the maximum speed in the loss of mass occurs.

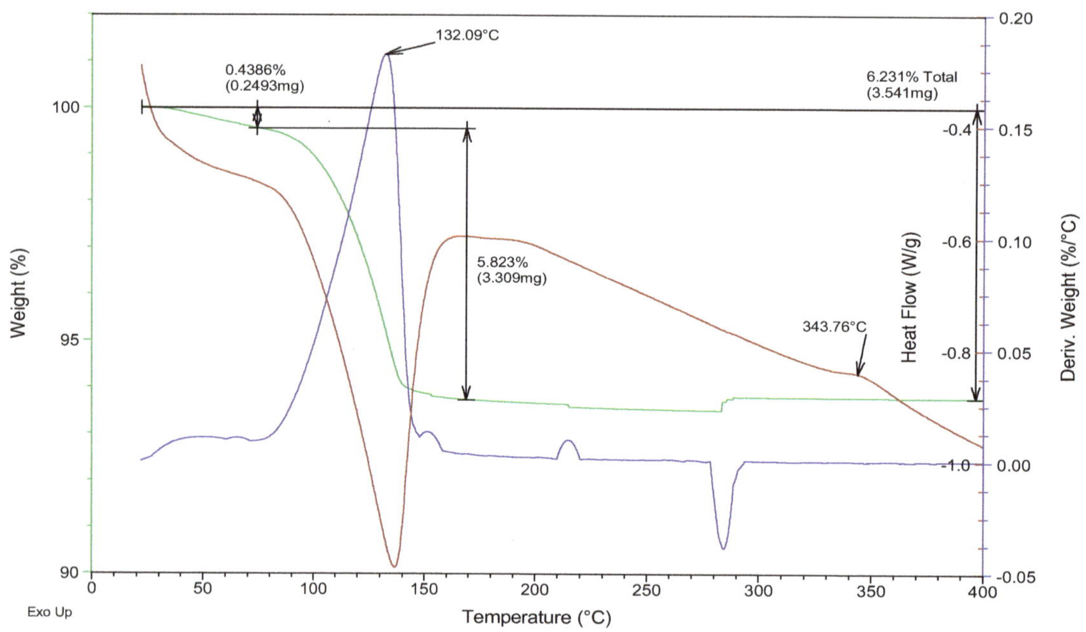

Figure 2. Thermogravimetric analysis of the E35 plaster used.

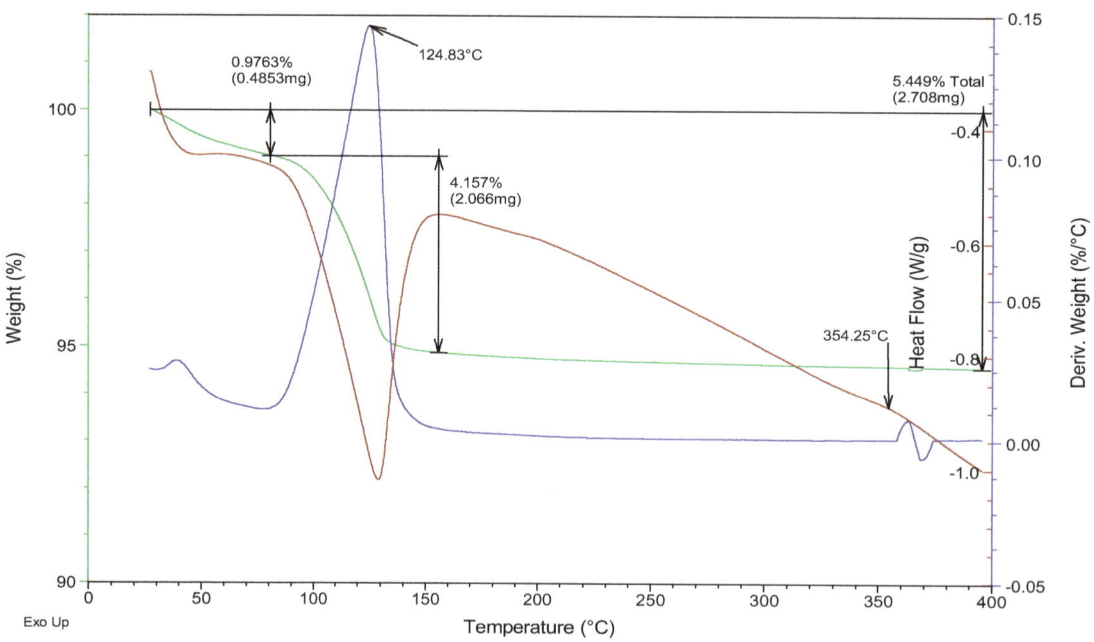

Figure 3. Thermogravimetric analysis of the fine gypsum YF used.

From the analysis of Figures 2 and 3, it was possible to observe that both the E35 plaster and the YF fine gypsum showed a first loss of mass (corresponding to 0.44% for plaster and 1% for gypsum) due to the surface humidity below 75 °C. Next, the second loss of mass was observed that took place between 75 and 175 °C as a consequence of the dehydration of the hemihydrate that gives rise to soluble anhydrite with a maximum of 131 °C for plaster and 125 °C for gypsum and associated mass losses of 5.8% and 4.2%, respectively; this event is endothermic as can be seen in the curve corresponding to heat flow. The exothermic phase transformation from soluble anhydrite to insoluble anhydrite did not lead to associated weight loss and was observed in plaster at 344 °C and less clearly in gypsum at 357 °C. These temperatures indicate that in both cases the sample contains β-hemihydrate.

The amount of hemihydrate (HH) present in the raw material samples on both a dry and wet basis is presented in Table 1. For the calculation, the initial mass of the sample subjected to thermal analysis, the loss in mass of the second thermal event between approximately 75 and 175 °C, and the stoichiometry of the reaction Equation (2) have been taken into account.

Table 1. Results of the thermal analysis of raw materials.

Sample	% H_2O	% HH Wet Base	% HH Dry Base
E35	0.4	93.85	94.27
YF	1.0	67.00	67.66

As can be seen in Table 1, the purity of the E35 plaster sample was much higher with a hemihydrate content close to 94%, which was reduced to 67% in the fine gypsum sample (wet base). This was due, on the one hand, to the presence of anhydrite in fine gypsum as shown in the X-ray diffractogram and, on the other hand, to the additive content of fine gypsum raw materials which decomposes at high temperatures (not shown in the thermogravimetric analysis).

On the other hand, Table 2 shows the dosages used for the preparation of the RILEM (The International Union of Laboratories and Experts in Construction Materials, Systems, and Structures) standardized 4 cm × 4 cm × 16 cm test specimens. It should be noted that high water/conglomerate ratios were chosen to show the effect that the variation in the content of mixing water and humidity has on the setting process of the samples.

Table 2. Dosages used for mixing the test specimens.

Denomination	Water (g)	Gypsum (g)	Denomination	Water (g)	Plaster (g)
YF-0.6	600	1000	E35-0.7	700	1000
YF-0.8	800	1000	E35-0.9	900	1000

The preparation of all the batches was carried out with the same technique and equipment following the recommendations of the UNE-EN 13279-1: 2009 [34] standard, always under the same environmental conditions (65% relative humidity and 22 °C ambient temperature) and without performing an energetic kneading that could break the internal microstructure of the gypsum composite materials.

2.2. Design and Development of Low-Cost Measurement Equipment

2.2.1. Arduino Resistive Sensor

The low-cost Arduino sensor used in this work consists of the following electronic components: Arduino UNO board with ATmega328P microprocessor with 14 digital inputs/outputs and 6 analog inputs, with the possibility of a USB connection to the computer, 8 GB MicroSD memory module to store the information without saturating the internal memory of the Arduino controller, and a humidity sensor, the Arduino Grove-Moisture Sensor HL-69 model (RS component, Madrid, Spain). Figure 4 shows the sensor connection diagram.

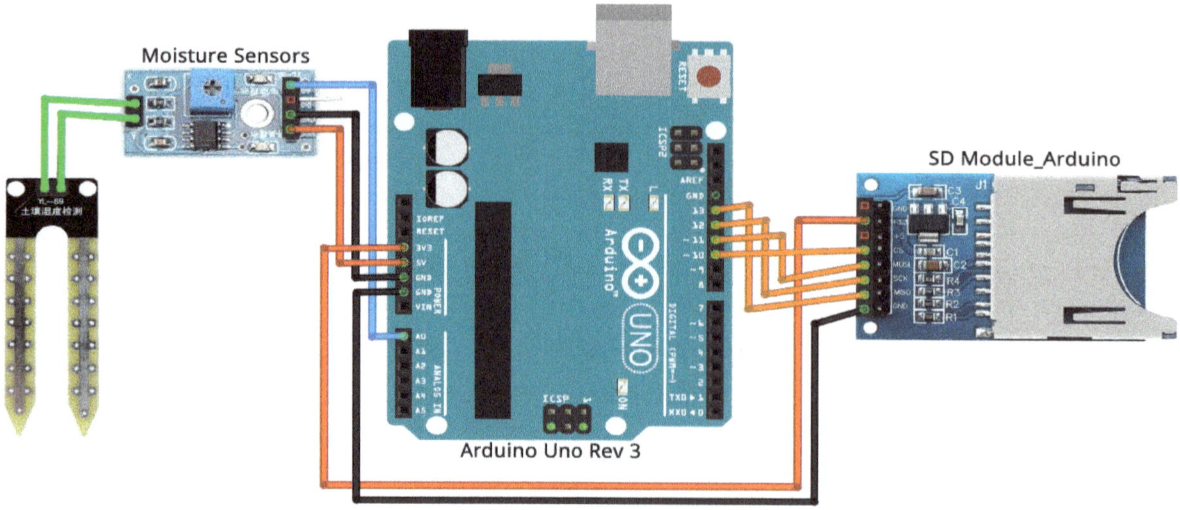

Figure 4. Arduino sensor connection diagram.

It is worth highlighting the versatility that Arduino has to connect the physical world with the digital world, the possibility of automating the samplings, and the large existing free software community that allows quick access to the information flows about the sensors used [35]. On the other hand, the resistive sensor HL-69 has been used previously in other research works to determine the moisture content of different types of soils [36], basing its operation on the electrical conductivity of the medium in which it is introduced according to the Equation (3):

$$I = \sigma \cdot E \cdot S \quad (3)$$

where I is the current intensity, E is the intensity of the electric field generated by the potential difference between the sensor tips, S is the section, and σ is the conductivity of the mixture that decreases as the mixing water is lost in the test specimens.

In Figure 5, an explanatory flow diagram of the program developed for monitoring the moisture content inside the gypsum and plaster specimens during the setting process is shown.

As can be seen in Figure 5, the Arduino programming code is divided into three blocks. The first consists of the initialization of variables and constants that will be used in it as well as the recognition of all the devices that intervene in the process and the verification of the sending of information from the sensor to the MicroSD card or the computer's database. Second, is the automation of the sampling through a previously predetermined data capture period, which for the case of this work was considered five minutes during the seven days established by the UNE-EN 13279-2 standard for gypsum and plaster. The third and last was the creation of a single plain text file per test with the data: number of samples being carried out, percentage of humidity, and time at which the capture was carried out.

Finally, it is necessary to highlight that the total cost of the designed Arduino humidity sensor was approximately 35 euros, this being a competitive price compared to other commercial humidity sensors on the market, such as hygrometers or inductive sensors, which have higher prices [37]. Also, the device used had the advantage of being able to carry out measurements in the variation of the moisture content of the samples from when the gypsum material was poured into the mold once it was kneaded; this presents great industrial advantages since data were obtained from the beginning of the setting of the paste and without the need to unmold the specimens to perform the measurements [28].

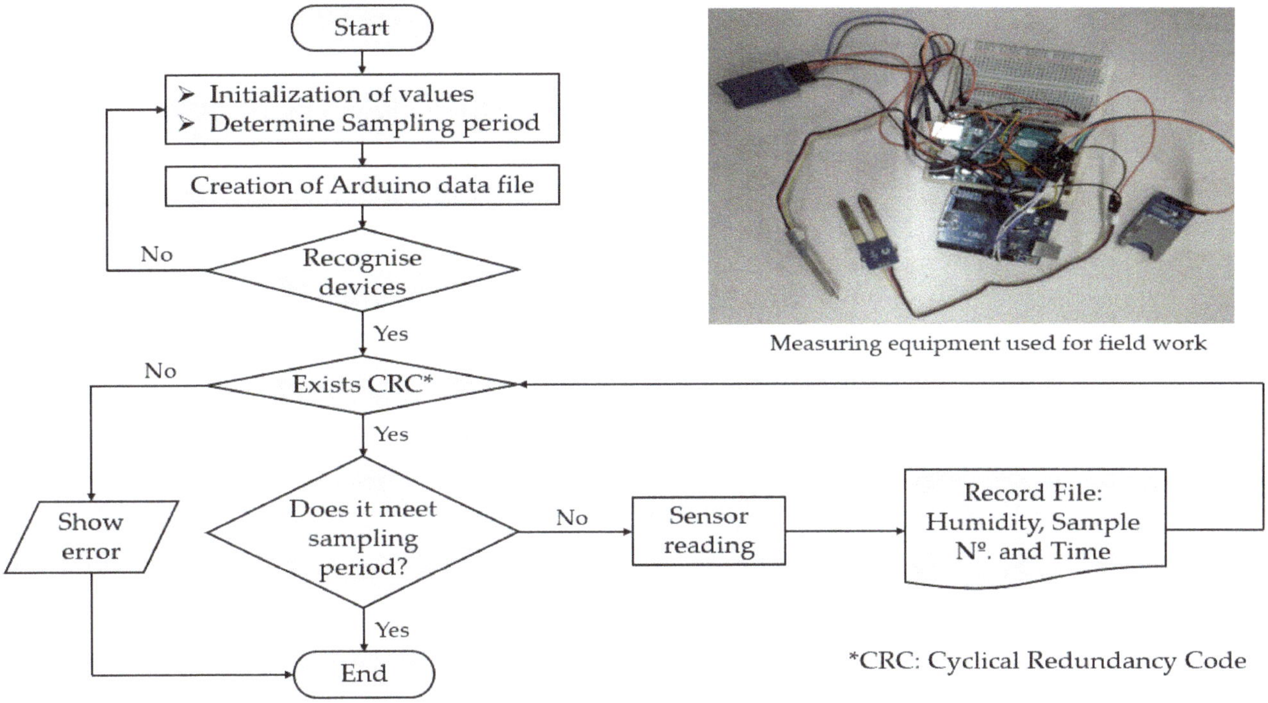

Figure 5. Explanatory flow diagram of the program developed for data capture.

2.2.2. Capacitive Sensor

Capacitive sensors that base their operation on the action of an electric field have been used successfully in industry to measure physical parameters. Some authors have used this type of device to measure humidity in construction elements such as wood or cement mortars [23]. This type of sensor has the advantage of being easy to use and economically viable. In this work, a capacitive sensor linked to a self-oscillating circuit, like the one shown in Figure 6, was developed. Using this system, it was possible to measure small variations in the loss of mixing water contained in the gypsum and plaster specimens as they hardened.

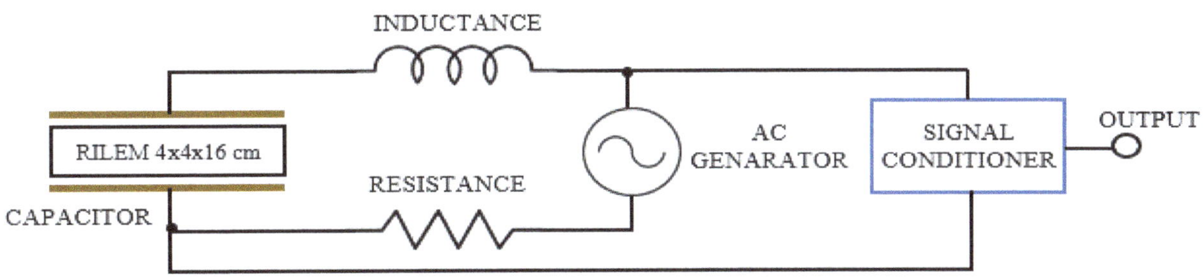

Figure 6. Diagram of the capacitive sensor developed for taking measurements. RILEM: (The International Union of Laboratories and Experts in Construction Materials, Systems, and Structures)

In this way, the implemented sensor consists of a flat-parallel plate capacitor with a surface area of 4×16 cm^2 corresponding to the side of the test specimens and which is part of a self-oscillating RLC (resistance–coil–capacitor) circuit. Copper plates that are conventionally used in the elaboration of printed circuits have been used, with a conductive copper face and the outer face insulated with bakelite to avoid measurement errors that could falsify the experiment. The operation of the measuring equipment was based on finding the resonance frequency of the circuit, that is, the one in which the current intensity that circulates was maximum and the pulsation was zero $(X_L - X_c = 0)$. This resonant frequency was found by viewing the Lissajous figure corresponding to the composition of two signals that are in phase on the oscilloscope. In this way, according to the determination of said resonance frequency measured with the help of an oscilloscope, it is possible to know the variation in the moisture content of the materials tested employing Equation (4):

$$L\omega = \frac{1}{C\omega} \Longrightarrow f = \frac{1}{2\pi\sqrt{LC}}\ [Hz] \qquad with\ C = \epsilon \cdot \frac{A}{d}\ [F] \tag{4}$$

where f is the resonance frequency in Hz, C is the capacitor capacity in F, L is the inductance of the sensor in H, Єis the permittivity of the dielectric (in this work of the wet specimen), A is the surface of the condenser plates m^2, and d is the distance between the plates of 0.04 m corresponding to the thickness of the standard specimens.

However, sometimes it is necessary to take into account factors that can influence the capacitance measurement. This is the case of edge effects that appear when the distance between plates is excessive; for this reason, it is recommended to use the corrected equation to calculate the capacitance:

$$C = \varepsilon \cdot \frac{l \cdot a}{d} \cdot \left[1 + \frac{l}{\pi}\left[\ln\left(\frac{\pi \cdot a}{d} + 1\right) + 1 + \ln 2\right]\right] \tag{5}$$

where l and a are the length and width, respectively, in m of the capacitor plates. Besides, it is advisable to always maintain the same ambient humidity in laboratory conditions to avoid the appearance of parasitic resistances that could harm the measurement. The importance of using shielded cables to avoid capacitive interference in the measuring equipment, and the need to ensure good contact between the copper plate and the gypsum or plaster material, should also be highlighted, thus avoiding the effect of air as a dielectric that could get in the way of taking measurements.

Therefore, it is through the variation in the electrical permittivity of the material caused during the setting that the measurements related to the humidity of the sample are obtained. It is a very useful method that allows knowing the variation in the percentage of humidity in the specimens just before carrying out the mechanical tests, to subsequently establish relationships that allow the two variables to be related. In addition to being an alternative method to the traditional gravimetric tests, the system has a low cost, is easy to use, and its measurement is not only on the surface as are most commercial hygrometers [38].

2.3. Methodology

The experimental program carried out in this investigation is presented visually in Figure 7. A total of 24 specimens from the same batch were made for each type of gypsum material and dosage. On the one hand, a test piece of each type was used for monitoring, using the resistive sensor in Arduino, the variation in the mixing water content (100%) until the time of testing according to the UNE-EN 13279-2 standard at 7 days (0%) with a sampling frequency of five minutes. A series of three specimens of each type were tested daily to determine the evolution of the mechanical properties (bending and compression), and the mean value of the three values is presented in this work and their relationship with the moisture content obtained with the help of a capacitive sensor. Finally, complementary tests were also carried out employing gravimetric thermal analysis; these have been carried out on the same specimen at ages of 1, 4, and 7 days in dosages YF-0.8 and E35-0.7 to monitor the composition of the sample and quantify the proportion of dihydrate that is formed at the expense of the hemihydrate as the gypsum sets.

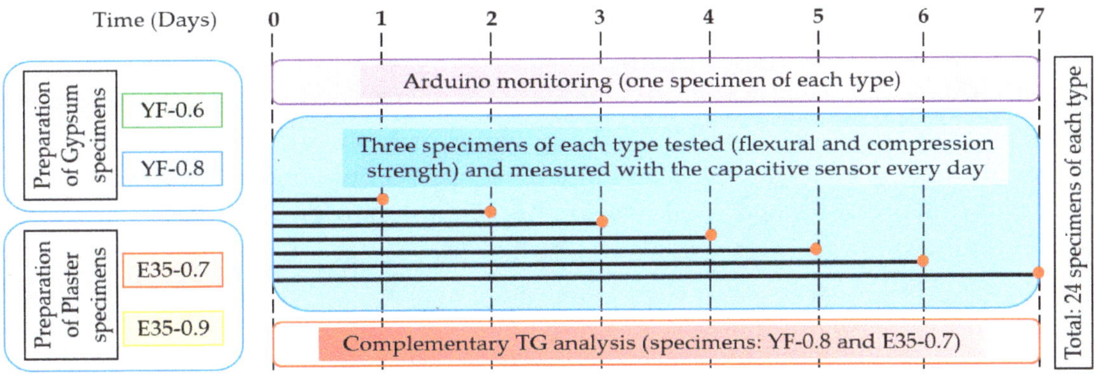

Figure 7. Experimental program and tests carried out.

3. Results and Discussion

3.1. Arduino Sensor Monitoring

Figure 8a shows the measurement equipment implemented in Arduino and responsible for monitoring the variation in the mixing water content of the test specimens, expressed in terms of percentage of humidity. The sampling time was five minutes for seven days, the time specified by the UNE-EN 13279-2 standard as the time necessary to allow the gypsum composite materials to set before testing their mechanical resistance. The sensors were duly calibrated by setting the value of 100% in the measure corresponding to the pouring of the material in the mold and replaced in the measure of the different specimens so as not to distort the results. In addition, and in order to avoid measurement errors, two sensors per specimen were used to counteract the validity of the results, and the mean value of both measurements is reflected in this work.

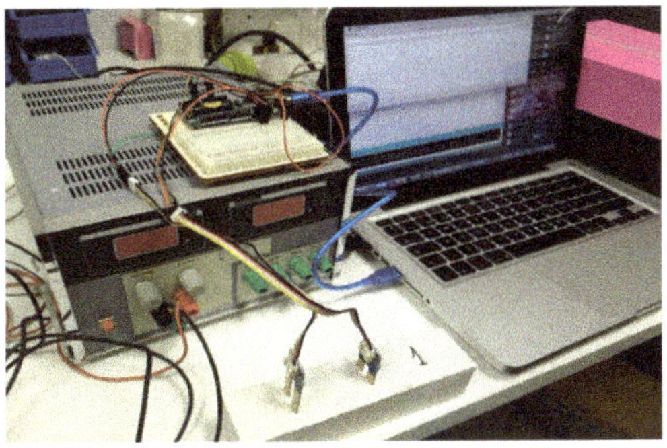

(a)

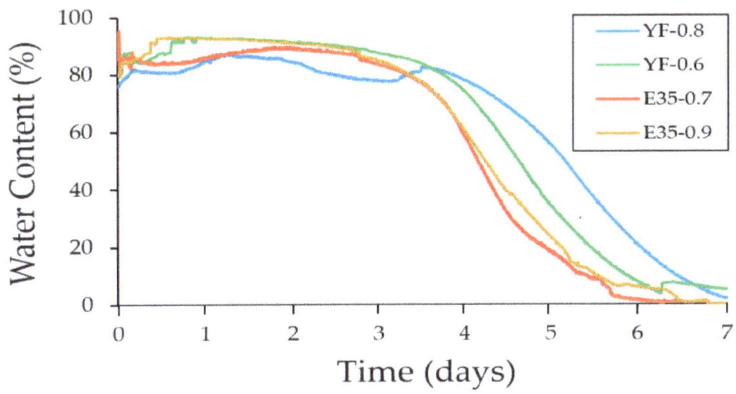

(b)

Figure 8. Testing using the Arduino sensor. (**a**) Measurement equipment monitoring a plaster specimen; (**b**) results obtained from the variation in the mixing water content against time for each of the tested specimens.

In parallel, the results obtained for each of the samples tested are shown in Figure 8b. From the analysis of the figure, it can be seen how the variation in the content of mixing water collected by the sensor in the first three days is very scarce, being from that moment on where there is a sharp decrease in the percentage of water in the test specimens. On the seventh day collected by the UNE-EN 13279-2 standard as the age to carry out the mechanical tests, it can be seen how all the samples have reached zero or values close to it; this indicates variations in humidity not measurable by the sensor and the end of hardening of the studied samples. It can also be observed that the higher the mixing water content in the samples, the longer it takes to reach zero, and the specimens made with fine gypsum material took longer to lower their moisture content than the test specimens made with plaster.

In this way, the Arduino system presented was very useful to carry out simple and fast monitoring of the setting process and variation in the mixing water content of the test specimens made with gypsum materials. This monitoring finds an immediate application in industrial companies destined to the elaboration of precast since it allows to check if the hardening process of the samples is developing normally.

In addition to this monitoring, a thermogravimetric analysis was carried out with the help of the same equipment and techniques for powder samples at YF-0.8 and E35-0.7 dosages for test specimens aged 1, 4, and 7 days. As an example, Figures 9–12 show the results of the thermogravimetric analysis of samples E35-0.7 at 1 and 7 days of age, and of samples YF-0.8 at 1 and 7 days of age, respectively. In the thermal analysis of all the specimens, three mass losses were observed, all corresponding to endothermic events. The first would correspond to the loss of free water, which occurred at a temperature below 100 °C, and which in the younger specimens, both plaster and gypsum, presents a maximum in the derivative of the mass at approximately 58 °C. This first loss of water is of much lower magnitude in the seven-day-old specimens and did not present a maximum in the curve derived from the mass.

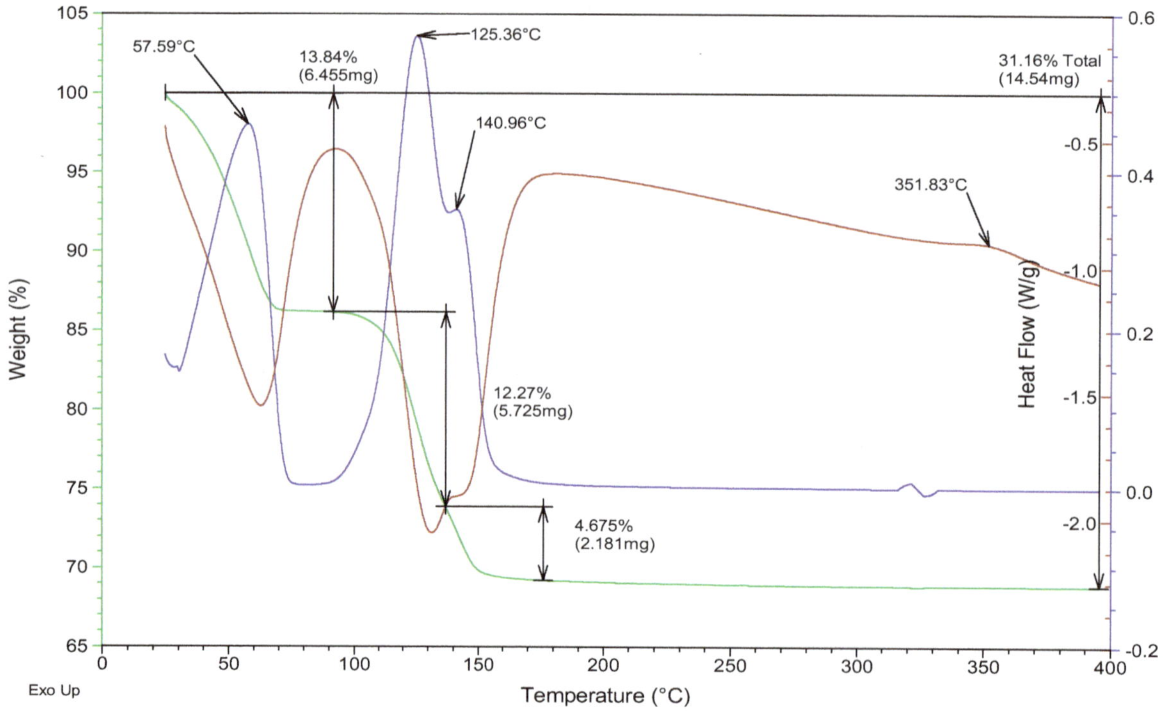

Figure 9. Thermogravimetric analysis of the test tube E35-0.7 aged 1 day.

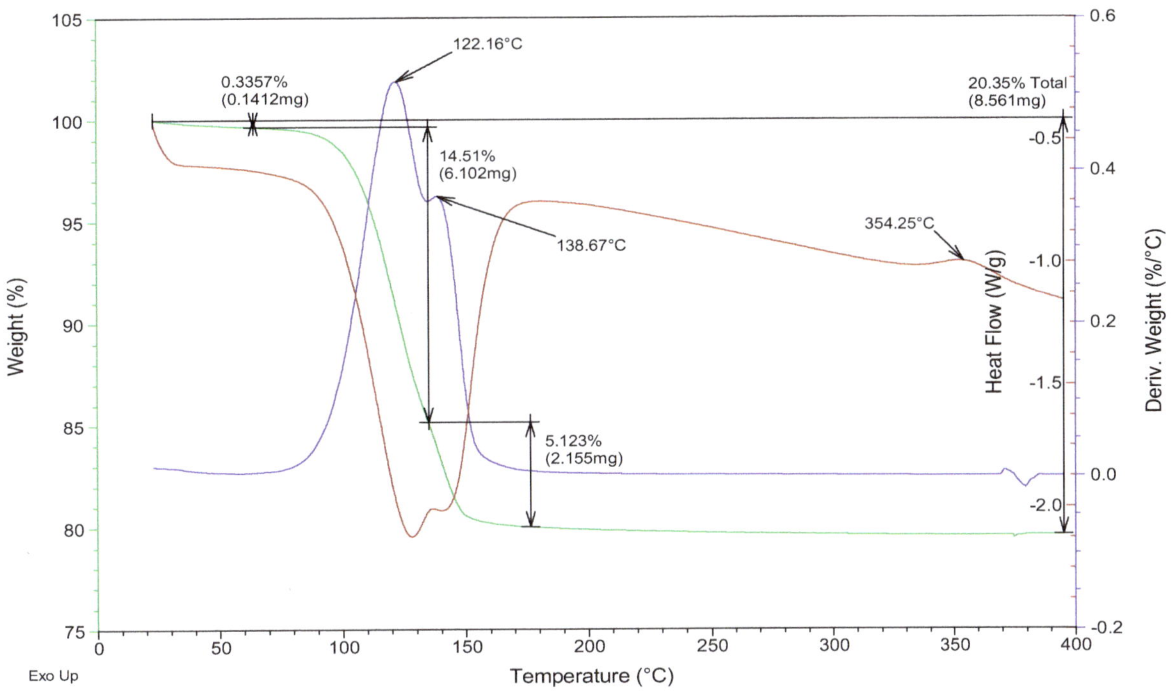

Figure 10. Thermogravimetric analysis of the E35-0.7 test tube aged 7 days.

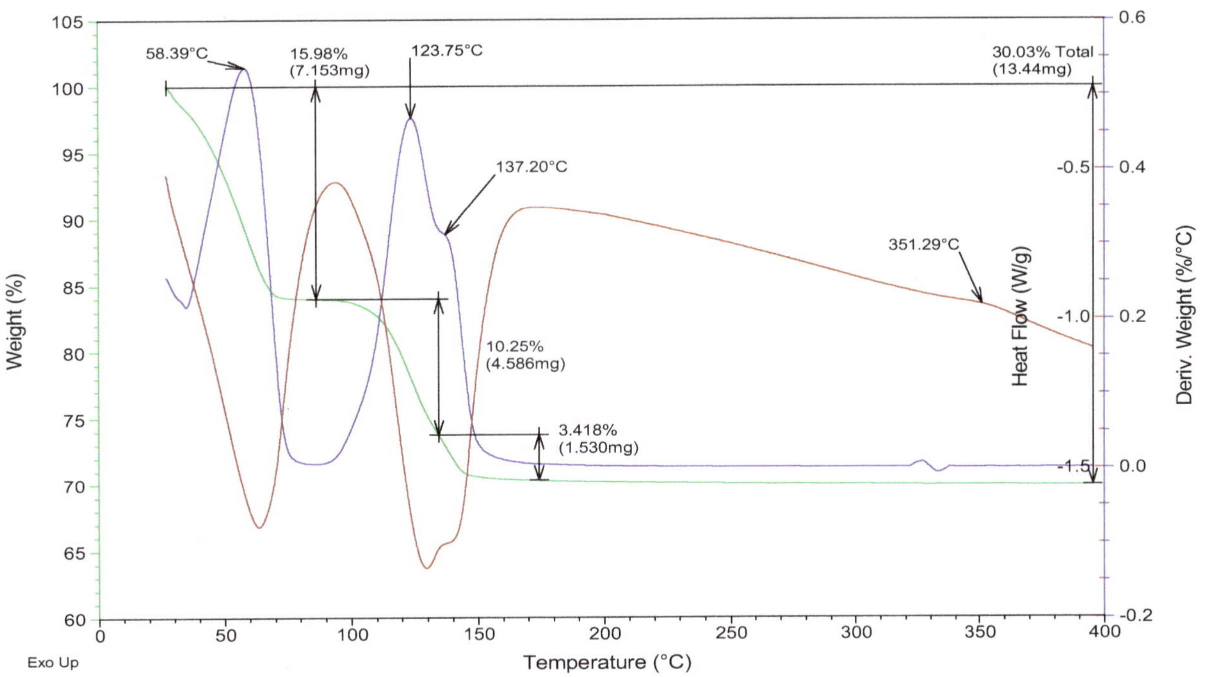

Figure 11. Thermogravimetric analysis of the test tube YF-0.8 aged 1 day.

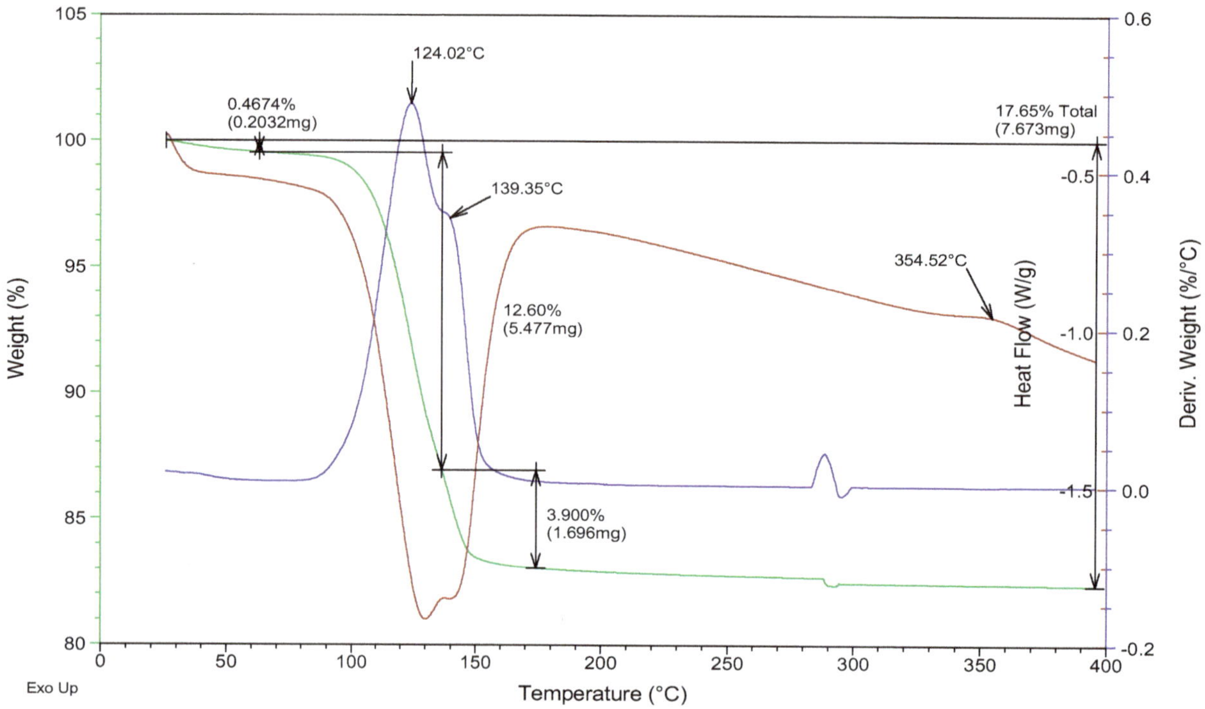

Figure 12. Thermogravimetric analysis of the YF-0.8 test tube aged 7 days.

In the temperature interval between 75 and 175 °C, the second mass loss occurs with a maximum in the derivative of the mass around 122–125 °C. This second loss presents a peak with a maximum in the derivative of the mass between 137 and 141 °C which constitutes the third mass loss of the sample. The second loss of mass was due to the dehydration of the dihydrate gypsum formed in the reaction of the hemihydrate with the mixing water; the third loss was due both to the dehydration of the hemihydrate formed in the second mass loss and to the dehydration of the original hemihydrate remaining in the test tube and from the original raw material which gives rise to soluble anhydrite. The exothermic phase change from soluble to insoluble anhydrite can be seen in all the test specimens with a maximum temperature of 351–355 °C.

Table 3 shows the mass loss, as a percentage concerning the original mass, of each of these three events, for the specimens of different ages as well as the total mass loss.

Table 3. Results of the thermogravimetric analysis for samples YF-0.8 and E35-0.7.

Sample *	Mass Loss				Composition	
	Total Δm *(%)*	Δm H_20 *(%)*	Δm *DH (%)*	Δm *HH (%)*	*%DH*	*%HH*
E35-0.7-1	31.2	13.8	12.3	4.7	78.2	9.8
E35-0.7-4	20.4	0.3	14.4	5.2	91.9	6.7
E35-0.7-7	20.4	0.3	14.5	5.1	92.4	5.1
YF-0.8-1	30.0	16.0	10.3	3.4	65.3	0.4
YF-0.8-4	18.1	0.5	12.3	4.2	78.6	2.3
YF-0.8-7	17.7	0.5	12.6	3.9	80.3	0.0

* The numbers 1, 4, and 7 indicate the age in days that the sample was tested.

It was observed that the percentage of free water in the plaster specimen decreased from 13.8% in the youngest specimen to a value of 0.3% in the oldest specimen; in the gypsum specimen, these values decreased from 16% to 0.5%. This decrease was due both to the evaporation of the mixing water during drying and to the progressive hydration of the hemihydrate to form dihydrate.

Accordingly, the loss of mass associated with the dehydration of the hemihydrate present in the specimen (second loss in the thermogram) increased in plaster samples from 12.3% to 14.5% with the increasing age of the specimen, while in the gypsum samples, the values evolved from 10.3% to 12.6% at the age of seven days. These values reflect the hydration of the hemihydrate that constitutes the raw materials to produce an increasing amount of dihydrate with the age of the specimen as the setting process occurs.

The amount of dihydrate (DH) and hemihydrate (HH) present in the test specimens were calculated from the second and third mass loss, taking into account the initial mass of each sample, the stoichiometry of reactions (1) and (2), and discounting the mass of hemihydrate formed in reaction (1) in calculating the amount of hemihydrate. The results are presented in the last two columns of Table 3.

In the plaster specimens, the dihydrate represents 78.2% of the youngest specimen and 92.4% in the oldest specimen. These values were, respectively, 65.3% and 80.3% for the fine gypsum specimens. The amount of hemihydrate in the plaster specimens decreased from 9.8% to 5.1% as the set progressed; these values were 0.4% and 0%, respectively, for the fine gypsum specimens.

These results show the evolution in the composition of the specimens throughout the setting process, with the progressive formation of dihydrate, responsible for the mechanical resistance of the gypsum compound as well as the progressive decrease in the free water of the sample (humidity) throughout the process [39]. Also, the results indicate that the plaster specimens present a higher amount of dihydrate than the plaster samples, according to the composition of the raw materials indicated in Table 1, and that it accounts for the higher mechanical strengths for these specimens.

3.2. Tests Using the Capacitive Sensor

Figure 13 shows the non-intrusive capacitive sensor developed for taking moisture measurements in the gypsum materials used in this investigation. This sensor allows us to know the variation in the moisture content in the test specimens over time and its determination at the exact moment of the mechanical tests.

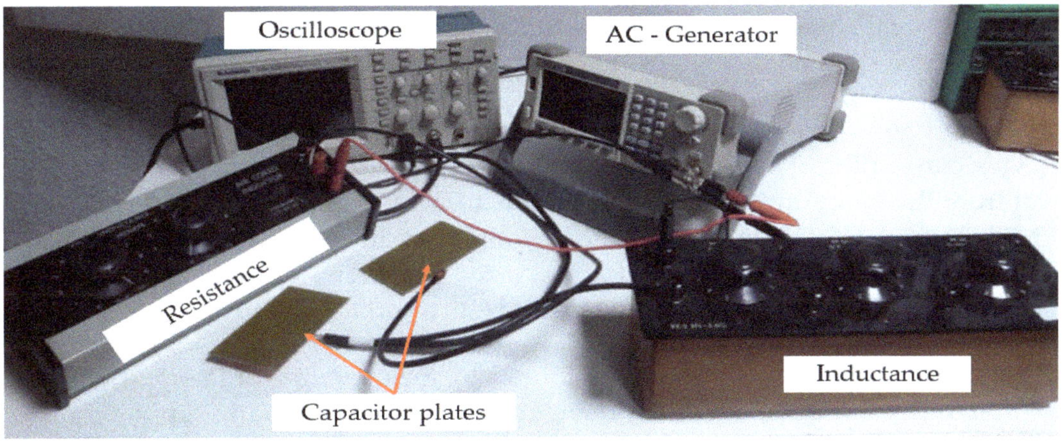

Figure 13. Capacitive measurement equipment used.

To do this, first, and as shown in Figure 14, a sensor calibration has been carried out to determine its behavior and relate the obtained capacitance values with the moisture content of the test piece. Thus, the initial measurement of capacitance collected with the sensor after removing the specimens was taken as a value of 100% in humidity and as a value of 0% the measurement of the specimen dried in an oven at a temperature of 50 ± 2 °C for 48 h. Said capacitances have been related to the determination of the humidity obtained by gravimetric methods according to the following Equation (5):

$$M = \frac{W_{nd} - W_0}{W_0} \cdot 100\ [\%] \tag{6}$$

where W_{nd} is the humidity on day n studied and W_0 is the weight of the test piece in an anhydrous state dried in an oven.

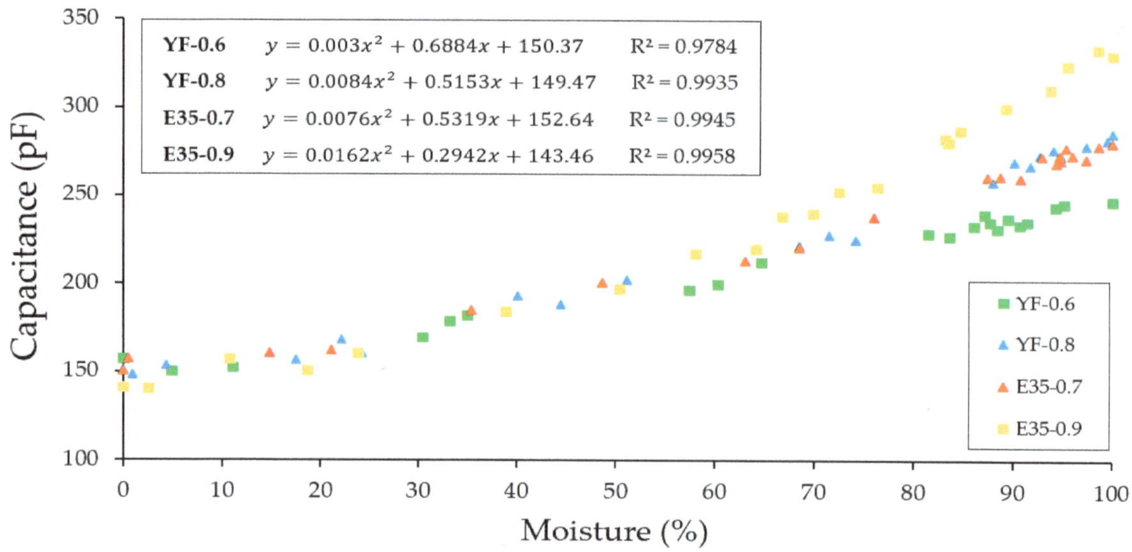

Figure 14. Capacitance values versus the moisture content of the test specimens.

As can be seen in Figure 14, the capacitance values oscillate around 150 and 350 pF for the gypsum materials tested. You can see how there is a strong positive correlation when fitting the values using second-degree polynomial equations. At high humidity values, there are differences between the capacitance values among all the tested samples, observing higher capacitance values for the samples with the highest water/conglomerate ratio of each typology. For humidity values lower than approximately 50%, a single curve was obtained regardless of the material treated. Furthermore, it can be seen that the higher the water/conglomerate ratio, the higher the capacitance values are obtained regardless of the material treated.

Figures 15 and 16 show the results derived from the flexural and compression rupture tests at different ages concerning the moisture content of the specimen, carried out daily from day one after removal from the mold until reaching seven days. The techniques and recommendations of the UNE-EN 13279-2 [40] standard have been followed and a hydraulic press of the Autotest-200/10-SW model have been used to carry out the mechanical tests. Furthermore, all laboratory tests have been carried out under the same conditions of relative humidity and temperature (50 ± 5% and 22 ± 2 °C).

As can be seen after the analysis of Figures 15 and 16, the evolution of the mechanical strengths of the gypsum material test specimens tested develops, as there is a variation in the moisture content measured with the help of the capacitive sensor developed. It can be seen how the maximum levels of resistance are reached at the age of seven days set by the reference regulations and corresponding to the minimum levels of humidity determined with the equipment implemented. In turn, there is a significant change in the evolution of resistance in the transition between the fourth and fifth days, this period being in which the variations in water content of the test specimens have been more pronounced. Finally, and following other previous studies [41], the plaster material specimens have shown better performance in terms of mechanical resistance than their homologs made with gypsum, presenting higher resistance values than those that contained less amount of mixing water in origin.

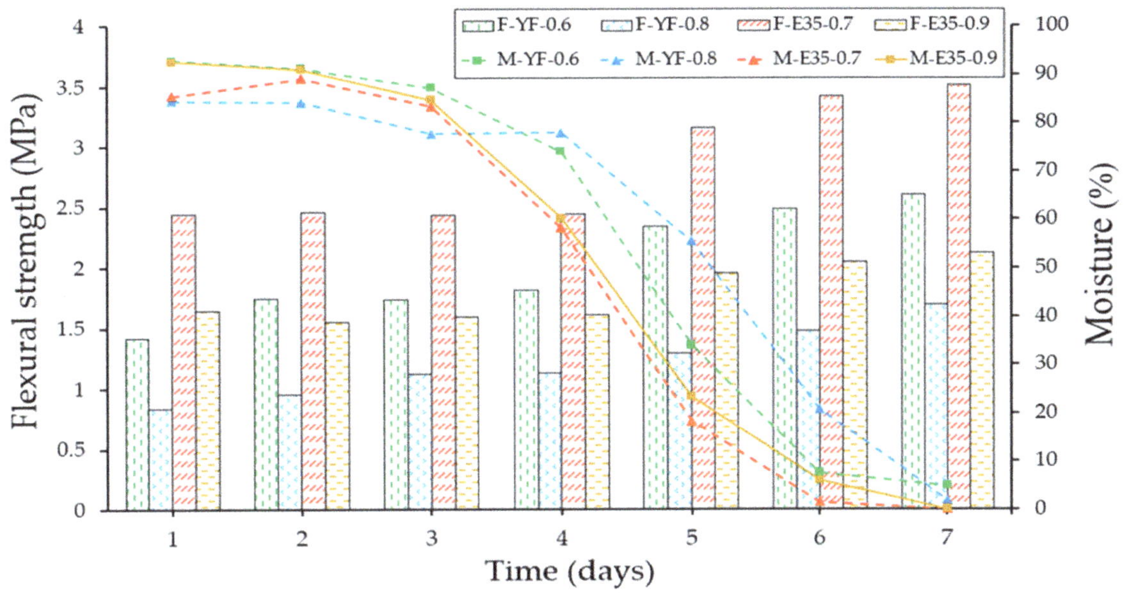

Figure 15. Evolution of flexural strength (F) and moisture content (M) of the different specimens over time.

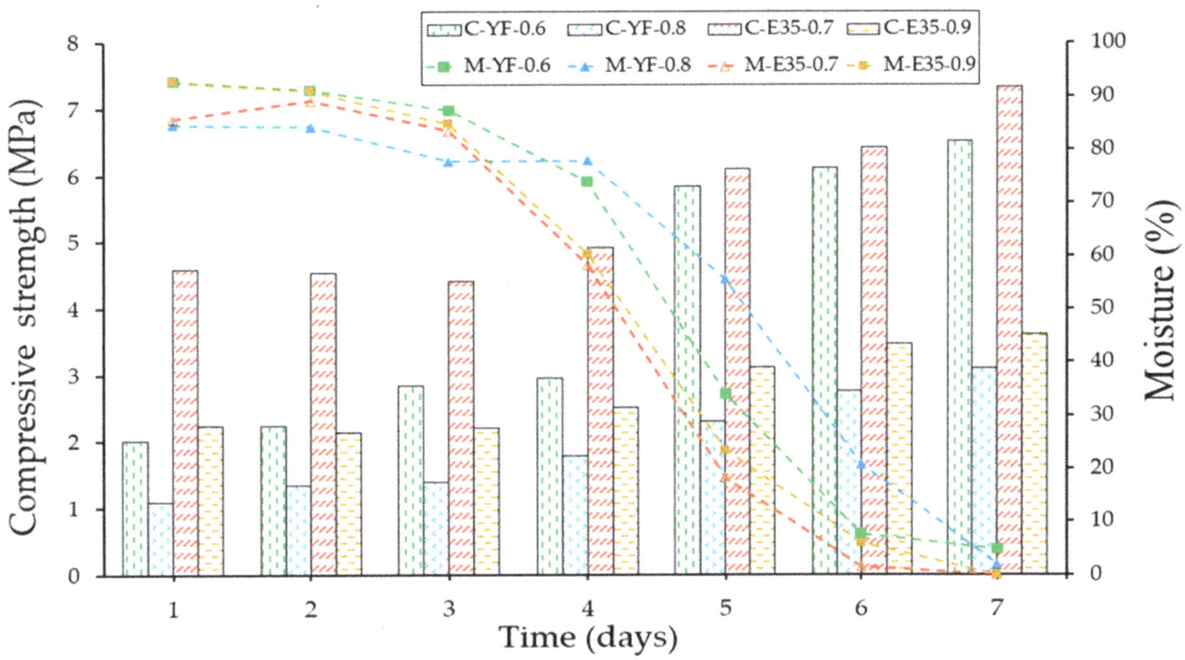

Figure 16. Evolution of the compressive strength (bars) and the moisture content (lines) of the different specimens against time.

In addition to the tests carried out, the analysis by X-ray diffraction was carried out on the test pieces YF-0.8 and E35-0.7. The results obtained are shown in Figure 17.

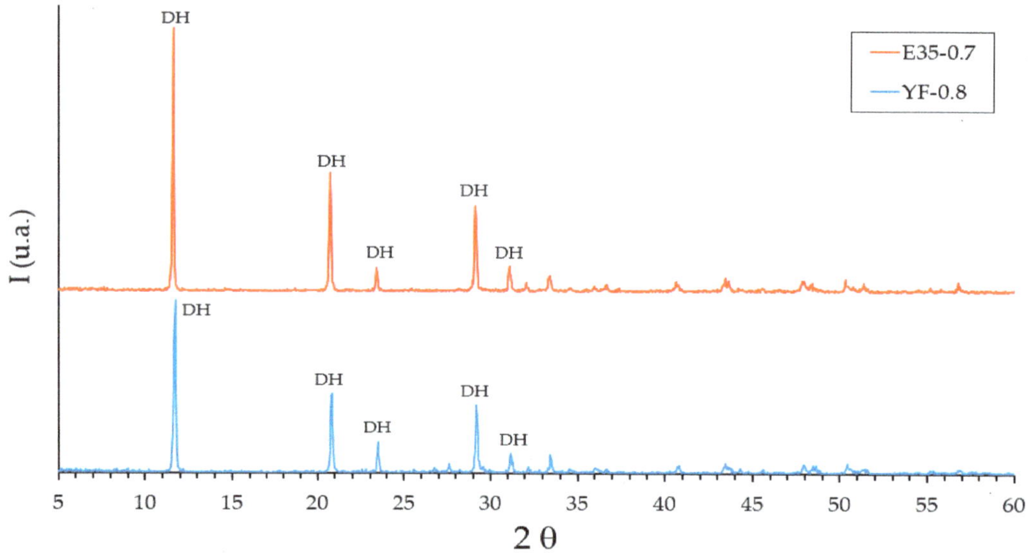

Figure 17. Results of the analysis by X-ray diffraction of the test specimens YF-0.8 and E35-0.7 at the age of 7 days. DH: dihydrate.

X-ray diffraction analysis for the specimens hardened at the age of seven days (Figure 17) shows diffraction peaks located at angles of 11.6°, 20.7°, 29.1°, 23.4°, and 31.1° corresponding to dihydrate (DH), and did not show peaks corresponding to hemihydrate (HH). The diffractograms are practically coincident between the YF-0.8 and E35-0.7 specimens with the only exception that the diffraction peaks were more intense for the specimen made with plaster material.

Finally, in Table 4, the confidence intervals for humidity, capacitance, flexural strength, and compressive strength were calculated for each day of measurements. These intervals were calculated to contain at least 75% of the data because the hypothesis of normality cannot be verified.

Table 4. Confidence intervals for moisture, capacitance, flexural strength, and compressive strength measured daily on each sample.

Intervals for Humidity Expressed as a Percentage				
Day	**YF-0.6**	**YF-0.8**	**E35-0.7**	**E35-0.9**
1	(90.10, 100)	(96.05, 100)	(96.07, 100)	(93.27, 100)
2	(88.56, 92.49)	(91.58, 96.04)	(94.1, 96.5)	(80.1, 98.3)
3	(86.57, 88.97)	(86.21, 93.65)	(92.02, 95.98)	(72.86, 89.14)
4	(79.00, 88.40)	(65.83, 77.03)	(85.53, 92.41)	(64.08, 75.32)
5	(53.44, 68.16)	(34.15, 56.31)	(56.24, 82.16)	(43.79, 71.27)
6	(28.23, 37.51)	(14.36, 28.24)	(7.67, 62.47)	(6.19, 48.15)
7	(0, 16.56)	(0, 6.42)	(0, 21.9)	(0, 15.75)
Ranges for capacitance expressed in pF				
Day	YF-0.6	YF-0.8	E35-0.7	E35-0.9
1	(242.29, 248.73)	(274.69, 289.65)	(266.52, 286.80)	(320.25, 338.33)
2	(231.96, 239.04)	(269.63, 278.75)	(269.47, 280.11)	(276.26, 322.5)
3	(227.68, 234.44)	(252.94, 276.82)	(267.02, 274.70)	(242.49, 303.69)
4	(223.35, 236.15)	(219.29, 230.89)	(260.09, 261.85)	(228.20, 259.56)
5	(186.64, 219.92)	(180.48, 208.68)	(198.34, 249.46)	(187.22, 236.38)
6	(164.48, 189.72)	(150.41, 173.54)	(144.03, 221.31)	(130.75, 199.75)
7	(145.97, 161.01)	(145.57, 156.21)	(145.64, 166.64)	(127.74, 165.10)

Table 4. *Cont.*

Intervals for Humidity Expressed as a Percentage				
Day	**YF-0.6**	**YF-0.8**	**E35-0.7**	**E35-0.9**
Intervals for flexural strength expressed in MPa				
Day	YF-0.6	YF-0.8	E35-0.7	E35-0.9
1	(1.26, 1.58)	(0.75, 0.95)	(2.32, 2.55)	(1.56, 1.72)
2	(1.67, 1.83)	(0.91, 0.99)	(2.42, 2.50)	(1.49, 1.61)
3	(1.68, 1.80)	(0.96, 1.28)	(2.35, 2.51)	(1.41, 1.77)
4	(1.70,1.94)	(1.03, 1.23)	(2.32, 2.56)	(1.55, 1.67)
5	(2.20, 2.48)	(1.21, 1.37)	(3.08, 3.24)	(1.85, 2.05)
6	(2.38, 2.58)	(1.40, 1.56)	(3.26, 3.58)	(1.99, 2.11)
7	(2.54, 2.66)	(1.61, 1.73)	(3.43.3.59)	(2.04, 2.20)
Intervals for compressive strength expressed in MPa				
Day	YF-0.6	YF-0.8	E35-0.7	E35-0.9
1	(1.94, 2.10)	(0.91, 1.27)	(4.41, 4.77)	(2.12, 2.35)
2	(2.04, 2.44)	(1.05, 1.65)	(4.30, 4.78)	(1.95, 2.35)
3	(2.56, 3.12)	(1.21, 1.57)	(4.63, 4.49)	(2.09, 2.41)
4	(2.80, 3.14)	(1.58, 1.98)	(4.69, 5.13)	(2.25, 2.77)
5	(5.69, 6.01)	(2.17, 2.45)	(5.94, 6.28)	(3.02,3.22)
6	(5.92, 6.32)	(2.64, 2.88)	(6.28, 6.56)	(3.21, 3.70)
7	(6.36, 6.36)	(3.00, 3.20)	(7.14,7.54)	(3.49, 3.77)

Note that at least 75% of the data are within the calculated range.

4. Conclusions

In this work, two low-cost sensors were developed capable of determining in real-time the variation in mixing water content experienced by gypsum materials during their hardening. In this way, it was possible to delve into the study of the setting of this type of construction material and its relationship with the evolution of their mechanical resistance to bending and compression.

On the one hand, an Arduino resistive sensor was used that made it possible to monitor in real-time the percentage variation in the moisture content of the test tube, since the fresh mix material was poured into the mold. This inexpensive system is easily usable and can be used to determine the state of the materials located in situ at the site, as well as to study the evolution of the humidity that may appear in rehabilitation projects. The effect of increasing the water/conglomerate ratio during the hardening process was also verified, since it not only translates into an increase in the workability of the mixture but also produces slower hardening of the samples. Complementary to this study, XRD analyzes and thermogravimetric analyses were performed at different ages to visualize the evolution of the hemihydrate and dihydrate compounds typical of gypsum materials.

On the other hand, a capacitive sensor based on obtaining the resonance frequency of a self-oscillating RLC circuit was implemented. The results showed how the designed sensor allows for measuring the moisture content of the gypsum and plaster samples in a reliable and non-intrusive way. These characteristics have made it possible to relate the variation in mixing water content of the test specimens with the evolution of mechanical resistance. As has been proven, both in the case of flexural and compressive strength, these increase as the moisture content of the test tube decreases until reaching the maximum value at the age of seven days, a date set by the UNE-EN standard, of which UNE-EN standard 13279-2 is the optimal one for conducting mechanical tests. It was verified that the plaster samples showed higher resistance values than the gypsum specimens. This can be attributed, in part, to the higher proportion of gypsum hemihydrate present in E35 compared to YF, which was revealed by thermogravimetric analysis. Thus, this sensor has a strong application for the determination of the evolution in the moisture content and its relationship with the mechanical resistance in different construction elements, as long as the positioning of the condenser plates and the separation of these is not possible or greater than 20 cm (e.g., interior partitions of dry partition or plates for false ceilings).

As future lines of research, it would be good to test the operation of this type of sensor for measuring humidity in situ. Moreover, it would be convenient to carry out tests with other types of construction materials such as lime, mortar or concrete. On the other hand, in materials, such as gypsum and plaster, an exothermic reaction occurs during the setting. Also, it is intended to monitor this internal temperature of the samples and relate it to the variations in the content of mixing water in the specimens to improve understanding of the phenomenon studied.

Author Contributions: Conceptualization, P.S. and D.F.; methodology, C.M.; software, E.Y.; validation, D.F., E.A.-S. and P.S.; formal analysis, E.A.-S. and D.F.; investigation, D.F. and C.M.; resources, C.M.; data curation, P.S. and E.A.-S.; writing—original draft preparation, D.F. and E.A.-S.; writing—review and editing, E.Y.; visualization, D.F.; supervision, E.A.-S. and C.M.; project administration, P.S.; funding acquisition, C.M. All authors have contributed equally to the development of the definitive manuscript. All authors have read and agreed to the published version of the manuscript.

Acknowledgments: The authors would like to acknowledge the collaboration provided by the Laboratories of the Higher Technical School of Engineering and Industrial Design of the Universidad Politécnica de Madrid for carrying out the chemical analyses presented.

References

1. Regueiro, M.; Calvo, J.R. El yeso: Geología y yacimientos en España. *Boletín Soc. Española Cerámica Vidr.* **1997**, *36*, 563–571.
2. Vidales, A. Caracterización Fisicoquímica y Aplicaciones de Yeso con Adición de Residuo Plástico de Cables Mediante Criterios de Economía Circular. Ph.D. Thesis, E.T.S. de Edificación (Universidad Politécnica de Madrid), Madrid, Spain, 2019. [CrossRef]
3. Vega, D.F.; Martinez, P.S.; Fernandez, C.M.; Sanchez, E.A. Comparative analysis of fibre-reinforced plasters for the production of precast elements. *DYNA Ing. E Ind.* **2020**, *95*, 333–338. [CrossRef]
4. Del Rio, M. Aplicaciones del yeso y la escayola en la edificación. Nuevas aplicaciones. *Inf. Constr.* **2004**, *56*, 53–60.
5. Brunello, V.; Bersani, D.; Rampazzi, L.; Sansonetti, A.; Tedeschi, C. Gypsum based mixes for conservation purposes: Evaluation of microstructural and mechanical features. *Mater. Constr.* **2020**, *70*, 1–14. [CrossRef]
6. García, A. *Material de Construcción de Yeso o Escayola Aligerado y su uso en Placas y Paneles*; ES 2 293 309 A1; Oficina Española de Patentes y Marcas: Madrid, Spain, 2008.
7. Villoria, P.; del Rio, M.; Sorrentino, M.; Porras, C.; Santa Cruz, J.; Viñas, C. Mechanical Characterization of Gypsum Composites Containing Inert and Insulation Materials from Construction and Demolition Waste and Further Application as A Gypsum Block. *Materials* **2020**, *13*, 193. [CrossRef]
8. Zhang, Y.; Yang, J.; Cao, X. Effects of several retarders on setting time and strength of building gypsum. *Constr. Build. Mater.* **2020**, *240*, 117927. [CrossRef]
9. Erbs, A.; Nagalli, A.; De Carvalho, K.Q.; Mymrin, V.; Passig, F.H.; Mazer, W. Properties of recycled gypsum from gypsum plasterboards and commercial gypsum throughout recycling cycles. *J. Clean. Prod.* **2018**, *183*, 1314–1322. [CrossRef]
10. Herrero, S.; Mayor, P.; Olivares, F.H. Influence of proportion and particle size gradation of rubber from end-of-life tires on mechanical, thermal and acoustic properties of plaster–rubber mortars. *Mater. Des.* **2013**, *47*, 633–642. [CrossRef]
11. Botero, J.C.O.; Bonilla, M.; Borrachero, M.V.; Payá, J. Efecto de un aditivo extraído de la planta Agave americana sobre las propiedades físicas y mecánicas de un yeso. *Mater. Constr.* **2013**, *63*, 79–92. [CrossRef]
12. Nicoleau, L.; Van Driessche, A.E.; Kellermeier, M. A kinetic analysis of the role of polymers in mineral nucleation. The example of gypsum. *Cem. Concr. Res.* **2019**, *124*, 105837. [CrossRef]
13. Ding, X.; Shan, Z.; Long, Z.; Chen, Z. Utilization of collagen protein extracted from chrome leather scraps as a set retarders in gypsum. *Constr. Build. Mater.* **2020**, *237*, 117584. [CrossRef]

14. Liu, C.; Gao, J.; Tang, Y.; Zhao, Y. Early hydration and microstructure of gypsum plaster revealed by environment scanning electron microscope. *Mater. Lett.* **2019**, *234*, 49–52. [CrossRef]
15. Li, Z.; Zhang, J.; Li, S.; Gao, Y.; Liu, C.; Qi, Y. Effect of different gypsums on the workability and mechanical properties of red mud-slag based grouting materials. *J. Clean. Prod.* **2020**, *245*, 118759. [CrossRef]
16. Chen, J.; Chen, C. Uncertainty Analysis in Humidity Measurements by the Psychrometer Method. *Sensors* **2017**, *17*, 368. [CrossRef]
17. Papez, J.; Kic, P. Wood moisture of rural timber construction. *Agron. Res.* **2013**, *11*, 505–512.
18. Rodríguez-Abad, I.; Martínez-Sala, R.; Lladró, R.C.; Barra, R.D.; Garcia-Garcia, F. Estudio de la variación del contenido de humedad en el Pinus pinaster Ait. por medio de la técnica no destructiva del georradar. *Mater. Constr.* **2010**, *61*, 143–156. [CrossRef]
19. Nielsen, A. Gamma-ray-attenuation used for measuring the moisture content and homogeneity of porous concrete. *Build. Sci.* **1972**, *7*, 257–263. [CrossRef]
20. Kopinga, K.; Pel, L. One-dimensional scanning of moisture in porous materials with NMR. *Rev. Sci. Instrum.* **1994**, *65*, 3673–3681. [CrossRef]
21. Then, Y.L.; You, K.; Dimon, M.N.; Lee, C.Y. A modified microstrip ring resonator sensor with lumped element modeling for soil moisture and dielectric predictions measurement. *Measurement* **2016**, *94*, 119–125. [CrossRef]
22. Guilbaud, J.P.; Carvalho, H.; Baroghel-Bouny, V.; Raharinaivo, A. Estudio del contenido de humedad en un material en base cemento midiendo su impedancia y su gamma-densitometría. *Mater. Constr.* **2000**, *50*, 5–13. [CrossRef]
23. Ferrández, D.; Martínez, P.S.; Ferrández, D.; Garcia-Fuentevilla, L. New System of Shrinkage Measurement through Cement Mortars Drying. *Sensors* **2017**, *17*, 522. [CrossRef]
24. Ferrández, D. Estudio de la Transmisión de Vibraciones por Impacto en Losas de Hormigón y Mortero. Ph.D. Thesis, E.T.S. de Edificación (Universidad Politécnica de Madrid), Madrid, Spain, 2016. [CrossRef]
25. Janetti, M.B.; Colombo, L.; Ochs, F.; Feist, W. Determination of the water retention curve from drying experiments using infrared thermography: A preliminary study. *Int. J. Therm. Sci.* **2017**, *114*, 271–280. [CrossRef]
26. Yang, S.; Liu, Y.; Wu, N.; Zhang, Y.; Svoronos, S.; Pullammanappallil, P. Low-cost, Arduino-based, portable device for measurement of methane composition in biogas. *Renew. Energy* **2019**, *138*, 224–229. [CrossRef]
27. Fernandez, C.M.; Vega, D.F.; Martinez, P.S.; Martinez, F.F. Comportamiento frente a la humedad de los morteros de albañilería elaborados con áridos reciclados. *DYNA Ing. E Ind.* **2019**, *94*, 442–446. [CrossRef]
28. Yedra, E.; Vega, D.F.; Martínez, P.S.; Fernández, C.M. Low cost system for measuring the evolution of mechanical properties in cement mortars as a function of mixing water. *Constr. Build. Mater.* **2020**, *244*, 118127. [CrossRef]
29. Spinelli, G.M.; Gottesman, Z.L.; Deenik, J. A low-cost Arduino-based datalogger with cellular modem and FTP communication for irrigation water use monitoring to enable access to CropManage. *HardwareX* **2019**, *6*, e00066. [CrossRef]
30. Calva, A.A.; Duarte, C.; Amilton, J.; Florentino, J.E.; Gomes, B.; Nascimiento, F.H. Accuracy of capacitive sensors for estimating soil moisture in northeastern Brazil. *Soil Tillage Res.* **2019**, *195*, 104413. [CrossRef]
31. de Villanueva Domínguez, L.; García Santos, A. *Manual del Yeso*; CIE Inversiones Editoriales-Dossat 2000: Madrid, Spain, 2001.
32. Strydom, C.; Potgieter, J. Dehydration behaviour of a natural gypsum and a phosphogypsum during milling. *Thermochim. Acta* **1999**, *332*, 89–96. [CrossRef]
33. Fleck, W.E.P.; Jones, M.H.; Kuntze, R.A.; McAdie, H.G. The differential thermal analysis of natural and synthetic hydrates of calcium sulphate. *Can. J. Chem.* **1960**, *38*, 936–943. [CrossRef]
34. UNE-EN 13279-1:2009. *Yesos de Construcción y Conglomerantes a Base de Yeso Para la Construcción. Parte 1: Definiciones y Especificaciones*; AENOR: Madrid, Spain, 2009.
35. Velilla, D.; Pablo, J. Desarrollo de un Prototipo Mecatrónico de Seguidor Solar Parabólico Para la Edificación. Ph.D. Thesis, E.T.S. de Edificación (UPM), Madrid, Spain, 2018. [CrossRef]
36. Noreña, L.; Cortés, D. Characterization and calibration of low-cost humidity sensor to optimize ther performance in a WSN network in indoor settings. *Actas Ing.* **2015**, *1*, 169–173.
37. Pallás, R. *Sensores y Acondicionadores de Señal*, 4th ed.; S.A. Marcombo: Barcelona, Spain, 2005; ISBN 9788426713445.

38. Puertolas, R.F.; Gonzalez, J.C.L.; Castro, S.S.; Benito, P.A.; Arrebola, C.V. A bioclimatic building in Madrid: Analysis of the thermal response and long-term comfort indices review. *Dev. Built Environ.* **2020**, *3*, 100015. [CrossRef]
39. Mettler Toledo, Determination of calcium sulfate dihydrate and hemihydrate in cement. Available online: https://www.mt.com/cn/zh/home/supportive_content/matchar_apps/MatChar_UC264.html (accessed on 23 September 2020).
40. UNE-EN 13279-2:2014. *Yesos de Construcción y Conglomerantes a Base de Yeso Para la Construcción*; AENOR: Madrid, Spain, 2014.
41. García, M.; Sánchez, J.L.; García-Fuentevilla, L.; Saiz, P. Influence of moisture content on the physical and mechanical properties of gypsum with a water/powder ratio of 0.9. In Proceedings of the 2nd International Conference on Technological Innovation in Building, CITE2017, Madrid, Spain, 6–8 March 2017; pp. 149–150, ISBN 978-84-16397-56-3.

2

Mechanical, Thermal and Moisture Buffering Properties of Novel Insulating Hemp-Lime Composite Building Materials

Yaser Abdellatef [1,2], Mohammad Amil Khan [3], Asif Khan [3], Mehdi Iftekharul Alam [3] and Miroslava Kavgic [1,3,*]

1 Civil Engineering Department, University of Ottawa, 161 Louis Pasteur, Ottawa, ON K1N 6N5, Canada; yabde079@uottawa.ca
2 Mechanical Power Department, Faculty of Engineering, Cairo University, Giza 12613, Egypt
3 Civil Engineering Department, University of Manitoba, 15 Gillson St., Winnipeg, MB R3T 5V6, Canada; mohammadamilkhan@gmail.com (M.A.K.); asifkhantupu@gmail.com (A.K.); mehdi17uofm@gmail.com (M.I.A.)
* Correspondence: mkavgic@ottawa.ca

Abstract: Hempcrete is a sustainable biocomposite that can reduce buildings' embodied energy while improving energy performance and indoor environmental quality. This research aims to develop novel insulating hemp-lime composites using innovative binder mixes made of recycled and low-embodied energy pozzolans. The characterization of composites' mechanical and hygrothermal properties includes measuring compressive strength, splitting tensile strength, thermal conductivity, specific heat capacity, and moisture buffer capacities. This study also investigates the impact of sample densities and water content on compressive strength at different ages. The findings suggest that mixes with a 1:1 binder to hemp ratio and 300–400 kg/m^3 density have hygrothermal and mechanical properties suitable for insulating infill wall applications. Hence, compressive strengths, thermal conductivity, and specific heat capacity values range from 0.09 to 0.57 MPa, 0.087 to 0.10 W/m K, and 1250 to 1557 J/kg K, respectively. The average moisture buffer value for all hempcrete samples of 2.78 (gm/m^2 RH%) indicates excellent moisture buffering capacity. Recycled crushed brick pozzolan can enhance the hygrothermal performance of the hemp-lime composites. Thus, samples with 10% crushed brick have the lowest thermal conductivity considering their density and the highest moisture buffer capacity. The new formulas of hydrated lime and crushed brick have mechanical properties comparable to metakaolin and hydraulic lime formulas.

Keywords: hemp-lime composites; mechanical properties; thermal properties; moisture buffering capacity; crushed brick; metakaolin; hydrated lime; hydraulic lime

---

1. Introduction

Hempcrete, made of the hemp plant's inner woody core mixed with a lime-based binder, is a promising lightweight biocomposite material that holds the potential to considerably improve energy and indoor environmental performance and sustainability of buildings [1,2]. Nevertheless, the utilization of hempcrete in the construction industry remains low, mainly due to the high variability of hemp-lime composites and lack of formulas that use widely available and locally sourced ingredients. For example, hempcrete's production cost and embodied energy could be considerably higher in Canada than in Europe due to imported materials [3]. The use of innovative binders and a better understanding of the material's performance are indispensable steps for addressing hempcrete underutilization.

Previous studies have shown that hempcrete hygrothermal and mechanical properties depended on various factors. The binder to hemp ratio, compaction level, water content, and binder mix design have been considered to be critical [4–7]. For instance, an increase in the proportion of porous hemp hurd in hempcrete mixture reduced the dry density and the thermal conductivity [6,8]. In contrast, there was a positive correlation between density and thermal conductivity of the hempcrete [9,10]. Thus, studies have reported thermal conductivity in the range of 0.06−0.11 W/m K for hempcrete mixtures with dry densities ranging from 200 to 400 kg/m^3 and hemp/binder ratios of 1:1, 1:1.5, and 1:2 [11,12]. In comparison, hempcrete samples with higher dry density and hemp/binder ratio in the range of 450−800 kg/m^3 and 1:2−1:4, respectively, had higher thermal conductivity ranging from 0.12 to 0.18 W/m K [11,12]. Previous findings have also indicated a negative correlation between hempcrete's density and moisture buffering capacity. For example, Latif et al. (2015) reported a moisture buffer value (MBV) of 3.47 gm/m^2%RH with a binder to hemp ratio of 1.2:1, and density of 290 kg/m^3 [13], while Collet et al. (2013) reported an MBV of 2.14 gm/m^2%RH, with a binder to hemp ratio of 2:1, and density of 430 kg/m^3 [14].

Furthermore, an increase in water content tends to increase the thermal conductivity of the hempcrete mixes. Rahim et al. (2016) reported a rise in thermal conductivity of approximately 8% and 16% at 10 °C and 40 °C, respectively, due to an increase in the water content of 0.08 kg/kg from the dry state [15]. The variation in specific heat capacity and thermal diffusivity under different humidity conditions was considerably less apparent than the variation in thermal conductivity [8]. Thus, specific heat capacity in the range of 1000−1590 J/kg K was reported for hempcrete with dry densities ranging from 381 to 627 kg/m^3 and hemp to binder ratio 1:2 [5,10]. Thermal diffusivity in the range of 1.48×10^{-7} m^2/s in the dry state to 0.98×10^{-7} m^2/s in the fully saturated condition and high specific heat capacity can provide a better thermal performance than suggested by hempcrete's thermal transmittance [16]. Moreover, hempcrete integration with other innovative materials can modify its thermal energy storage properties [17].

The typical value of the compressive strength of hempcrete mixes is between 0.2 and 1.15 MPa, whereas their flexural and splitting tensile strength values are in the range of 0.06−1.3 MPa and 0.02−0.15 MPa, respectively [9,18]. A high content of hemp hurd in the design mix reduces the compressive and flexural strength of hemp-lime composites [9,19]. However, an increase in the density through compaction of the hempcrete can enhance its mechanical characteristics while reducing the binder's quantity, which is the major contributor to embodied carbon [18,20,21]. In this respect, Dinh (2014) reported approximately 15% higher compressive strength of a 556.0 kg/m^3 hempcrete sample than a 439.7 kg/m^3 sample with a compressive strength of 0.65 ± 0.06 MPa [6]. Likewise, the hempcrete mixes' splitting tensile strength increased with their density [18]. For instance, Elfordy et al. (2008) reported flexural strength in the range of 0.749−1.209 MPa, with a 2:1 binder to hemp ratio and a high-density range of 430−607 kg/m^3 [18].

Calcium hydroxide (i.e., hydrated lime) is particularly suitable for hemp-lime composites' production due to advantages such as breathability, high water vapour permeability, durability, mould resistance, and pest deterrence [22,23]. Furthermore, adding hydraulic lime and metakaolin can improve hempcrete's performance due to their fast setting, high reactivity, and lower embodied energy as compared with Portland cement [22,24]. Crushed brick is another promising pozzolan that belongs to an aluminosilicate group and is reactive towards lime, creating interfacial surface alkaline [25]. Therefore, utilizing reclaimed bricks from demolished buildings as pozzolanic lime mortars could extend their lifetime and reduce hempcrete's embodied energy.

This research study aims to develop innovative insulating hemp-lime composites made of recycled and locally sourced materials. To the best of the authors' knowledge, this is the first study that has used recycled crushed brick as a pozzolan to increase hemp-lime composites' availability and performance, while reducing their environmental impact. Unlike most previous studies that focused on hempcrete's mechanical, thermal, or moisture properties [10,23], this study experimentally estimated the mechanical strength (compressive and splitting tensile strength), thermal performance, and moisture regulation capabilities of eight hempcrete design mixes. Furthermore, in contrast to the existing studies that utilized a hemp/binder ratio of 1:2 or higher for wall formulas [18,26], this research focused on maximizing the hemp hurd ratio to improve the thermal properties of the hempcrete mixtures. Therefore, a hemp/binder ratio of 1:1 with a targeted density of 300 to 400 kg/m^3 was applied to create an insulating "infill wall" formula. The effect of water content on the mechanical properties was also studied. Consequently, this research provides valuable information about insulating hemp-lime composites' critical performance required for material characterization that is likely to interest audiences in academia and industry, focusing on sustainable low-carbon composite building materials.

2. Materials and Methods

2.1. *Raw Materials*

Local hemp producer "Plains Hemp" provided the hemp hurd used as aggregate in this study. The bulk density of the hurd at room temperature of 20 °C was 110 ± 5 kg/m^3, whereas its dry density after oven drying at 105 °C for 24 hours was 101 ± 2 kg/m^3. Sieve analysis was performed on an oven-dried batch of 100 g, according to ASTM C136–06 [27], to determine the distribution of fine and coarse aggregates. Approximately 87% of the particles were less than 2.36 mm, and 10% were between 2.36 and 6.3 mm. The hydrated lime, metakaolin, and hydraulic lime, used in this study complied with ASTM C207–06 [28], ASTM C618–12 [29], and ASTM 1707–09 [30], respectively. Utilized recycled pozzolanic crushed brick had a density of ~1355 kg/m^3, firing temperature between 600 and 900 °C, and it was primarily composed of silicon dioxide (50–60%) and aluminum oxide (20–30%).

2.2. *Compositions of Hempcrete Samples*

This study focused on maximizing the hemp hurd ratio within the hempcrete mixture to improve its thermal properties while reducing its environmental impact. Hence, the hemp hurd to binder ratio used in the hempcrete sample preparation was 1:1 with a wall application density of 300 and 400 kg/m^3 [5,31]. Table 1 summarizes eight hempcrete design mixes with varying pozzolan concentrations and fifteen batches with varying water contents. The amount of water added to hemp-lime composites affects their physical properties. For instance, low water content may cause incomplete hydration and unreacted binder in powder form, leading to reduced compressive strength [23,32]. In contrast, high water content can result in higher density, leading to an increase in thermal conductivity, issues with a setting for the lime binder, and excessively long drying times for the hempcrete [6,33]. Furthermore, there is a lack of tests for measuring the workability of hemp-lime concretes, and experience is used to determine the appropriate water content. Therefore, we investigated different water contents that allowed the suitable workability of the developed hempcrete mixes.

Table 1. Composition of hempcrete samples.

Design Mix	Name	Mix Ratio by Mass Hemp/Lime/Pozzolan/Water *						Hemp/Lime/Pozzolan/Water (by Mass)
		H	L	CB	MK	NHL	Water	
Hydrated lime-crushed brick	LCB10	1	0.9	0.1	-	-	2.5	22%/21%/2%/55%
	LCB20	1	0.8	0.2	-	-	2.5	22%/18%/4%/56%
	LCB20A	1	0.8	0.2	-	-	3	20%/16%//4%/60%
Hydrated lime-metakaolin	LMK10	1	0.9	-	0.1	-	2.6	22%//20%/2%/56%
	LMK20	1	0.8	-	0.2	-	3	20%/16%/4%/60%
	LMK20A	1	0.8	-	0.2	-	2.5	22%/18%/4%/56%
	LMK50	1	0.5	-	0.5	-	3	20%/10%/10%/60%
	LMK50A	1	0.5	-	0.5	-	2.5	22%/11%/11%/56%
	LMK70	1	0.3	-	0.7	-	2.5	22%/7%/15%/56%
	LMK70A	1	0.3	-	0.7	-	2.75	21%/6%/15%/58%
	LMK70B	1	0.3	-	0.7	-	3	20%/6%/14%/60%
Hydrated lime-natural hydraulic lime	LNHL50	1	0.5	-	-	0.5	2.5	22%/11%/11%/56%
	LNHL50A	1	0.5	-	-	0.5	3	20%/10%/10%/60%
	LNHL70	1	0.3	-	-	0.7	2.5	22%/7%/15%/56%
	LNHL70A	1	0.3	-	-	0.7	3	20%/6%/14%/60%

* H, hemp hurd; L, hydrated lime; MK, metakaolin; NHL, natural hydraulic lime; CB, crushed brick.

2.3. Mixing, Moulding, and Curing

Mixing binders and water to create a slurry was the first step in creating hempcrete samples. Then, hemp hurd was added to the slurry and blended in an industrial mixer for approximately 5 min until a uniform mixture was obtained. For the thermal properties' tests and moisture buffering tests, rectangular wooden moulds (26 cm wide × 26 cm long × 5.5 cm high) were filled with the mixture and tamped until reaching the desired wet density. The cylindrical moulds (10 cm diameter × 20 cm long) for compression tests were prepared according to ASTM C39/C39M–12 [34]. Furthermore, for splitting tensile tests, the cylindrical moulds (15 cm diameter × 30 cm long) were produced by following ASTM C496/C496M–17 [35]. Then, a one-quarter portion of the mould at a time was filled with the mixture and tamped until the desired wet density was reached (i.e., 750–800 kg/m^3). The samples were demolded after seven days and cured at room temperature of 22 ± 1 °C, with RH ~50% for another 21–28 days until they reached a constant mass. In total, approximately 220 samples were tested for their mechanical, thermal, and moisture buffering properties.

2.4. Experimental Tests

2.4.1. Mechanical Tests

The compression and splitting tensile tests were performed by following the ASTM D4832 [36] and ASTM C496 [35] standards, respectively. The compressive strength was measured after 28, 60, and 90 days, while splitting tensile strength was measured after 90 days. The load was applied continuously without any shock at a 5 mm/min rate and run for approximately 4–5 min per specimen. The mode of failure was determined by the drop observed in the load-displacement curve after the initial increase. For each design mix and age, we tested four samples and calculated compressive strength as their average. Data obtained from the compression test were used to calculate Young's modulus, according to ASTM C469 [37]. For splitting tensile tests, we tested two samples for each design mix and computed their averages.

2.4.2. Thermal Tests

A FOX 314 Heat Flow Meter Apparatus (HFMA) was used for thermal conductivity and specific heat capacity measurements of the hempcrete samples, according to ASTM C518 [38] and ASTM

C1784 [39] standards, respectively. Thermal tests included "wet" samples at room temperature of 22 ± 1 °C, with RH ~50% and "dry" samples, oven-dried at 100 ± 5 °C until a constant mass was reached (i.e., changes in weight readings of less than 0.1%). The HFMA created steady one-dimensional heat flux through the specimens by setting both plates at constant but different temperatures with an accuracy of ± 0.03 °C. The HFMA applied Fourier's law of heat conduction to obtain hempcrete's thermal conductivity with an accuracy of approximately 1% at a specific mean temperature of the plates. Additionally, we calibrated the HFMA to convert transducers' voltage signals to heat fluxes using NIST 1450b SRM (Standard Reference Material of the National Institute of Standards and Technology) [40]. The HFMA was modified to measure hempcrete samples' specific heat capacity by first maintaining both plates at the equal temperature until reaching a steady-state condition, then changing the plates' temperatures by the same step, and holding them until reaching a new steady-state condition. By assuming transient heat conduction in a finite body for a significant time, the specific heat capacity was determined at each temperature step using Equation (1), according to Tleoubaev and Brzezinski (2007) [41]:

$$C_p = \frac{k}{\rho L^2 \, \pi^{-2}\left(\frac{\Delta ln(Q_i)}{\Delta t}\right)} \tag{1}$$

where L is sample thickness (m), ρ is density (kg/m^3), k is thermal conductivity (W/m K), and $(\Delta lnQ_{i/}\Delta t)$ is the slope of the curve of the natural logarithm of HFMA output values versus time.

The uncertainty in specific heat capacity was calculated at each temperature setpoint to be approximately ± 2.5%, using Equation (2):

$$\mathrm{U}_{C_p} = \sqrt{\left(\frac{\partial C_p}{\partial k} U_k\right)^2 + \left(\frac{\partial C_p}{\partial \rho} U_\rho\right)^2 + \left(\frac{\partial C_p}{\partial L} U_L\right)^2 + \left(\frac{\partial C_p}{\partial (slope)} U_{Slope}\right)^2} \tag{2}$$

where U_k is uncertainty in thermal conductivity (± 1%), U_ρ is uncertainty in density (±1.75%), U_L is uncertainty in thickness (±0.025 mm), and U_{slope} is uncertainty in slope (±0.4% with 95% confidence level).

Furthermore, the hempcrete samples' thermal diffusivity was calculated using HFMA measurements of thermal conductivity and specific heat capacity, as shown in Equation (3):

$$\alpha = \frac{k}{\rho \, C_p} \tag{3}$$

where k is the thermal conductivity (W/m K), ρ is density (kg/m^3), and C_p is the specific heat capacity (J/kg K).

2.4.3. Moisture Buffering Capacity Tests

According to the NORDTEST protocol [42], the measurements of hempcrete's moisture buffer capacity included sealing all specimens' sides except the upper surface (26 cm × 26 cm), exposed to humidity cycles. The air temperature in the climate chamber was kept constant at 23 ± 1 °C, while the relative humidity (RH) was kept constant at 75 ± 2% for 8 hours and 33 ± 2% for 16 hours. Due to the importance of the initial conditions of the samples' moisture contents [43,44], their effect was studied using two different initial states, high moisture content (75% RH) and low moisture content (33% RH). The mass of each sample mass was measured using an electronic scale with an accuracy ± 0.01 gm to monitor the mass change during the uptake and release phases. The humidity cycles were repeated until the steady-state condition was reached, and the mass change in the cycles' uptake and release phases was less than 5% [42]. The moisture buffer value (MBV) was calculated as the average of the last three steady cycles, using Equation (4):

$$MBV = \frac{\Delta m}{A\ (\Delta RH\ \%)} \tag{4}$$

where Δm in the mass change (gm), A is the exposed surface area (m^2), and $\Delta RH\%$ is the difference between the high and low relative humidity.

3. Results

3.1. Density Results

Figure 1 shows that approximately 91% of the specimens are within the density range from 300 to 350 kg/m^3. The average density and the amount of variation are 323 kg/m^3 and 12 kg/m^3, respectively, the minimum is 290 kg/m^3, and the maximum is 372 kg/m^3. These results suggest consistency between the average densities of hempcrete samples produced in this study. Furthermore, Table 2 summarizes descriptive statistics of compression, tensile, and thermal samples. The samples prepared for compressive tests have the highest average density (~340 kg/m^3), followed by the splitting tensile samples (~324 kg/m^3), and thermal samples (~315 kg/m^3). Similarly, the specimens created for the compressive tests exhibit the highest standard deviation of 17.2 kg/m^3 (~5%), followed by tensile specimens with a standard deviation of 12.1 kg/m^3 (~3.8%). Thermal samples share similar variations in density (~2.7%), for both wet and dry tests.

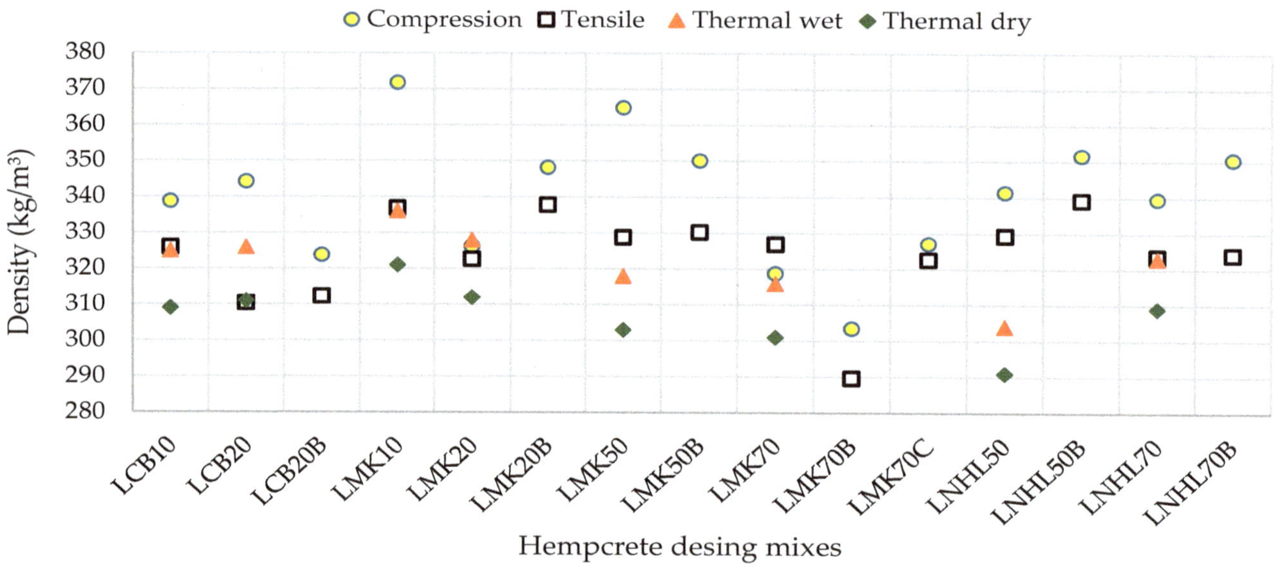

Figure 1. Average densities of all hempcrete samples.

Table 2. Descriptive statistics of samples.

Density (kg/m^3)	Compression	Splitting Tensile	Thermal Wet	Thermal Dry
Minimum	304	290	304	291
Average *	340 ± 9	324 ± 6	322 ± 6	307 ± 6
Maximum	372	339	336	321
Standard deviation	17.2	12.1	8.9	8.3

* Confidence intervals (for 95% confidence level.).

3.2. Compression and Splitting Tensile Test Results

Table 3 summarizes the average compressive strength, splitting tensile strength, modulus of elasticity, and hempcrete samples' densities. The samples' average compressive strength is 0.28 MPa, with a standard deviation of 0.098 MPa (35%), and it ranges from 0.11 MPa (LMK70B) to 0.47 MPa (LMK50B). Due to less variation in the samples' density for splitting tensile tests than compressive tests (see Table 2), tensile strength dispersion is slightly lower than compression values. Thus, the average splitting tensile strength of all samples is 0.024 MPa, with a standard deviation of 0.0065 MPa (27%), and it ranges from 0.0101 MPa to 0.0348 MPa.

Table 3. Average compression strength, tensile strength, Young's modulus, density, and water content.

Name	Water (kg)	Density (CT [1]) (kg/m^3)	Compressive Strength (MPa)	E (MPa)	Density (STT [2]) (kg/m^3)	Splitting Tensile Strength (MPa)
LCB10	2.5	339	0.32	10.42	326	0.0244
LCB20	2.5	344	0.19	6.38	310	0.0219
LCB20B	3	324	0.26	3.16	312	0.0223
LMK10	2.6	372	0.36	10.84	329	0.0219
LMK20	2.5	326	0.21	7.20	330	0.0270
LMK20B	3	348	0.28	9.00	323	0.0222
LMK50	2.5	365	0.33	8.05	338	0.0307
LMK50B	3	350	0.47	8.31	337	0.0342
LMK70	2.5	319	0.14	7.64	327	0.0160
LMK70B	2.75	304	0.11	4.27	290	0.0101
LMK70C	3	327	0.18	4.00	323	0.0183
LNHL50	2.5	341	0.26	9.01	329	0.0307
LNHL50B	3	352	0.42	5.21	339	0.0348
LNHL70	2.5	340	0.30	10.22	324	0.0260
LNHL70B	3	351	0.38	8.22	324	0.0229

[1] Compression test; [2] splitting tensile test.

Overall, an increase in hempcrete density leads to a rise in its compressive and tensile strengths. These findings are consistent with the results of previous studies [18]. Thus, the hempcrete samples of 350 kg/m^3 and denser have an average compressive strength of 0.33 MPa and higher. In contrast, the specimens with the lowest density, including LMK70B (303.5 kg/m^3), LMK70 (318.8 kg/m^3), and LMK70C (327.1 kg/m^3), have the lowest average compressive strength ranging from 0.11 MPa to 0.18 MPa. Similarly, the highest splitting tensile strength of 0.0348 MPa has the LNHL50B sample with the highest density of 339.15 kg/m^3, whereas the least dense LMK70B (~289 kg/m^3) has the lowest splitting tensile strength of 0.01 MPa. Nevertheless, the two samples with the highest compressive strengths, LMK50B, and LNHL50B, are not the most compacted hempcrete samples. These results suggest that water content, binder type, and content also significantly impact hemp-lime concretes' mechanical properties.

Hence, on the one hand, an increase in the water content increases the compressive strength, even of the samples with lower densities. For example, the LCB20 sample with 2.5 kg of water and 344.1 kg/m^3 density has a compressive strength of 0.19 MPa, while the LCB20B sample with 3 kg of water and 323.8 kg/m^3 density has a compressive strength of 0.26 MPa. Metakaolin samples with percentages above 20%, and all LNHL samples exhibit a similar trend. On the other hand, the design mixes with smaller water content (i.e., 2.5:1 water to hemp ratio) show inferior compressive and tensile strengths compared with other hempcrete specimens. These results echo Walker and Pavia's (2014) observations regarding hemp's high suction ability that undermined the binder's hydration and adversely affected the strength of hemp-lime composites [23]. The sample produced from LMK70 (lime 30% and metakaolin 70%) show the highest amount of unreacted binder and has the lowest compressive and splitting tensile strengths. The LNHL50 sample, cast in two batches, shows an average compressive strength of only 0.26 MPa using a water to hemp ratio of 2.5:1 and approximately 60% higher (0.42 MPa) using a water to hemp ratio of 3:1. Similarly, the LNHL70 sample prepared with a water to hemp ratio of 2.5:1 exhibits an average compressive strength of 0.30 MPa, which is around 21% below the compressive strength of the sample with water to hemp ratio of 3:1.

Furthermore, an increase in the percent of metakaolin and crushed brick above a certain amount reduces the compressive strength of hempcrete. This effect is most pronounced in the LMK samples in which compressive strength decreases from 62% to 76%, with an increase in the percent share of metakaolin from 50% to 70%. Similarly, an increase in the percent share of metakaolin and NHL hurts the tensile strength of hempcrete. Thus, an increase in the percentage of metakaolin from 50% to 70% results in a 16% to 62% decrease in hempcrete tensile strength, depending on the batch. An increase in hydraulic lime from 50% to 70% decreases the tensile strength for approximately 15% to 34%, whereas an increase in the crushed brick from 10% to 20% exhibits the lowest decreases in the tensile strength of 8% to 10%. On the one hand, these results support Dinh (2014) findings regarding the adverse effect of an increase in metakaolin share above 70% on the compressive strength of hempcrete [6]. On the other hand, Eires et al. (2006) stated that 75% of metakaolin and 25% of lime were the best binder mix [45]. However, they also reported that this design mix required higher curing temperatures to increase strength gain.

As presented in Table 3, the modulus of elasticity shows a wide variation between design mixes and a weak relationship with the samples' density. Thus, Young's modulus ranges from 3.16 to 10.84 MPa, and its average value of all samples is 7.46 MPa, with a standard deviation of 2.327 MPa (31%). Increasing the metakaolin and crushed brick content reduces Young's modulus, whereas increasing hydraulic lime does not have the same effect. The LCB20 sample shows the lowest value of Young's modulus (3.16 MPa), followed by LMK70C (4.00 MPa). In contrast, the LMK10 specimen has the highest Young's modulus (10.84 MPa), followed by LCB10 (10.42 MPa) and LNHL70 (10.22 MPa).

Figure 2 compares densities and compressive strengths on the 28th, 60th, and 90th day among the developed samples and against previous studies with similar hempcrete densities [18,26]. The results show that specimens, on the one hand, with the lowest densities of approximately 303 kg/m^3 and 318 kg/m^3 have 30–50% lower compressive strength than the results in the literature [18,26]. The likely reason is a higher content of binder used in these studies. In this regard, Elfordy et al. (2008) [18] and Williams et al. (2017) [26] used a 2:1 and 2.25:1, respectively, binder to hemp ratio. On the other hand, samples with a density of approximately 340 kg/m^3 and above have compressive strength comparable with Elfordy et al. (2008) [18] and Williams et al. (2017) [26]. Some studies have also differentiated between testing in the perpendicular (⊥) and the parallel (||) direction to the direction of compaction. In this respect, Williams et al. (2017) reported slightly higher compressive strengths of the samples with similar density tested in perpendicular than parallel direction (see Figure 2) [26]. In this research, the compaction and testing directions (i.e., perpendicular or parallel) were designed to present the worst-case scenario for mechanical properties. Therefore, parallel testing to the compaction direction might be another reason for differences between our results and values reported by Williams et al. (2017) [26].

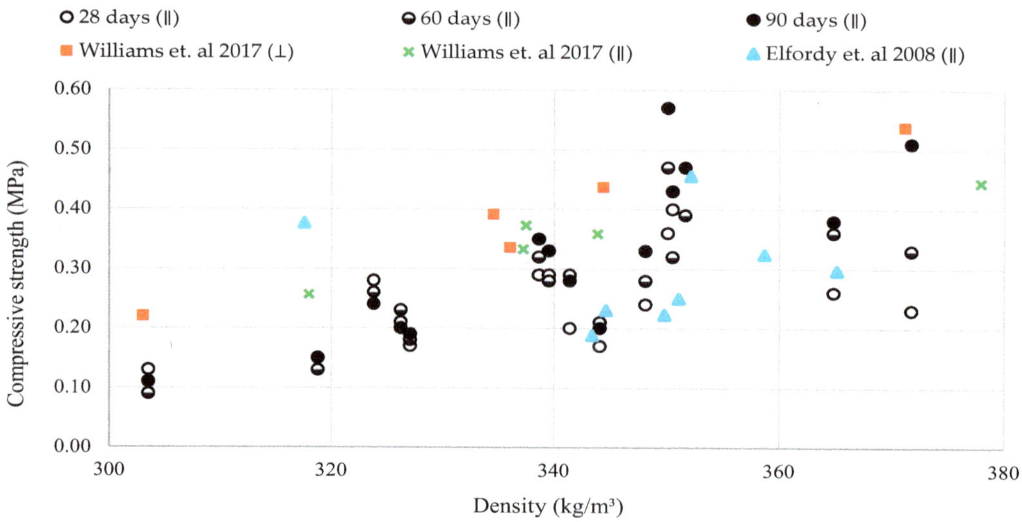

Figure 2. Comparison of density, compressive strength, and age.

3.3. Thermal Properties Results

Table 4 presents the dry and wet hempcrete samples' thermal properties, including their density, thermal conductivity, specific heat capacity, and thermal diffusivity. The wet hempcrete samples have higher density, thermal conductivity, and specific heat capacity than their dry counterparts due to the higher moisture content. Thus, the wet density ranges from 304 kg/m^3 to 336 kg/m^3, and the average is 322 kg/m^3. In comparison, the dry density is approximately 4.5% lower, ranging from 291 kg/m^3 to 321 kg/m^3 with an average of 307 kg/m^3. The thermal conductivities of wet and dry samples vary from 0.091 to 0.101 W/m K and from 0.087 to 0.096 W/m K, with averages of 0.096 and 0.0913 W/m K, respectively. Furthermore, the specific heat capacities of wet and dry samples range from 1398 to 1557 J/kg K and from 1250 to 1421 J/kg K, with averages of 1508 and 1365 J/kg K, respectively. These results range from specific heat capacity values reported in the literature, between 1000 and 1560 J/kg K [5,46].

Table 4. Thermal properties of the hempcrete samples.

Sample	State	Density (kg/m^3)	Thermal Conductivity (W/m K)	Specific Heat Capacity (J/kg K)	Thermal Diffusivity (m^2/s) $\times 10^{-7}$
LCB10	wet	325	0.0962	1504	1.968
	dry	309	0.0921	1350	2.209
LCB20	wet	326	0.0943	1398	2.069
	dry	311	0.0902	1250	2.321
LMK10	wet	336	0.1007	1536	1.951
	dry	321	0.0959	1397	2.140
LMK20	wet	328	0.0981	1530	1.955
	dry	312	0.0937	1385	2.169
LMK50	wet	318	0.0955	1523	1.972
	dry	303	0.0910	1379	2.178
LMK70	wet	316	0.0941	1498	1.988
	dry	301	0.0898	1352	2.206
LNHL50	wet	304	0.0907	1520	1.962
	dry	291	0.0868	1383	2.156
LNHL70	wet	323	0.0952	1557	1.893
	dry	309	0.0912	1421	2.076

In addition, the wet samples' thermal diffusivity is lower as compared with their dry counterparts. In this respect, thermal diffusivity of the wet samples ranges from 1.893 to 2.069 (m^2/s) $\times 10^{-7}$, whereas dry specimens range from 2.076 to 2.321 (m^2/s) $\times 10^{-7}$, with averages of 1.970 (m^2/s) $\times 10^{-7}$ and 2.182 (m^2/s) $\times 10^{-7}$, respectively. These findings indicate that, in the wet state, the total increase in heat storage $\left(\rho \times c_p\right)$ is higher than the increase in thermal conductivity. Thus, the wet state's thermal diffusivity $\left(\frac{k}{\rho\, C_p}\right)$ is lower than that of the dry state (i.e., have higher thermal inertia than the dry samples.) Our results are slightly higher than the range of thermal diffusivity values reported in the literature, between 0.98 and 1.68 (m^2/s) $\times 10^{-7}$ [5,47]. A possible explanation might be the higher density values that previous studies used to calculate thermal diffusivity.

The results presented in Table 4 indicate that the hempcrete samples with crushed brick have the lowest thermal conductivity and specific heat capacity, considering their density and the highest thermal diffusivity. For example, the average density, thermal conductivity, specific heat capacity, and thermal diffusivity of all hempcrete samples with crushed brick (wet and dry) are approximately

318 kg/m^3, 0.93 W/m K, 1376 J/kg K, and 2.142 (m^2/s) × 10^{-7}, respectively. In comparison, the metakaolin hempcrete samples have a similar average density (~317 kg/m^3) but higher thermal conductivity (~0.95 W/m K) and specific heat capacity (~1450 J/kg K). The LNHL hempcrete samples have the lowest average density of approximately 307 kg/m^3, and therefore the lowest thermal conductivity of 0.091 W/m K. Nevertheless, they have the highest average specific heat capacity of 1470 J/kg K, and the lowest thermal diffusivity of 2.020 (m^2/s) × 10^{-7}.

Figure 3 illustrates the relation between thermal conductivity and density in the dry and wet states for all design mixes. Similar to previous studies [9,26], both the dry and wet samples show a positive linear relationship between the samples' conductivities and densities with R^2 values of 0.907. Thus, the LNHL50 samples with the lowest dry and wet densities of 291 and 304 kg/m^3 have the lowest thermal conductivity of approximately 0.087 and 0.90 W/m K, respectively. In contrast, the LMK10 samples with the highest dry and wet densities of 321 and 336 kg/m^3 have the highest thermal conductivity of approximately 0.096 and 0.1 W/m K, respectively.

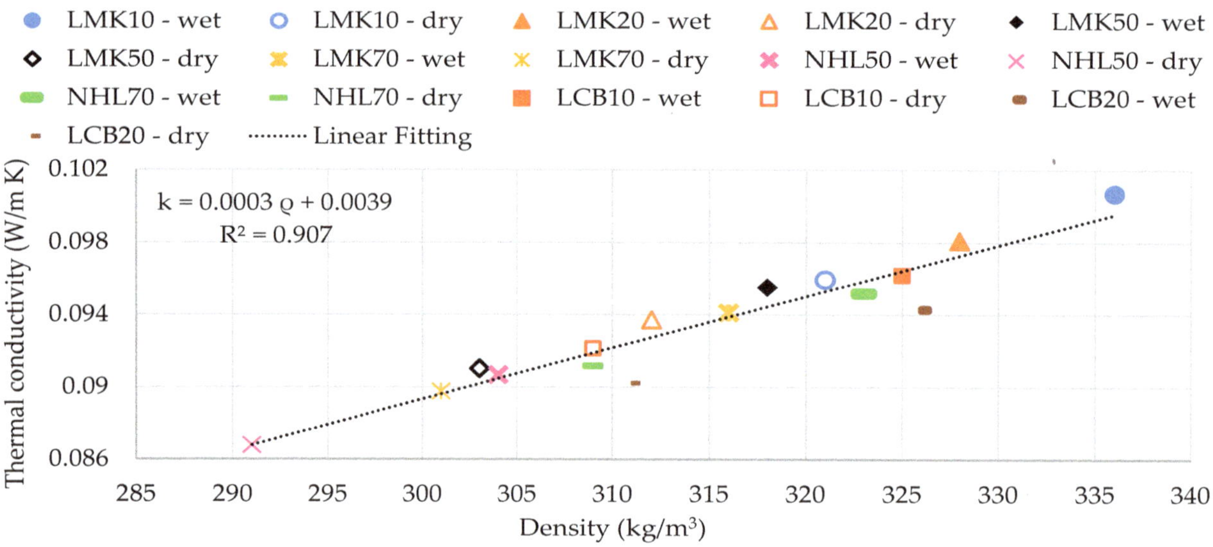

Figure 3. Thermal conductivity as a function of density.

Figure 4 compares the experimental average thermal conductivities of hempcrete samples against the results of previous studies. The results show that the thermal conductivities and their increase with density are comparable to the values reported by previous studies for the specimens with similar densities. Furthermore, some studies differentiate between testing in the perpendicular (⊥) and parallel (||) direction (i.e., heat flux direction) to the compaction of the samples. In this respect, Williams et al. (2017) [26] and Nguyen et al. (2010) [48] reported higher thermal conductivity of the samples with similar density tested in perpendicular than parallel direction (see Figure 4). Moreover, Nguyen et al. (2010) [48] reported a higher difference between the parallel and perpendicular conductivities than Williams et al. (2017) [26]. The likely reason is the higher density range of the samples in the first study compared with that of the second study. In this research, the compaction and testing directions (i.e., perpendicular or parallel) represent the worst-case scenario for the thermal properties. In this respect, the HFMA created heat flux across the samples, perpendicular to the compaction direction. Thus, our results show a good match with experiments conducted in the perpendicular direction [26], and a more significant discrepancy with those performed in the parallel orientation [26].

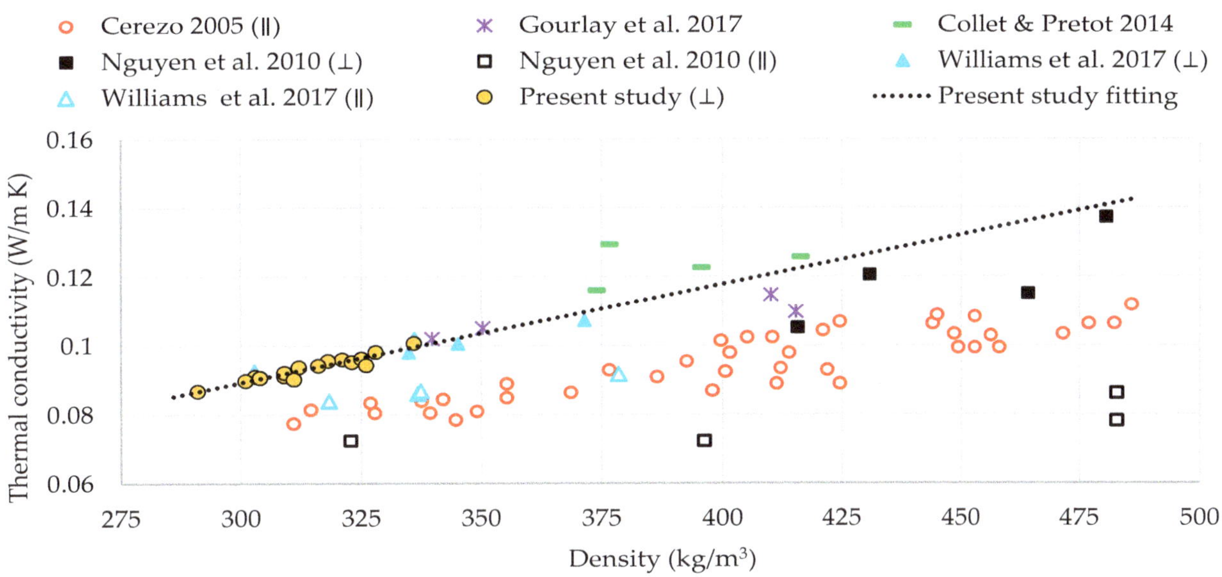

Figure 4. Thermal conductivity comparison with other studies.

3.4. Moisture Buffering Results

Figure 5 presents a comparison of moisture buffer values (MBV) for each cycle's uptake and release phases until reaching the steady-state of three selected hempcrete samples, LMK50, LNHL50, and LCB10, with two initial moisture conditions, high and low. As shown for the high moisture initial condition, the MBV for the release phase starts with a higher value (2.4–2.6 gm/m^2 RH%) than that of the uptake phase (1.3–1.7 gm/m^2 RH%). The hempcrete samples tend to lose the high moisture content obtained from the initial condition more than absorbing more moisture. However, after the initial moisture content reduces, the MBV for the uptake phase increases again. In contrast, the MBV for the release phase slightly decreases until both reach steady-state starting from the fourth cycle with less than 5% difference between MBV for uptake and release. For the low moisture initial condition case, the MBV for the uptake phase starts with a higher value (3–4 gm/m^2 RH%) than that of the release phase (2–2.8 gm/m^2 RH%). The hempcrete samples are capable of absorbing more moisture due to the low initial moisture content. The MBV for the uptake phase decreases again, and the MBV for the release phase slightly increases until both reach steady-state starting from the fourth cycle, with a difference of less than 5% between MBV for uptake and release.

The three design mixes express different moisture buffering behaviors. For example, the LNHL50 with high initial moisture condition starts with 14% and 26% higher MBV as compared with LCB10 and LMK50, in the uptake phase and 2.6% and 13.8% higher in the release phase, respectively. Furthermore, the starting MBV values of the LCB10 with low initial conditions are 11.2% and 39% higher than those of the LNHL50 and LMK50 counterparts in the uptake phase. Similarly, in the release phase, the starting MBV values of the LCB10 are 14% and 42% higher than those of LNHL50 and LMK50, respectively. Moreover, the LCB10 shows approximately 1.2% higher steady-state MBV than LNHL50, and 18% higher than LMK50.

Figure 6 shows the percentage mass change relative to the initial mass during the test for the selected samples subjected to the humidity cycles. As presented, all specimens show the same mass change profile with a gradual increase or decrease, depending on the initial condition, until they reach a steady-state starting from the fourth cycle. Moreover, the LCB10 sample has the highest mass gain and release in the uptake and release portions of the cycle, respectively, regardless of the initial condition. The LNHL50 and LMK50 samples show comparable rates of mass gain and release.

Figure 7 presents the average steady-state MBV values of the hempcrete samples. The average steady-state MBV for all hempcrete samples of 2.78 gm/m^2 RH%, with a standard deviation of 0.24 gm/m^2 RH% (8.6%), falls in the range reported in the literature. For example, Collet et al. (2013) reported an MBV of 2.14 gm/m^2 %RH, with a binder to hemp ratio of 2:1, and density of 430 kg/m^3 [14], while Latif et al. (2015) reported an MBV of 3.47 gm/m^2 %RH with a binder to hemp ratio of 1.2:1, and density of 290 kg/m^3 [13]. The findings suggest that the initial moisture content condition does not significantly affect the final steady-state average MBV for each design mix (i.e., the difference is less than 1%).

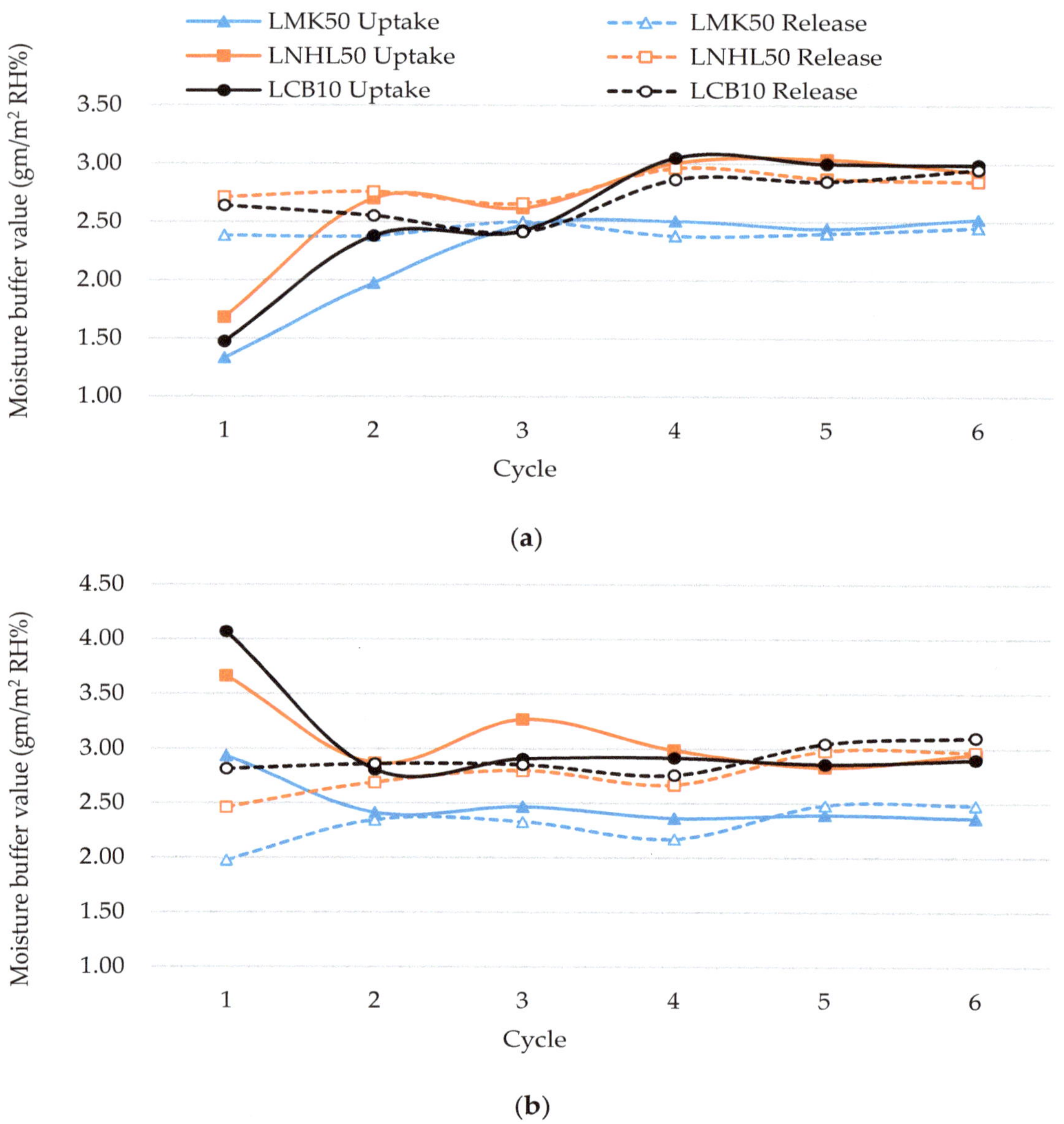

Figure 5. Moisture buffer value with cycles with different initial conditions. (**a**) High moisture content initial condition; (**b**) Low moisture content initial condition.

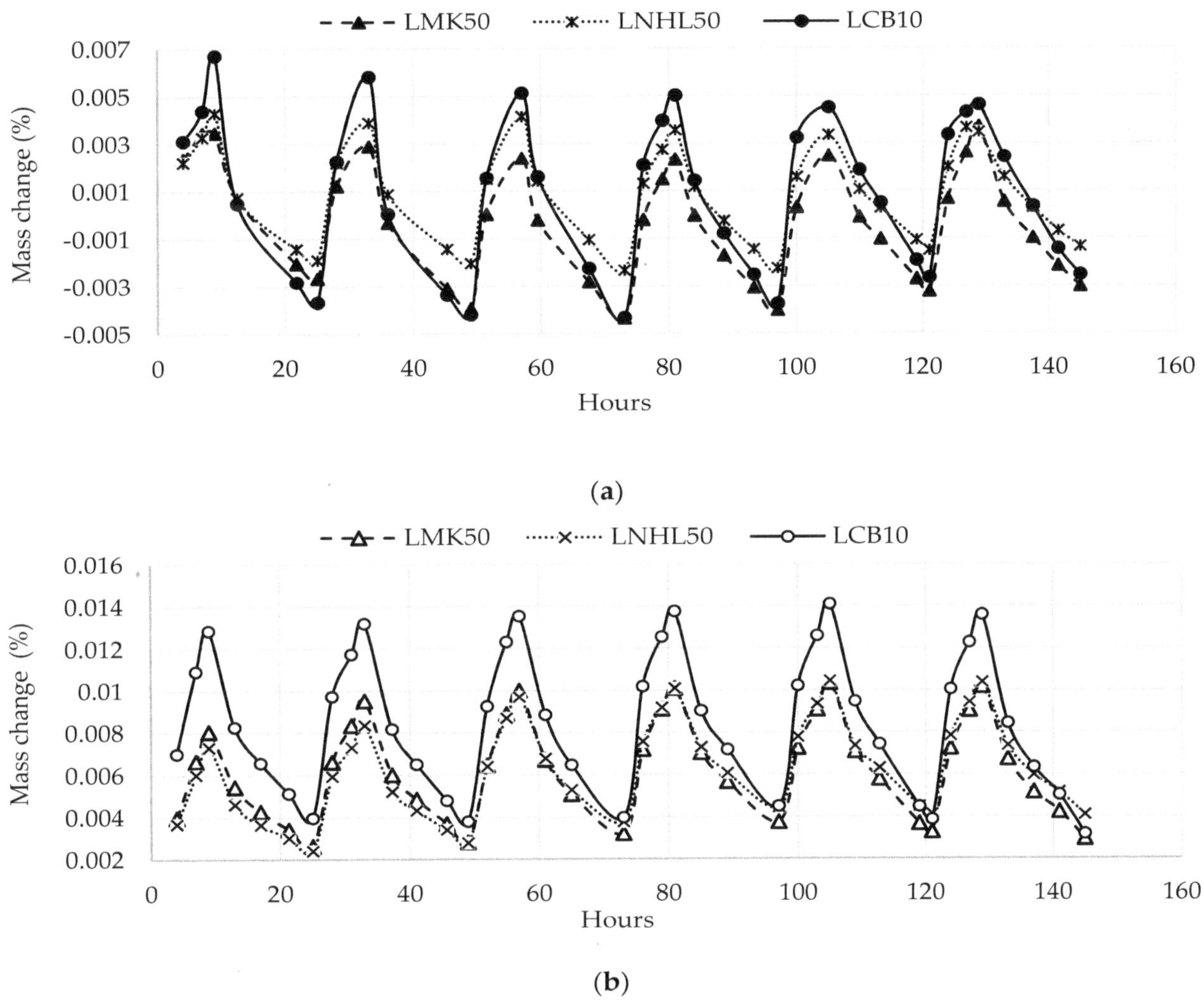

Figure 6. Typical steady cycle with different boundary conditions. (**a**) High moisture content initial condition; (**b**) Low moisture content initial condition.

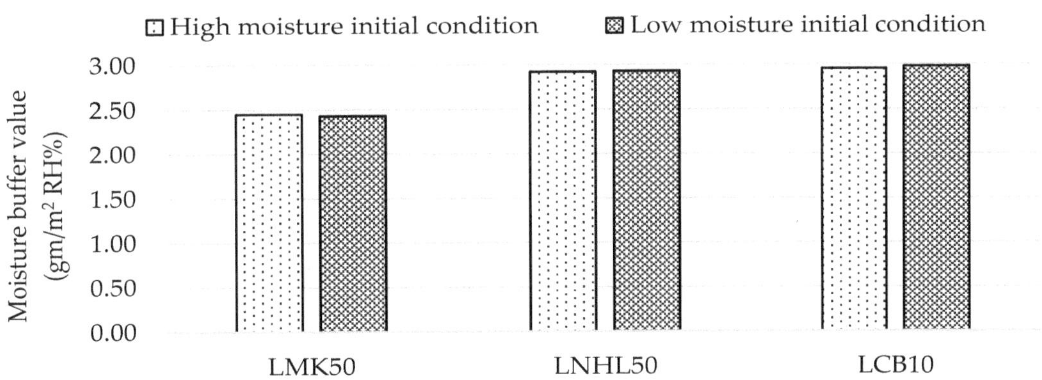

Figure 7. Average steady-state moisture buffer value (MBV).

4. Discussion

The overall findings suggest that hempcrete mixes with 1:1 binder to hemp hurd ratio and 300–400 kg/m^3 density have mechanical and hygrothermal properties suitable for infill wall applications. For instance, the compressive strengths ranging from 0.09 to 0.57 MPa are comparable to the values reported by the previous studies that used higher binder to hurd ratios (i.e., 2:1 and 2.25:1) [18,26]. Similarly, thermal conductivity and specific heat capacity values in the range of 0.087–0.10 W/m K and 1250–1557 J/kg K correspond to the values reported by the studies that developed hempcrete samples with similar densities [5,10,26,46]. Furthermore, the average steady-state moisture buffer value (MBV) for all hempcrete samples of 2.78 gm/m^2 RH indicates excellent moisture storage capabilities. The splitting tensile strength ranging from 0.0101 to 0.0348 MPa is the only measured parameter that is comparatively lower than the previous studies' results. The likely reasons are the lower density range and the binder to this study's hemp hurd ratio. In addition, hempcrete composites made from hydrated lime have significantly lower flexural strength than composites made from commercial composites [4,49].

The novel hempcrete mixes with hydrated lime and crushed brick (LCB) show comparable compressive and tensile strength to other design mixes, with the advantage of being composed of recycled material. For instance, the hempcrete samples with 10% crushed brick have compressive strengths higher than the average of all the design mixes. Additionally, the LCB samples have the lowest thermal conductivity considering their density and the highest average moisture buffer values for high and low initial moisture contents, which are approximately 18% higher than metakaolin formulas and comparable to design mixes with hydraulic lime.

The hempcrete's density is a vital design parameter due to its significant effects on the material's mechanical and thermal properties. For instance, on the one hand, hempcrete samples of approximately 340 kg/m^3 and above have 45%–75% higher compressive and tensile strengths than those with lower density (i.e., 300–330 kg/m^3). On the other hand, the low-density hempcrete samples (i.e., 290–310 kg/m^3) have 8–13% lower thermal conductivity than those in the range of 320–336 kg/m^3. Furthermore, as illustrated in Figure 8, an increase in thermal conductivity with a rise of compressive strength tends to be higher in the low than the upper range of compressive strength values. These findings indicate that the specific application should determine the density of the hempcrete biocomposites. The results also show that similar to previous studies [9,26], there is a strong positive linear correlation (R^2 = 0.91) between the dry and wet samples' conductivities and their densities. The relationship between compressive strength and density is weaker and increases over time from 0.36 at 28 days to 0.77 at 90 days.

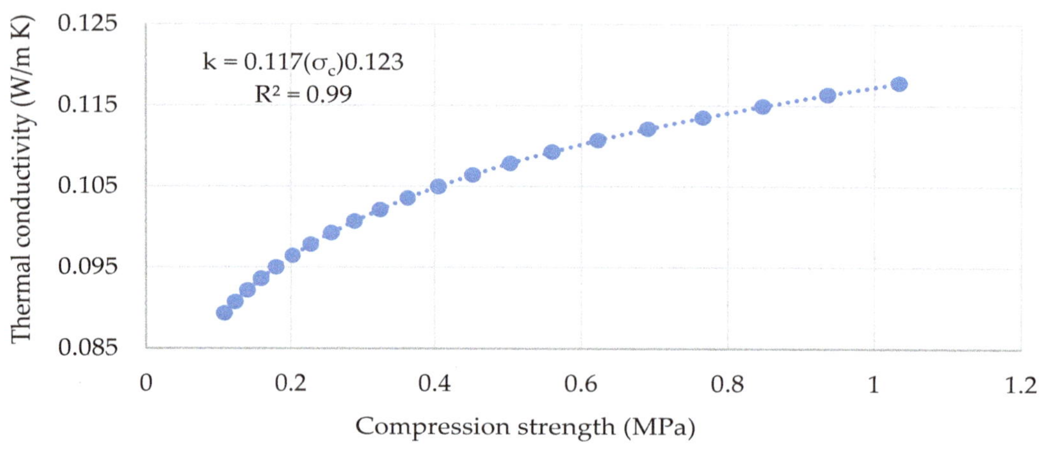

Figure 8. Thermal conductivity as a function of compressive strength.

The results also indicate that binder mix design and water content impact the prepared hempcrete samples' thermal and mechanical properties. For example, an increase in the percent share of metakaolin and crushed brick reduces thermal conductivity and specific heat capacity. In contrast, increasing the hydraulic lime percentage increases the conductivity and specific heat capacity. In addition, an increase in the percent share of pozzolans above a certain amount can harm the mechanical properties of hempcrete, particularly when combined with reduced water content (i.e., 2.5:1 water to hemp ratio). This hindering effect is most significant when metakaolin is increased above 50%, followed by crushed brick above 10%.

5. Conclusions and Future Research

Despite excellent hemp-lime composites' hygrothermal properties, their utilization in the construction industry remains low due to the high variability and lack of formulas with widely available ingredients. The use of innovative binders and a better understanding of the material's performance are indispensable steps for addressing hempcrete's underutilization. This is the first research study that applies recycled crushed brick as a pozzolan in hempcrete mixes to increase their availability and performance while reducing their environmental impact. Another contribution to the body of knowledge is the comprehensive mechanical and hygrothermal characterization of hemp-lime composites with different binder mix designs. Consequently, this study provides valuable information about insulating hemp-lime composites' critical performance required for material characterization that is likely to interest audiences, in academia and industry, who are focused on sustainable low-carbon composite building materials. The main conclusions of this study that point to the need for future work and investigation are as follows:

- The thermal and mechanical properties of hempcrete depend on its density. Hence, the large-scale utilization of hemp-lime composites in the construction industry requires the use of manufactured blocks, wall panels, or spraying techniques to enable the construction of consistent code complying envelope systems. Therefore, future work should focus on the processing methods for optimizing and standardization of hempcrete design mixes for different envelope applications and installations. Furthermore, future work should increase the number of tested samples for each design mix to ensure consistency.
- The higher content of hemp hurd combined with an application-orientated level and direction of compaction can lead to the development of products suitable for infill wall utilizations. A follow-up study should investigate mechanical properties in the perpendicular and thermal behavior in the parallel direction of hempcrete formulas with a 1:1 binder to hemp hurd ratio and 300–400 kg/m^3 density. Further research should also optimize compaction levels and orientations for infill wall installations of hemp-lime composites using blocks or wall panels.
- Although all the developed design mixes have mechanical and hygrothermal properties suitable for infill wall applications, in small amounts, recycled crushed brick is an excellent alternative to conventional pozzolans for creating more environmentally-friendly hempcrete composites. For example, the hempcrete samples with 10% crushed brick have the lowest thermal conductivity considering their density and the highest moisture buffer capacity. Furthermore, the new hydrated lime and crushed brick formulas have comparable mechanical properties to metakaolin and hydraulic lime formulas. Future research should combine and optimize the use of the investigated lime mortars and pozzolans in hemp-lime mixes. A follow-up study should also explore other locally available materials that can improve hemp-lime composites' performance and affordability while reducing their environmental impact.
- The water content is an essential design parameter due to its significant impact on the hempcrete samples' mechanical and thermal properties. This study indicates that a smaller amount of water (i.e., $\leq$2.5 kg) leads to some specimens' inferior mechanical properties due to incomplete hydration caused by the high suction ability of the hemp hurd. A follow-up study should optimize water content for different infill wall application approaches such as blocks, wall panels, or spray-in. Further research should also develop a hemp pretreatment process and investigate water retainers' application to balance the high suction of the hurd and improve mechanical behavior and durability of hemp-lime composites. Future research should also investigate hemp-lime composites' durability and decay under different environmental conditions.

Author Contributions: Conceptualization, Y.A. and M.C.; methodology, Y.A., M.A.K., A.K., M.I.A., and M.K.; software, Y.A.; validation, Y.A., M.A.K., A.K.; M.I.A., and M.K.; formal analysis, Y.A., M.A.K., A.K., M.I.A., and M.K.; investigation, Y.A., M.A.K., A.K.; M.I.A., and M.K.; resources M.K.; data curation, Y.A. and M.K.; writing—original draft preparation Y.A.; writing—review and editing, M.K., M.A.K., A.K., and M.I.A.; visualization,

Y.A.; supervision M.K.; project administration, M.K.; funding acquisition, M.K. All authors have read and agreed to the published version of the manuscript.

Acknowledgments: The authors would like to acknowledge the in-kind contributions of "Plains Hemp" and "WhiteMud Resources".

References

1. Stanwix, W.; Sparrow, A. *The Hempcrete Book: Designing and Building with Hemp-Lime*; Green Books; UK Hempcrete: Matlock, UK, 2014.
2. Pochwala, S.; Makiola, D.; Anweiler, S.; Bohm, M. The Heat Conductivity Properties of Hemp-Lime Composite Material Used in Single-Family Buildings. *Materials* **2020**, *13*, 1011. [CrossRef]
3. Magwood, C. *Essential Hempcrete Construction: The Complete Step-by-Step Guide*; New Society Publishers: Gabriola Island, BC, Canada, 2016.
4. Jami, T.; Karade, S.R.; Singh, L.P. A review of the properties of hemp concrete for green building applications. *J. Clean. Prod.* **2019**, *239*, 117852. [CrossRef]
5. Evrard, A. Transient Hygrothermal Behavior of Lime-Hemp Materials. Ph.D. Thesis, Universite Catholique De Louvain, Ottignies-Louvain-la-Neuve, Belgium, 2008.
6. Manh, D.T. Contribution to the Development of Precast Hempcrete Using Innovative Pozzolanic Binder. Ph.D. Thesis, Université Toulouse III-Paul Sabatier, Toulouse, France, 2014.
7. Ntimugura, F.; Vinai, R.; Harper, A.; Walker, P. Mechanical, thermal, hygroscopic and acoustic properties of bio-aggregates—Lime and alkali—Activated insulating composite materials: A review of current status and prospects for miscanthus as an innovative resource in the South West of England. *Sustain. Mater. Technol.* **2020**, *26*, e00211. [CrossRef]
8. De Bruijn, P.; Johansson, P. Moisture fixation and thermal properties of lime–hemp concrete. *Constr. Build. Mater.* **2013**, *47*, 1235–1242. [CrossRef]
9. Cérézo, V. *Propriétés Mécaniques, Thermiques et Acoustiques d'un Matériau à Base de Particules Végétales: Approche Expérimentale et Modélisation Théorique*; Le grade de docteur L'Institut National des Sciences Appliquées de Lyon: Lyon, France, 2005.
10. Walker, R.; Pavía, S. Moisture transfer and thermal properties of hemp–lime concretes. *Constr. Build. Mater.* **2014**, *64*, 270–276. [CrossRef]
11. Arnaud, L.; Boyeux, B.; Hustache, Y. *Hemp and the Construction Industry*; Hemp: Industrial Production and Uses; CABI: Wallingford, UK, 2013; pp. 239–259.
12. Lawrence, M.; Fodde, E.; Paine, K.; Walker, P. Hygrothermal Performance of an Experimental Hemp-Lime Building. *Key Eng. Mater.* **2012**, *517*, 413–421. [CrossRef]
13. Latif, E.; Lawrence, M.; Shea, A.; Walker, P. Moisture buffer potential of experimental wall assemblies incorporating formulated hemp-lime. *Build. Environ.* **2015**, *93*, 199–209. [CrossRef]
14. Collet, F.; Chamoin, J.; Pretot, S.; Lanos, C. Comparison of the hygric behaviour of three hemp concretes. *Energy Build.* **2013**, *62*, 294–303. [CrossRef]
15. Rahim, M.; Douzane, O.; Tran Le, A.D.; Langlet, T. Effect of moisture and temperature on thermal properties of three bio-based materials. *Constr. Build. Mater.* **2016**, *111*, 119–127. [CrossRef]
16. Yates, T. *Final Report on the Construction of the Hemp House at Haverhill, Suffolk*; Building Research Establishment: Watford, UK, 2002.
17. Abdellatef, Y.; Kavgic, M. Thermal, microstructural and numerical analysis of hempcrete-microencapsulated phase change material composites. *Appl. Therm. Eng.* **2020**, *178*, 115520. [CrossRef]
18. Elfordy, S.; Lucas, F.; Tancret, F.; Scudeller, Y.; Goudet, L. Mechanical and thermal properties of lime and hemp concrete ("hempcrete") manufactured by a projection process. *Constr. Build. Mater.* **2008**, *22*, 2116–2123. [CrossRef]
19. Barnat-Hunek, D.; Smarzewski, P.; Fic, S. Mechanical and Thermal Properties of Hemp-Lime Composites. *Compos. Theory Pract.* **2015**, *15*, 21–27.

20. Nguyen, T.-T.; Picandet, V.; Amziane, S.; Baley, C. Influence of compactness and hemp hurd characteristics on the mechanical properties of lime and hemp concrete. *Eur. J. Environ. Civ. Eng.* **2009**, *13*, 1039–1050. [CrossRef]
21. Tronet, P.; Lecompte, T.; Picandet, V.; Baley, C. Study of lime hemp composite precasting by compaction of fresh mix—An instrumented die to measure friction and stress state. *Powder Technol.* **2014**, *258*, 285–296. [CrossRef]
22. Walker, R.; Pavía, S. Behaviour and Properties of Lime-Pozzolan Pastes. In Proceedings of the 8th International Masonry Conference, Dresden, Germany, 4–7 July 2010.
23. Walker, R.; Pavia, S.; Mitchell, R. Mechanical properties and durability of hemp-lime concretes. *Constr. Build. Mater.* **2014**, *61*, 340–348. [CrossRef]
24. Sheridan, J.; Sonebi, M.; Taylor, S.; Amziane, S. Effect of linseed oil and metakaolin on the mechanical, thermal and transport properties of hemp-lime concrete. In Proceedings of the 2nd International Conference on Bio-Based Building Materials, Clermont Ferrand, France, 21–23 June 2017; pp. 125–133.
25. Silva, R.V.; De Brito, J.; Dhir, R.K. Properties and composition of recycled aggregates from construction and demolition waste suitable for concrete production. *Constr. Build. Mater.* **2014**, *65*, 201–217. [CrossRef]
26. Williams, J.; Lawrence, M.; Walker, P. The influence of the casting process on the internal structure and physical properties of hemp-lime. *Mater. Struct.* **2017**, *50*, 108. [CrossRef]
27. *ASTM C136–06, Standard Test Method for Sieve Analysis of Fine and Coarse Aggregates*; ASTM International: West Conshohocken, PA, USA, 2006.
28. *ASTM C207-06, Standard Specification for Hydrated Lime for Masonry Purposes*; ASTM International: West Conshohocken, PA, USA, 2006.
29. *ASTM C618, Standard Specification for Coal Fly Ash and Raw or Calcined Natural Pozzolan for Use in Concrete*; ASTM International: West Conshohocken, PA, USA, 2012.
30. *ASTM C1707, Standard Specification for Pozzolanic Hydraulic Lime for Structural Purposes*; ASTM International: West Conshohocken, PA, USA, 2009.
31. Nguyen, T.T. Contribution à l'étude de la Formulation et du Procédé de Fabrication d'éléments de Construction en Béton de Chanvre. Ph.D. Thesis, Université de Bretagne Sud, Morbihan, France, 2010.
32. Pinkos, J. The Effectiveness of Hempcrete as an Infill Insulation in the Prairies Compared to a Standard Building Based on Power Consumption. Master's Thesis, University of Manitoba, Winnipeg, MB, Canada, 2014.
33. Collet, F.; Pretot, S. Thermal conductivity of hemp concretes: Variation with formulation, density and water content. *Constr. Build. Mater.* **2014**, *65*, 612–619. [CrossRef]
34. *ASTM C39/C39M, Standard Test Method for Compressive Strength of Cylindrical Concrete Specimens*; ASTM International: West Conshohocken, PA, USA, 2012.
35. *ASTM C496/C496M-17, Standard Test Method for Splitting Tensile Strength of Cylindrical Concrete Specimens*; ASTM International: West Conshohocken, PA, USA, 2017.
36. *ASTM D4832-16e1, Standard Test Method for Preparation and Testing of Controlled Low Strength Material (CLSM) Test Cylinders*; ASTM International: West Conshohocken, PA, USA, 2016.
37. *ASTM C469/C469M-14, Standard Test Method for Static Modulus of Elasticity and Poisson's Ratio of Concrete in Compression*; ASTM International: West Conshohocken, PA, USA, 2014.
38. *ASTM C518–17, Standard Test Method for Steady-State Thermal Transmission Properties by Means of the Heat Flow Meter Apparatus*; ASTM International: West Conshohocken, PA, USA, 2017.
39. *ASTM C1784-14, Standard Test Method for Using a Heat Flow Meter Apparatus for Measuring Thermal Storage Properties of Phase Change Materials and Products*; ASTM International: West Conshohocken, PA, USA, 2014.
40. NIST SRM 1450. Energy and Environment Division, NIST. Available online: https://www.nist.gov/energy-andenvironment-division/nist-srm-1450 (accessed on 10 November 2018).
41. Tleoubaev, A.; Brzezinski, A. Thermal Diffusivity and Volumetric Specific Heat Measurements Using Heat Flow Meter Instruments. In Proceedings of the Thermal Conductivity 29/Thermal Expansion 17 Conference, Birmingham, Alabama, 24–27 June 2007.
42. Rode, C.; Peuhkuri, R.H.; Hansen, K.K.; Time, B.; Svennberg, K.; Arfvidsson, J.; Ojanen, T. NORDTEST Project on Moisture Buffer Value of Materials. In Proceedings of the AIVC 26th Conference: Ventilation in Relation to the Energy Performance of Buildings, Brussels, Belgium, 21–23 September 2005; pp. 47–52.
43. Ge, H.; Yang, X.; Fazio, P.; Rao, J. Influence of moisture load profiles on moisture buffering potential and moisture residuals of three groups of hygroscopic materials. *Build. Environ.* **2014**, *81*, 162–171. [CrossRef]

44. Li, Y.; Fazio, P.; Rao, J. An investigation of moisture buffering performance of wood paneling at room level and its buffering effect on a test room. *Build. Environ.* **2012**, *47*, 205–216. [CrossRef]
45. Eires, R.; Nunes, J.P.; Fangueiro, R.; Jalali, S.; Camões, A. New Eco-Friendly Hybrid Composite Materials for Civil Construction. In Proceedings of the European Conference on Composite Materials Biarritz, Biarritz, France, 29 August–1 September 2006.
46. Tran Le, A.D.; Maalouf, C.; Mai, T.H.; Wurtz, E.; Collet, F. Transient hygrothermal behaviour of a hemp concrete building envelope. *Energy Build.* **2010**, *42*, 1797–1806. [CrossRef]
47. Evrard, A.; Herde, A.D. Bioclimatic envelopes made of lime and hemp concrete. In Proceedings of the CISBAT2005 Conference—Renewables in a Changing Climate—Innovation in Building Envelopes and Environmental Systems at Lausanne, Lausanne, Switzerland, 27–28 September 2005.
48. Nguyen, T.T.; Picandet, V.; Carre, P.; Lecompte, T.; Amziane, S.; Baley, C. Effect of compaction on mechanical and thermal properties of hemp concrete. *Eur. J. Environ. Civ. Eng.* **2010**, *14*, 545–560. [CrossRef]
49. Murphy, F.; Pavía, S.; Walker, R. An assessment of some physical properties of hemp-lime concrete. In Proceedings of the BCRI Bridge Infrastructure Concrete Research Ireland, Cork, Ireland, 2 September 2010; pp. 431–438.

3

Machine Learning Techniques in Concrete Mix Design

Patryk Ziolkowski * and Maciej Niedostatkiewicz

Faculty of Civil and Environmental Engineering, Gdansk University of Technology, Gabriela Narutowicza 11/12, 80-233 Gdansk, Poland; mniedost@pg.edu.pl

* Correspondence: patziolk@pg.edu.pl

Abstract: Concrete mix design is a complex and multistage process in which we try to find the best composition of ingredients to create good performing concrete. In contemporary literature, as well as in state-of-the-art corporate practice, there are some methods of concrete mix design, from which the most popular are methods derived from The Three Equation Method. One of the most important features of concrete is compressive strength, which determines the concrete class. Predictable compressive strength of concrete is essential for concrete structure utilisation and is the main feature of its safety and durability. Recently, machine learning is gaining significant attention and future predictions for this technology are even more promising. Data mining on large sets of data attracts attention since machine learning algorithms have achieved a level in which they can recognise patterns which are difficult to recognise by human cognitive skills. In our paper, we would like to utilise state-of-the-art achievements in machine learning techniques for concrete mix design. In our research, we prepared an extensive database of concrete recipes with the according destructive laboratory tests, which we used to feed the selected optimal architecture of an artificial neural network. We have translated the architecture of the artificial neural network into a mathematical equation that can be used in practical applications.

Keywords: concrete; concrete mix design; concrete strength prediction; data mining; machine learning

1. Introduction

Concrete mix design is an essential and abstruse topic, which requires extensive knowledge of many expert issues. Obtaining concrete with appropriate strength, and other utility parameters, allows for the reliable use of the structure. The process of concrete hardening and hydration are irreversible. Therefore, any errors in the design of the concrete mix are incredibly costly for the investor, both at the construction stage and in the subsequent exploitation of the structure due to reduced durability. By definition, concrete mix is a mixture of cement, water, and coarse and fine aggregate, mostly enriched by additives and admixtures to improve some parameters, such as concrete strength, density, durability, or workability. The final product is in which the concrete mix is transformed into concrete. The concrete hardening is started by the cement hydration process, which is an exothermic chemical reaction between cement and water. Hydrated cement forms a tobermorite gel, hydroxide, and some secondary compounds that help with bonding between the fine and coarse aggregate. In the course of the hydration process, the hydration products gradually deposit on the original cement grains and fill the space occupied by water. The hydration process stops when there is no unreacted cement or the water molecules are retracted. The hardening of concrete continues further and ends around the twenty-eighth day, when the concrete reaches full compressive strength [1–3]. The necessary amount of water for full hydration of cement varies from 20% to 25% of its mass, without taking into account the water trapped in the pores [4,5]. According to Power's model, the water required to hydrate cement entirely is 42% by weight [6,7]. The issue of concrete mix design boils down to selecting the correct proportions of cement, fine and coarse aggregate, and water to produce concrete that has the specified properties. Progress in the design of concrete mixes is at a moderate level. The most popular method used to estimate the amount of main ingredients needed, in a little changed form, has been used for decades and consists of estimating the strength of concrete mortar for bending [8–10]. These methods have many disadvantages and are labour-intensive to use. We want to introduce a way to design a concrete mix based on a mathematical equation developed by the machine learning algorithm. In the following paper, we describe the adopted neural network architecture, which further will be feed by a large dataset of concrete mix recipes. Finally, we present a mathematical formula that allows estimating the compressive strength of concrete. The developed formula will evaluate the compressive strength of concrete based on four input variables, the amount of cement, water, fine, and coarse aggregate. The presented equation does not reflect the behaviour of the concrete perfectly and has boundary conditions. However, it is a step on the way to the introduction of machine learning techniques for concrete mix design. In its present form, it can be a tool for a rough estimation of the concrete class. In our further endeavours we would like to emphasise concrete mix design in terms of durability and service life estimation, then the influence of concrete admixtures, such as superplasticizers, would be essential [11–16].

2. The Contemporary Concrete Mix Design and Machine Learning Techniques

2.1. Concrete Mix Design in European Corporate Practice

The primary goal of concrete mix design is to estimate the proper quantitative composition and proportion of concrete mixture components. We should use a composition which allows us to achieve the best possible concrete performance. Concrete performance is characterised by several features, from which the most significant are compressive strength and durability. Both concrete strength and durability should play an essential role in the concrete mix design. The issue of durability is essential in the case of an aggressive environment [17–22]. Based on industry experience, we found out that, in European corporate engineering practice, there are a few most used methods for designing a concrete mix. These methods are the Bukowski method, the Eyman and Klaus method, and the Paszkowski method. All the solutions mentioned above are derived from the so-called "Three Equations Method", or Bolomey Method, which is a mixed experimental-analytical procedure [10,23,24]. It means that collected laboratory data should confirm that mathematical approach. We calculate a volume of required components by analytical measures and validate the results by destructive laboratory testing. In this method, we use a fundamental equation of strength, consistency, and tightness to determine the three searched values, as follows, the amount of aggregate, cement, and water, expressed in kilograms per cubic meter.

The first equation is the compressive strength equation or Bolomey formula (Equation (1)), which expresses the experimentally determined dependence of the compressive strength of hardened concrete on the grade of cement used, the type of aggregate used, and the water-cement ratio characterising the cement paste [23,24]. In this method, the concrete grade is assumed as input data.

$$f_{cm} = A_{1,2}\left(\frac{C}{W} \pm 0,5\right) \text{[MPa]}, \tag{1}$$

where, f_{cm} is a medium compressive strength of concrete, expressed in kilograms. The value $A_{1,2}$ means coefficients, depending on the grade of cement and the type of aggregate; C is an amount of cement in 1 m^3 of concrete, expressed in kilograms; and W corresponds to the amount of water in 1 m^3 of concrete, expressed in kilograms. A second consistency Equation (2), is included in the water demand formula necessary to make a concrete mix with the required consistency.

$$W = C \cdot w_c + K \cdot w_k \ [\text{dm}^3], \tag{2}$$

where W is the amount of water in 1 m^3 of concrete, expressed in kilograms; C corresponds to the amount of cement in 1 m^3 of concrete, expressed in kilograms; K means the amount of aggregate in 1 m^3 of concrete, expressed in kilograms; w_c is the cement water demand index in dm^3 per kilogram; and w_k is the aggregate water demand index in dm^3 per kilogram. The water-tightness of concrete Equation (3) is included in the simple volume formula, which indicates that a watertight concrete mix is obtained if the sum of the volume of the individual components is equal to the volume of the concrete mix.

$$\frac{C}{\rho_c} + \frac{K}{\rho_k} + W = 1000\ [\mathrm{dm}^3], \tag{3}$$

where W is the amount of water in 1 m^3 of concrete, expressed in kilograms, C corresponds to the amount of cement in 1 m^3 of concrete, expressed in kilograms, K means the amount of aggregate in 1 m^3 of concrete, expressed in kilograms; ρ_c is the cement density in kilograms per dm^3; and ρ_k is the aggregate density in kilograms per dm^3.

The system of equations presented above, with three unknowns variables, allows for calculating the sought amounts of cement (C), aggregate (K), and water (W) in one cubic meter of concrete mix. The system is valid, assuming that there are no air bubbles in the concrete. Another method used in the construction industry is "the double coating method" [25]. The methods above are ones that are used to determine the quantitative composition of the concrete mix. However, the actual process of creating a concrete mix is much broader, including the following steps: The first step is to determine the data needed to design the mix, such as the purpose of the concrete use, the compressive strength of the concrete, and the consistency of the concrete mix. Next, the qualitative characteristics of the components should be determined, namely the type and class of cement and the type and granularity of the aggregates. Subsequent steps include an examination of the properties of the adopted ingredients; a check of their compliance with the standard requirements; determining the characteristics of the components that will be needed to determine the composition of the concrete mix; and a projection of the aggregate pile. The successive step is the actual adoption of the design method and a calculation per unit of volume. The final stage is to make a trial sample and examine both the concrete mix and the hardened concrete with design assumptions [26].

2.2. *Machine Learning Techniques*

2.2.1. The Overall Concept of Machine Learning

Machine learning is an area of knowledge which is developing dynamically in recent times. This technology is a part science dealing with artificial intelligence and refers to scientific fields such as computer science, robotics, and statistics [27]. In practice, machine learning aims to use various state-of-the-art achievements in computer science to build upon a system that will be able to learn from data sets and, thus, seek patterns and relationships between variables and groups of variables, which would be challenging to conduct with conventional methods. Learning, in this case, can be considered as the instantiation of the sophisticated algorithm. One of the most popular methods of machine learning is artificial neural networks (ANN).

2.2.2. Artificial Neural Networks (ANN)

ANN are clusters of neurons, which are also its basic unit. We can consider an artificial neuron as a specific signal converter. The behaviour of artificial neurons, in a sense, imitates the behaviour of neurons in the human brain [28]. A primary example of ANN consist of three layers, called as follows:

- The input layer;
- The hidden layer;
- The output layer.

The input layer consists of input variables and combines them with neurons from the hidden layer. On the contrary, the output layer contains the target data to be obtained by the hidden layer [27]. Therefore, the whole process of learning happens in the hidden layer, where connections between neurons are sought. Vast numbers of neurons can build a complex model, which would be unattainable with simple architecture and so unobvious that it would be difficult to create a purely empirical formula. An essential thing that neural networks do is a search for patterns, which is why examples best teach neural networks. To teach a neural network how to solve a given problem, one must enter the input data into it using the first layer and put data in the output layer as a given target to which the network is to strive. Moreover, the input data can be adjusted by assigning weight to them, which can potentially represent the importance of a given variable. The weight control mechanism is also part of the neural network and is called the "learning rule". One artificial neuron has miserable problem-solving capabilities. Many neurons can be combined into more hidden layers, where layers pass the results to one another, looking to reach the target value [29,30].

2.3. *Use of Machine Learning in Concrete Compressive Strength Prediction*

Designing a concrete mix consists of selecting components and their amount to achieve specific parameters of the concrete. One of the most significant parameters for concrete performance is the compressive strength of concrete, which defines the class of concrete. Other important parameters that contribute to good concrete performance are durability and even the manufacturing process itself. Poor durability may contribute to lowering the service quality of building in time. With a wrong manufacturing process, for example, poor concrete care can cause excessive cracks and reduce concrete tightness [31].

The issue of machine learning applications, more precisely ANN, to predict the strength of concrete is present in the scientific discourse and is continuously evolving, making this topic very progressive.

The topic was first discussed in 1998 by Yeh et al. [32], which used linear regression and ANN to try to predict the strength of high-performance concrete using seven input variables. In the research, Yeh et al. used an extensive database, but in our opinion, they did not take into account the specificity of concrete and used samples in their database that were still in the maturing phase, even three days old, which, in our opinion, could seriously misrepresent the results.

Subsequently, the topic was taken up by Seung-Chang Lee [33], which used a modular network structure consisting of five ANN. In the presented solution, the author used the weighting technique of input neurons to improve the accuracy of predictions. To estimate the number of input neurons, he used the parameter condensation technique. The author concludes that the methods he uses, namely condensation and weighting techniques, are efficient in looking for the optimal performance network.

Another interesting approach in this matter is to use a neural-expert system, which was suggested further by Gupta et al. [34] to predict the compressive strength of high-performance concrete. The neural expert system architecture, in theory, allows for constructing the database automatically by learning from example inferences. In general, this architecture assumes the use of a multi-layered neural network, which is consequently trained with generalised backpropagation for interval training patterns. However, this may allow for the learning of patterns with irrelevant inputs and outputs. What is more, in the study by Gupta et al. [34], the input variables have very different input metrics and instead of the amounts of concrete mix components, the input variables refer to such parameters as curing time. In our opinion, the selected input parameters have no unambiguous effect on the strength of concrete and can imply false results. The topic of neural-expert systems was also undertaken by Dac-Khuong Bui et al. [35], which focused entirely on the practical application of the mentioned expert approach.

Fangming Deng et al. [36] practised deep learning architecture to predict the compressive strength of concrete. In this study their used recycled concrete with five input variables as follows, water-cement ratio, recycled coarse aggregate replacement ratio, recycled fine aggregate replacement ratio, and fly ash replacement. They used so-called deep features that refer to ratio rather than the individual amount of concrete mix components. We used a similar approach in our study by introducing feature scaling. To find out the proper prediction model they used a Softmax regression. In the results section of their paper, they state that the deep learning architecture they applied gives a higher efficiency, generalisation ability, and precision, in comparison with standard ANN. However, they do not present sufficient proofs to support their statements. Convolution networks are computationally expensive. This seems to be confirmed by a significantly lower number of samples (74 exactly) than in our study (741 records). However, such a small dataset might result in underfitting, which means that the model does not fit the data well enough to such an extent that it reduces the efficiency of the model. Moreover, Hosein Naderpour et al. [37] shows a comparable degree of precision between ANN and Deep Neural Networks (DNN).

3. Materials and Methods

3.1. Essentials

In our study, we want to implement machine learning for concrete mix design. Based on a large number of tested concrete mix recipes, we would like to build an ANN which will be able to estimate the compressive strength of the concrete mix. The ANN estimates the strength of the concrete based on the amount of the four main components of a concrete mix, more precisely cement, fine and coarse aggregate, and water. We translated the constructed ANN into the source code and simplified to one equation, defining the twenty-eight-day strength of concrete as a function of the four parameters. The equation can be used for concrete compressive strength estimation and can serve as a tool for a concrete mix recipe check. The practical application of this method in the concrete mix design process, required to adopt the approach, is presented in Figure 1.

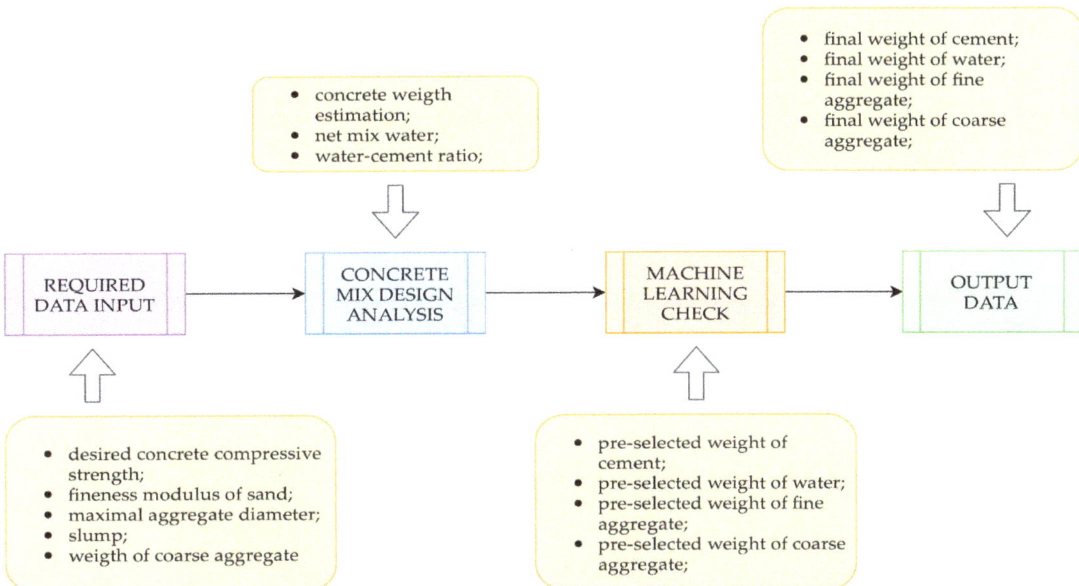

Figure 1. Block diagram of the practical application of machine learning in the concrete mix design.

It seems reasonable to set a boundary condition for this method. However, the ANN was trained on a limited number of samples so it may be difficult to predict how it will behave for amounts of material higher than in the considered ranges. It is essential to strictly control the water-cement ratio since the proper proportion is necessary for the full hydration of the cement. We have not analysed the influence of plasticisers.

3.2. The Database of Concrete Mix Recipes

In our research, we intend to teach the neural network the relationships between the number of individual components in a concrete mix and the compressive strength of concrete with a large number of examples. Thanks to this, the potential user of our solution will be able to design the right composition of ingredients and try to predict the compressive strength of concrete. To handle that task, we need a wide-ranging database containing a variety of concrete recipes with according data of their destructive laboratory tests. We prepared the database, which has many records from numerous sources, including literature, companies, institutions, and laboratories. The concrete mix recipes that we used for the analysis were designed for concrete structures of different dimensions, functions, and destinations. Therefore, there may be some differences between them, the sources of which we will not be able to predict. What is more, many of the recipes we have, besides the essential ingredients, have additives that have different functions. The most popular concrete additives are binding retardants, plasticisers, and workability boosters. The samples tested are standardised concrete cylinders with a diameter of 15 cm. Samples that were not cylindrical were converted into cylindrical ones according to valid norms [38]. The size of the aggregate in the dataset did not exceed 20 mm. The samples were made from normal Portland cement. We have carried out extensive consultations with experts and have adopted four components that have a significant impact on the compressive strength of concrete. The adopted input parameters are presented in Table 1.

Table 1. The parameters adopted in the dataset.

Parameter	Compressive Strength after 28 days	Cement	Water	Sand 0–2 mm	Aggregate above 2 mm
Codename Type	cs_28 target	cement input	water input	fine_aggregate input	coarse_aggregate input
Description	The compressive strength of concrete 28 days after hydration. Considered as full strength.	The weight of cement added to the mixture	The weight of water added to the mixture	The weight of sand added to the mixture	The weight of aggregate, which have more than 2 mm, added to the mixture

We divided the parameters from Table 1 into two groups, the inputs and target, which characterise input and output variables, respectively. After initiating the cement hydration process, concrete strength grows, progressively over time, to full strength. In our deliberations, we adopted a general assumption that concrete achieves its designed compressive strength in twenty-eight days. Prior to the twenty-eighth day, the concrete has a partial strength, but it cannot be considered full strength. We assumed in our research that the concrete reached its full strength because a mixture is designed for such strength. We removed all records for concrete of lower ages from the base. Many factors have an indirect effect on the obtained concrete strength, which has not been included in the analysis, such as the curing process. We assumed that quality control was sufficient to produce full strength concrete. The minimum, maximum, and average values for every input variable are presented in Table 2.

Table 2. Ranges of input features of database input features.

Input Features	Minimum (kg/m^3)	Maximum (kg/m^3)	Average (kg/m^3)
Cement	86.00	540.00	278.00
Water	121.80	247.00	182.42
Fine aggregate (sand 0–2 mm)	372.00	1329.00	768.55
Coarse aggregate (aggregate above 2 mm)	597.00	1490.00	969.08

3.3. *Results and Discussion*

To carry out the simulation, we divided our set into three subsets, as follows: The training dataset, the selection dataset, and the testing dataset. The training dataset is used to create a neural network, the selection dataset is used to adjust parameters of the neural network, and the testing dataset is used to evaluate the efficiency of the network. The database has 741 records, but we had to exclude 79 records (10.7%) from the analysis as univariate outliners. The training dataset has 395 records (53.3%), the selection dataset has 133 records (17.9%), and the testing dataset has 134 records (18.1%). The scatter plots of a target variable versus the input variables are presented in Figure 2.

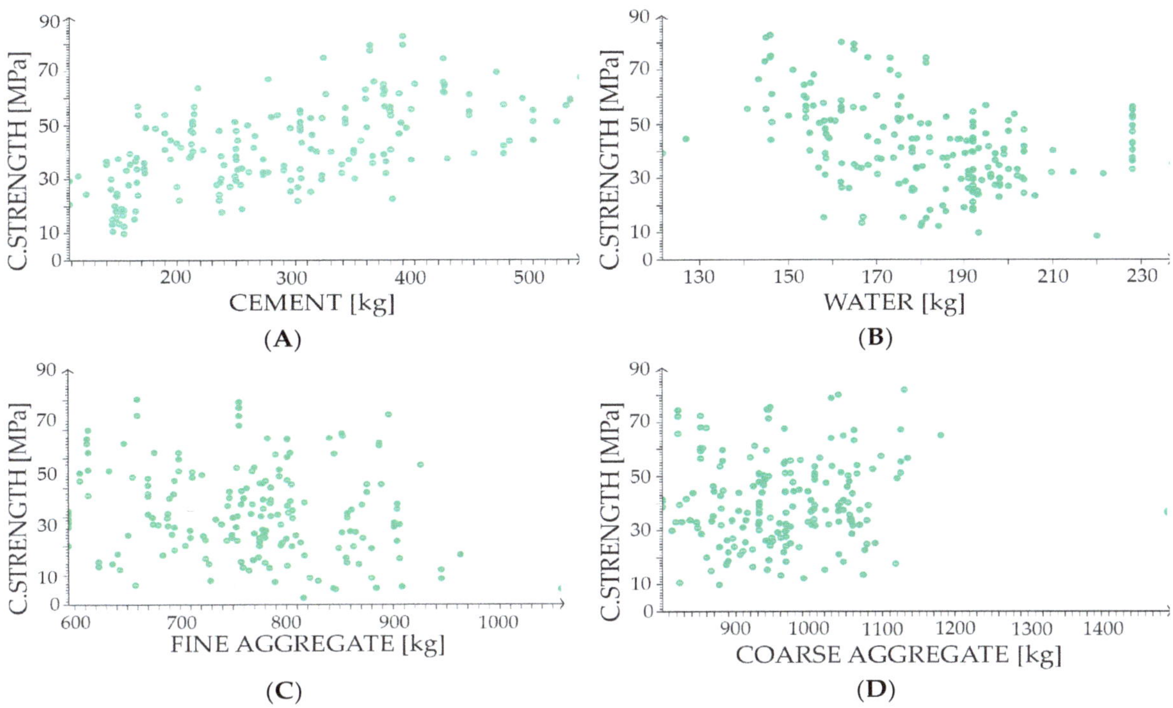

Figure 2. The scatter plots target versus input variables. The vertical axis is the full compressive strength of concrete expressed in megapascals. The horizontal axis is the amount of material in kilograms for cement and aggregate as well as in litres for water. (**A**) Cement; (**B**) water; (**C**) Fine aggregate (sand 0–2 mm); and (**D**) coarse aggregate (aggregate above 2 mm). The legend is in the left bottom corner.

Our neural network consists of four input variables, which refers to four principal components and generates one target output. The complexity of the model is expressed by the number of hidden layers, which in our case is three. The initial architecture that we prepared is shown in Figure 3, which consists of principal components (blue), perceptron neurons (red), and, because we used feature scaling, there are scaling and unscaling layers. The scaling and unscaling neurons are green and yellow, respectively.

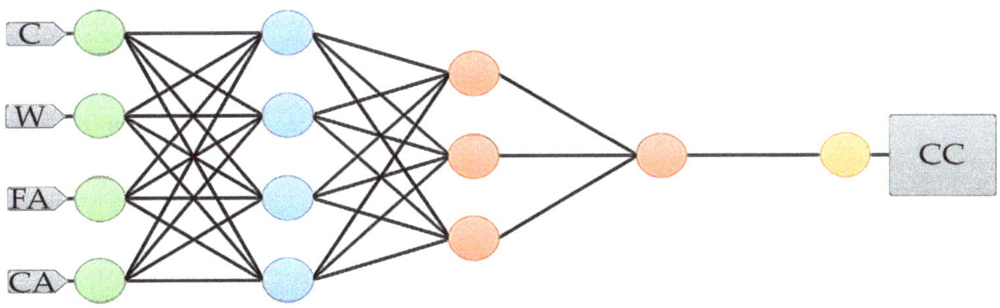

Figure 3. The initially used architecture of the ANN. The figure shows the network architecture, which includes the following parts, principal components (blue), perceptron neurons (red), and, because we use feature scaling, there are scaling and unscaling layers. The scaling and unscaling neurons are green and yellow, respectively. Abbreviations: C, cement; W, water; FA, fine aggregate; CA, coarse aggregate; CC, full compressive strength of concrete.

We want to point out that some input variables (cement, water, fine_aggregate, coarse_aggregate) correspond with some input neurons and target variable (cs_28) is associated with the output neuron. To obtain a proper training rate, we used the Broyden-Fletcher-Goldfarb-Shanno algorithm [39–44]. Then, to designate the quasi-Newton training direction step, we utilised the Brent method [45–48]. For the analysis, we calculated the linear correlation and determined a correlation matrix.

We have assessed the impact of individual variables on the final result, which is presented in Figure 4. We eliminated training input selectively and inspected the output results. An input contribution value 1.0 or lower than one denotes that the variable has less contribution to the results. Successively, a value higher than 1.0 means a more significant contribution. Our analysis indicates that the biggest contribution to the results have cement, which is in line with our assumptions that the water-cement ratio has the most significant impact on concrete strength. Literature findings also confirm that the cement content and type have a high influence on the compressive strength [49]. There are also other issues, including curing conditions and added admixture impact, that influence the compressive strength and concrete durability, especially an environments with a high risk of carbonation [50]. The detailed nature and the shape of the aggregate influence the workability and durability of concrete. The shape and texture of the aggregate affect the properties of fresh concrete more than hardened concrete [51]. Additionally, the grading or size distribution of aggregate is an important characteristic because it determines the paste requirement for workable concrete [52]. However, in our procedure, we did not make an exact distinction between the nature and shape of the aggregate. We only diversified the coarse and fine aggregates and sacrificed it for the sake of having larger data sample pools in these two categories. We also have not analysed the impact of environmental aggression and admixtures.

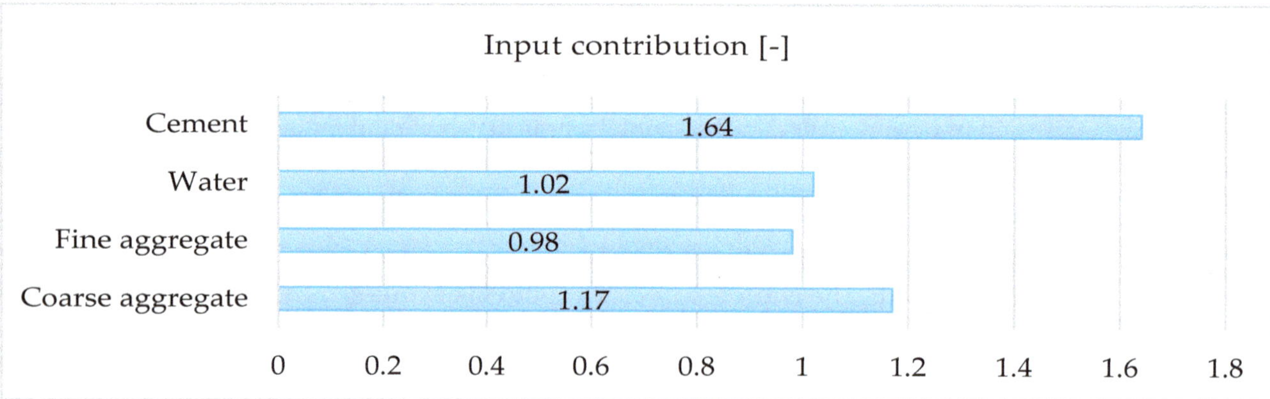

Figure 4. Input contribution.

We performed input selection by the growing inputs algorithm [53–56]. We found the optimal number of neurons by the order selection algorithm [57,58]. We carried out the output selection by the incremental order algorithm [59–61]. The loss history for the subsets used is presented in Figure 5.

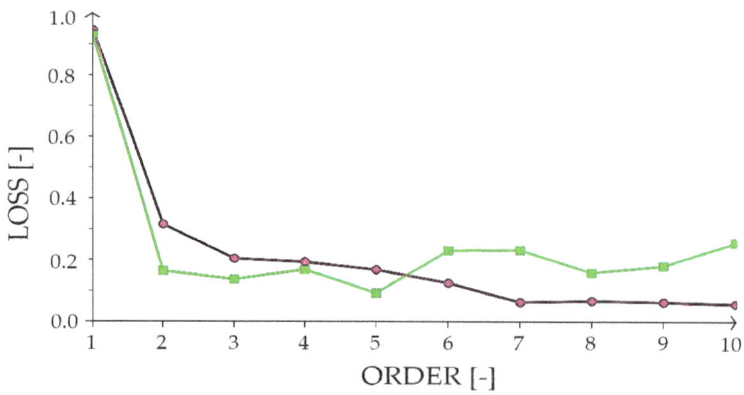

Figure 5. Incremental order algorithm performance. The chart presents a loss history, where the purple line is the training loss and the green one is the selection loss. The vertical axis is a loss and the horizontal axis is an order.

In Figure 6 we present a final architecture of the ANN, which consists of principal components (blue), perceptron neurons (red), and, because we used feature scaling there are scaling and unscaling layers. The scaling and unscaling neurons are green and yellow, respectively. We used a deep architecture with features scaling. Therefore it contains scaling and unscaling layers. Our final model, which is the most optimal for performing the given task, has four inputs, one output, and three hidden layers.

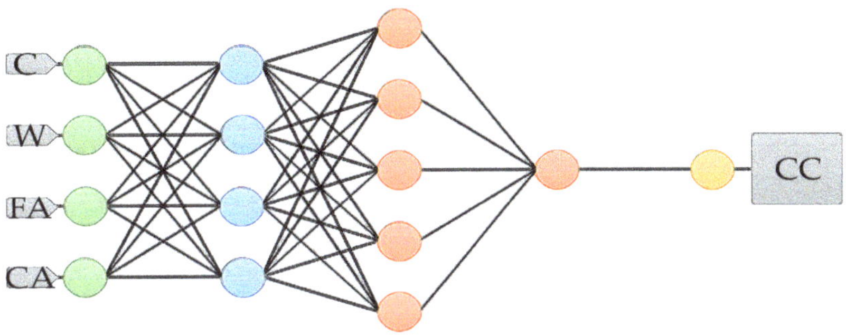

Figure 6. Finally used the architecture of ANN. The figure shows network architecture that includes following parts, principal components (blue), perceptron neurons (red), and, because we use feature scaling, there are scaling and unscaling layer. The scaling and unscaling neurons are green and yellow, respectively. Abbreviations: [C]—cement; [W]—water; [FA]—fine aggregate; [CA]—coarse aggregate; [CC]—full compressive strength of concrete.

In our study, we created an ANN which can be used for concrete mix design. The network targets the compressive strength of concrete with the four following input variables, cement, water, fine, and coarse aggregate. We can express our ANN by a mathematical Equation (4). The Equation (4) refers to the 28 day strength of concrete, which, as we mentioned, can be considered as full strength.

$$f_c^{full\ cs} = 0.5 \cdot \left(f_{ax}^{full\ cs} + 1.0\right) \cdot (62.62184 + 240.13674) - 240.13674\ [\text{MPa}], \tag{4}$$

Auxiliary mathematical formulas are presented in Equations (5)–(14). The variables C, W, FA, CA used in Equations (15)–(18) mean cement, water, fine aggregate, and coarse aggregate, respectively. Units are expressed in kilograms per cubic meter.

$$f_{ax}^{full\ cs} = tanh(0.246033 - 0.959961 \cdot ax_{11} + 0.816467 \cdot ax_{12} + 0.526611 \cdot ax_{13} + 0.73407 \cdot ax_{14} - 0.270081 \cdot ax_{15}) \tag{5}$$

$$ax_{11} = tanh(0.15979 - 0.00384521 \cdot ax_{21} + 0.837402 \cdot ax_{22} - 0.148804 \cdot ax_{23} - 0.569336 \cdot ax_{24}) \tag{6}$$

$$ax_{12} = tanh(0.1297 + 0.820862 \cdot ax_{21} + 0.0808105 \cdot ax_{22} + 0.206116 \cdot ax_{23} - 0.601257 \cdot ax_{24}) \tag{7}$$

$$ax_{13} = tanh(0.262573 - 0.0964355 \cdot ax_{21} + 0.610413 \cdot ax_{22} - 0.380981 \cdot ax_{23} + 0.150024 \cdot ax_{24}) \tag{8}$$

$$ax_{14} = tanh(0.0534668 + 0.3526 \cdot ax_{21} + 0.929932 \cdot ax_{22} - 0.734924 \cdot ax_{23} - 0.415405 \cdot ax_{24}) \tag{9}$$

$$ax_{15} = tanh(-0.927551 - 0.993103 \cdot ax_{21} - 0.202698 \cdot ax_{22} - 0.719788 \cdot ax_{23} - 0.637817 \cdot ax_{24}) \tag{10}$$

$$ax_{21} = 0.984539 \cdot C_{ax} - 0.144211 \cdot W_{ax} - 0.0968967 \cdot FA_{ax} + 0.0222878 \cdot CA_{ax} \tag{11}$$

$$ax_{22} = -0.117904 \cdot C_{ax} - 0.946362 \cdot W_{ax} + 0.249228 \cdot FA_{ax} + 0.168472 \cdot CA_{ax} \tag{12}$$

$$ax_{23} = 0.0747692 \cdot C_{ax} - 0.00246874 \cdot W_{ax} + 0.576173 \cdot FA_{ax} - 0.813897 \cdot CA_{ax} \tag{13}$$

$$ax_{24} = 0.105786 \cdot C_{ax} + 0.28913 \cdot W_{ax} + 0.772348 \cdot FA_{ax} + 0.555601 \cdot CA_{ax} \tag{14}$$

$$C_{ax} = \frac{2 \cdot (C - 86)}{454} - 1 \tag{15}$$

$$W_{ax} = \frac{2\cdot(W - 121.8)}{125.5} - 1 \tag{16}$$

$$FA_{ax} = \frac{2\cdot(FA - 372)}{957} - 1 \tag{17}$$

$$CA_{ax} = \frac{2\cdot(CA - 597)}{893} - 1 \tag{18}$$

We simplified the mathematical formula translated from the ANN source code and presented it in the form of Equation (19), $f_c^{full\ cs}$ with four variables C, W, FA, CA, which represent cement, water, fine aggregate, and coarse aggregate, respectively.

$$\begin{gathered} f_c^{full\ cs} = 151.37929\cdot tanh(0.24603 - 0.95996\cdot tanh(-0.00077\cdot C - 0.01524\cdot W - \\ 0.00066\cdot FA - 0.00012\cdot CA + 3.90060) + 0.816467\cdot tanh(0.00331\cdot C - 0.00588\cdot \\ W - 0.00085\cdot FA - 0.00105\cdot CA + 1.99855) + 0.526611\cdot tanh(-0.00079\cdot C - \\ 0.00828\cdot W + 0.00012\cdot FA + 0.00111\cdot CA + 0.78030) + 0.73407\cdot tanh(0.00061\cdot C - \\ 0.01672\cdot W - 0.00114\cdot FA + 0.00119\cdot CA + 2.67645) - 0.270081\cdot tanh(-0.00474\cdot \\ C + 0.00243\cdot W - 0.00180\cdot FA + 0.00039\cdot CA + 1.22858)\ ) - 88.757459\ [\text{MPa}], \end{gathered} \tag{19}$$

To illustrate how the equation works we presented the charts of the output variable and the single input variable, while the other input variable is fixed. The charts are shown in Figure 7. It should be noted that, as presented in Figure 7, the output charts do not correspond to the combined correlation of the variables, but only show a trend of a given variable concerning the target variable. It also should be noted that the parameters give a different contribution to the final results, as we have shown in Figure 4.

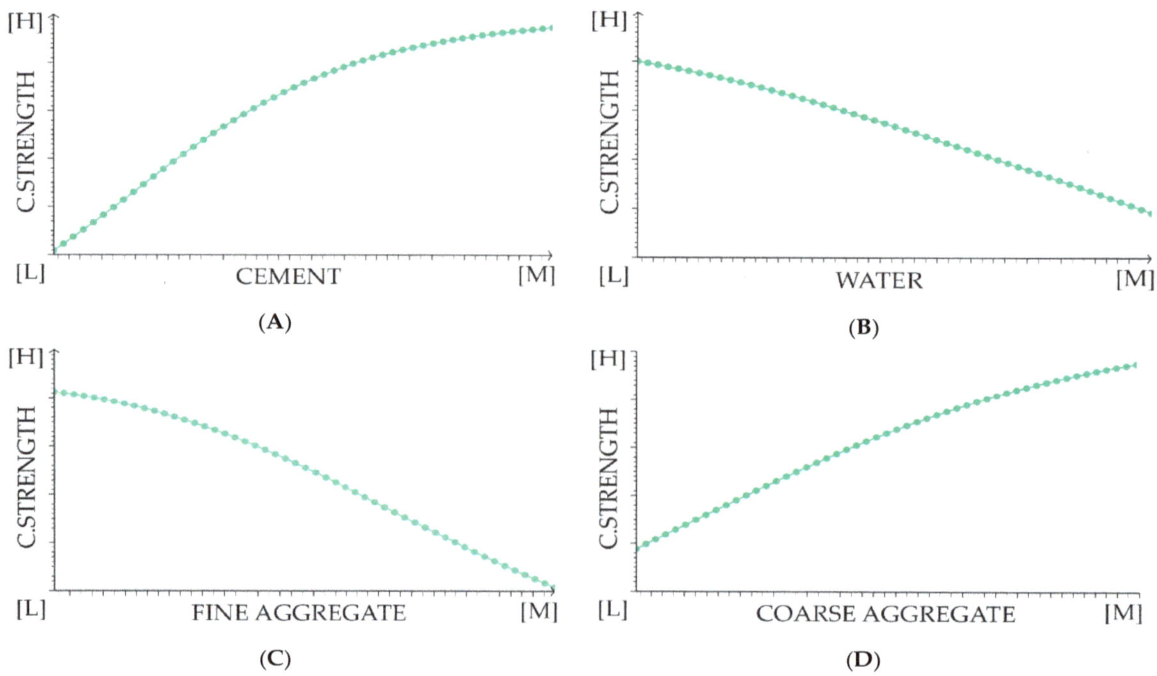

Figure 7. Output charts. The diagrams show variations in output variables for a single input variable, while the others are fixed. The vertical axis is the full compressive strength of concrete. The horizontal axis is the amount of material. (**A**) Cement; (**B**) water; (**C**) fine aggregate (sand 0–2 mm); (**D**) coarse aggregate (aggregate above 2 mm); [H], [M]—Higher/More; [L]—Lower/Less.

We compared the presented Formula (19) with a standard concrete mix design approach, based on the Bolomey design method. The comparison was prepared for 1 m^3 of concrete designed for the concrete slab, with direct pouring, plastic slump, no special desired finishing, no special ambient conditions when casting, and negligible environmental aggression. To design a concrete mix, we used the following materials: Portland cement; network water; natural sand; limestone gravel 4/10 mm; and limestone gravel 10/20 mm. The tested recipes are presented in Table 3. The gradings and fitting curves for the designed recipes are shown in Figure 8. The comparison is presented in Figure 9.

Table 3. Tested concrete mix recipes.

Designed CC of Concrete Mix	Cement (kg/m^3)	Water (L/m^3)	Natural Sand (kg/m^3)	Limestone Gravel 4/10 (kg/m^3)	Limestone Gravel 10/20 (kg/m^3)
10	190.42	61.89	1089.78	531.24	676.11
20	249.52	81.09	991.52	541.32	661.89
30	308.62	94.81	901.23	554.15	651.75
40	367.71	101.25	821.33	570.82	646.97
50	426.81	107.97	745.14	583.21	641.99

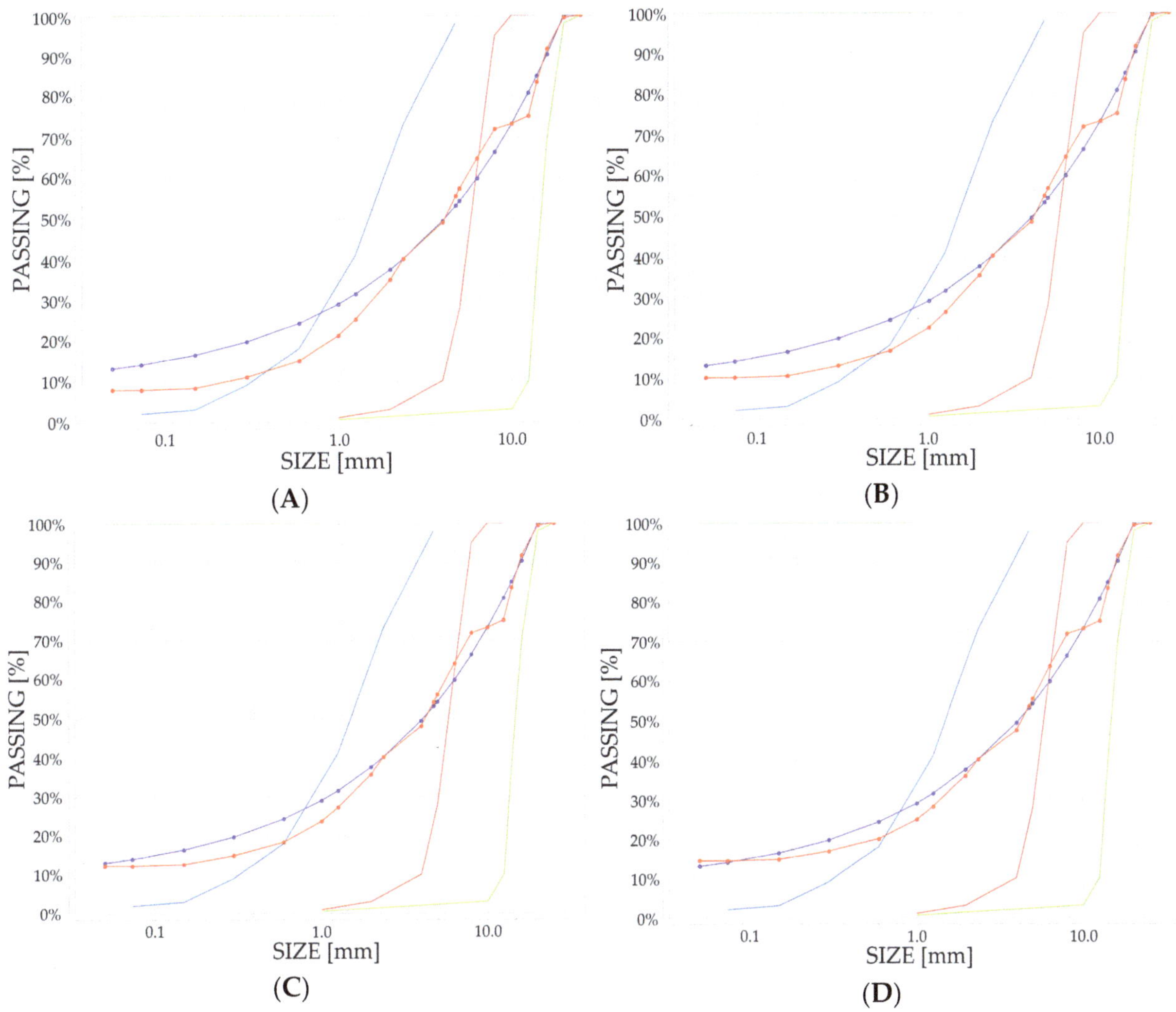

Figure 8. *Cont.*

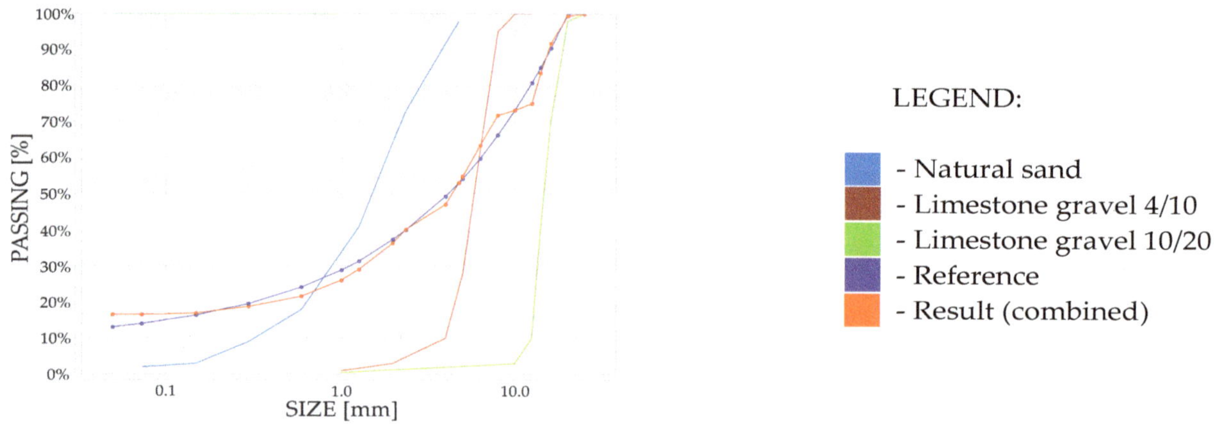

(E)

Figure 8. Gradings and fitting curves for designed concrete recipes: (**A**) 10 MPa; (**B**) 20 MPa; (**C**) 30 MPa; (**D**) 40 MPa; (**E**) 50 MPa; Legend in the left bottom corner.

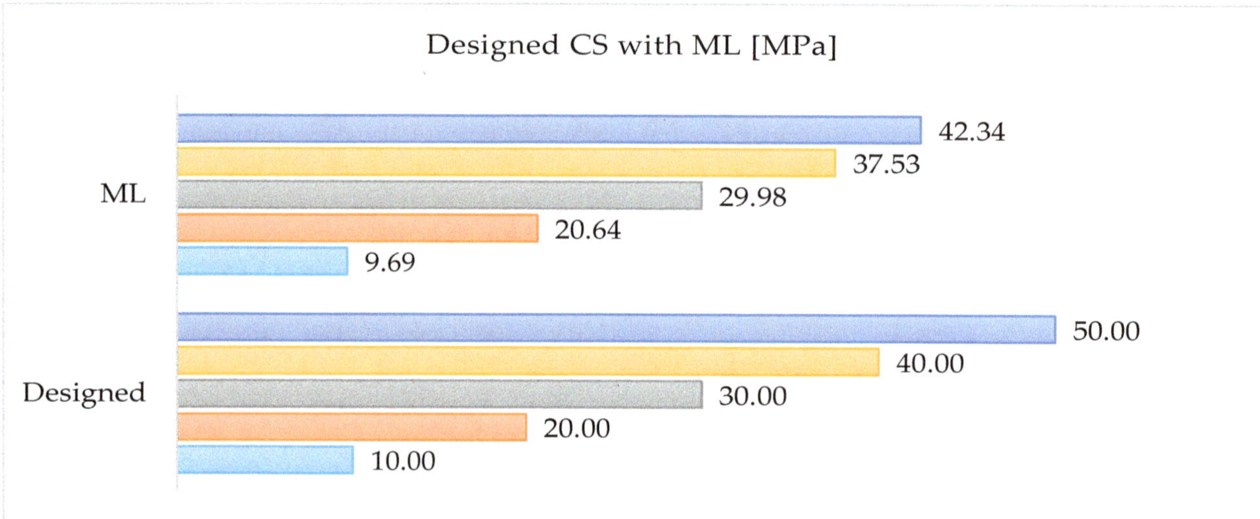

Figure 9. Comparison between designed Compressive Strength (CS) and calculated from Machine Learning (ML) formula.

We observed a low resilience of the ANN formula for recipes of high strength (50 MPa and above) concrete. It may be due to the smaller number of recipes on which we trained the ANN for these ranges. This behaviour of the ANN may be a sign of underfitting [27,28]. We must point out that the presented method is only an introduction to the broader use of machine learning in the design of concrete mixes and does not exhaust this topic. In particular, it does not take into account some crucial issues, such as durability and the technological process.

4. Conclusions

Our study is focusing on the application of machine learning in concrete mix design and building a practical tool that could be used in engineering practice. We designed the optimal ANN architecture and fed it with an extensive database of concrete mix recipes for the study. Every concrete mix recipe record has a corresponding laboratory destructive test. While building a neural network, the goal was to predict the compressive strength of the concrete resulting from a specific composition of concrete mix ingredients, or more precisely, what ratio of ingredients should be selected to obtain concrete with an adequate compressive strength. Our database has 741 records. We excluded 79 (10.7%) concrete samples from the dataset, as univariate outliners. The specificity of machine learning requires us to divide the database into three subsets, which we split up as follows: The training subset has 395 records (53.3%), the selection subset has 133 records (17.9%), and the testing subset has 134 records (18.1%). The initially adopted ANN model has four input variables, four principal components, four hidden neurons, and one target output. The suitable training rate and the step for the quasi-Newton training direction were calculated by the Broyden–Fletcher–Goldfarb–Shanno algorithm and the Brent method, respectively. Our input contribution analysis discloses that the most significant impact on the obtained results is the amount of cement that accurately points out the significance of the water-cement ratio to reach a higher concrete compressive strength. The finally adopted ANN model has four input variables, four principal components, six hidden neurons, and one target output. The pivotal point in making the machine learning techniques more applicable was a transformation of the ANN into an actual mathematical equation, which can be used in engineering practice. The initial conversion of the ANN into the mathematical formula had fifteen equations and required fourteen auxiliary variables. We simplified the expression into one general equation for the calculation of twenty-eight-day compressive strength of concrete. The equation we developed can be used as a rapid tool for concrete mix design check. The method allows checking the composition of four main concrete mix ingredients, cement, water, fine, and coarse aggregate, for achieving the desired concrete class. However, we would like to denote that the presented mathematical expression does not adequately reflect all the relationships between the components and have certain boundary conditions. We want to further develop the presented method. In the next step, to make this method more reliable, we would like to analyse the effect of admixtures and concrete durability.

Author Contributions: Conceptualisation, P.Z.; methodology, P.Z.; software, P.Z.; validation, P.Z.; formal analysis, P.Z.; investigation, P.Z.; resources, P.Z. and M.N.; data curation, P.Z.; writing—original draft preparation, P.Z.; writing—review and editing, P.Z.; visualisation, P.Z.; supervision, P.Z.; project administration, P.Z.; funding acquisition, M.N.

Acknowledgments: The author wishes to acknowledge every institution which provided data and tools to conduct this study.

References

1. Marchon, D.; Flatt, R.J. Mechanisms of cement hydration. In *Science and Technology of Concrete Admixtures*; Elsevier: Amsterdam, The Netherlands, 2016; pp. 129–145.
2. Scrivener, K.; Snellings, R.; Lothenbach, B. *A Practical Guide to Microstructural Analysis of Cementitious Materials*; CRC Press: Boca Raton, FL, USA, 2018; ISBN 1498738672.
3. Kurdowski, W. *Cement and Concrete Chemistry*; Springer Science & Business: Berlin, Germany, 2014; ISBN 9400779453.
4. Young, J.F.; Mindess, S.; Darwin, D. *Concrete*; Prentice Hall: Upper Saddle River, NJ, USA, 2002; ISBN 0130646326.
5. Hover, K.C. The influence of water on the performance of concrete. *Constr. Build. Mater.* **2011**, *25*, 3003–3013. [CrossRef]
6. Jensen, O.M.; Hansen, P.F. Water-entrained cement-based materials: I. Principles and theoretical background. *Cem. Concr. Res.* **2001**, *31*, 647–654. [CrossRef]
7. Jensen, O.M.; Hansen, P.F. Water-entrained cement-based materials: II. Experimental observations. *Cem. Concr. Res.* **2002**, *32*, 973–978. [CrossRef]
8. Jiménez Fernández, C.G.; Barra Bizinotto, M.; Valls del Barrio, S.; Aponte Hernández, D.F.; Vázquez Ramonich, E. Durability of recycled aggregate concrete designed with the Equivalent Mortar Volume (EMV) method: Validation under the Spanish context and its adaptation to Bolomey methodology. *Mater. Construcción* **2014**, *64*, e006-1–e006-12. [CrossRef]
9. Abdelgader, H.S. How to design concrete produced by a two-stage concreting method. *Cem. Concr. Res.* **1999**, *29*, 331–337. [CrossRef]

10. Abdelgader, H.S.; Suleiman, R.E.; El-Baden, A.S.; Fahema, A.H.; Angelescu, N. Concrete Mix Proportioning using Three Equations Method (Laboratory Study). In Proceedings of the UKIERI Concrete Congress Innovations in Concrete Construction, Jalandhar, India, 5–8 March 2013.
11. Cartuxo, F.; De Brito, J.; Evangelista, L.; Jiménez, R.J.; Ledesma, F.E. Increased Durability of Concrete Made with Fine Recycled Concrete Aggregates Using Superplasticizers. *Materials* **2016**, *9*, 98. [CrossRef] [PubMed]
12. Serralheiro, M.I.; de Brito, J.; Silva, A. Methodology for service life prediction of architectural concrete facades. *Constr. Build. Mater.* **2017**, *133*, 261–274. [CrossRef]
13. Plank, J.; Sakai, E.; Miao, C.W.; Yu, C.; Hong, J.X. Chemical admixtures—Chemistry, applications and their impact on concrete microstructure and durability. *Cem. Concr. Res.* **2015**, *78*, 81–99. [CrossRef]
14. Huang, H.; Qian, C.; Zhao, F.; Qu, J.; Guo, J.; Danzinger, M. Improvement on microstructure of concrete by polycarboxylate superplasticizer (PCE) and its influence on durability of concrete. *Constr. Build. Mater.* **2016**, *110*, 293–299. [CrossRef]
15. Arredondo-Rea, S.P.; Corral-Higuera, R.; Gómez-Soberón, J.M.; Gámez-García, D.C.; Bernal-Camacho, J.M.; Rosas-Casarez, C.A.; Ungsson-Nieblas, M.J. Durability parameters of reinforced recycled aggregate concrete: Case study. *Appl. Sci.* **2019**, *9*, 617. [CrossRef]
16. Andrade, C.; Gulikers, J.; Marie-Victoire, E. *Service Life and Durability of Reinforced Concrete Structures*; Springer: Berlin, Germany, 2018; Volume 17, ISBN 3319902369.
17. Aïtcin, P.C. The durability characteristics of high performance concrete: A review. *Cem. Concr. Compos.* **2003**, *25*, 409–420. [CrossRef]
18. Ahmad, S. Reinforcement corrosion in concrete structures, its monitoring and service life prediction—A review. *Cem. Concr. Compos.* **2003**, *25*, 459–471. [CrossRef]
19. Schabowicz, K.; Gorzelańczyk, T.; Szymków, M. Identification of the Degree of Degradation of Fibre-Cement Boards Exposed to Fire by Means of the Acoustic Emission Method and Artificial Neural Networks. *Materials* **2019**, *12*, 656. [CrossRef]
20. Godycki-Ćwirko, T.; Nagrodzka-Godycka, K.; Wojdak, R. Reinforced concrete thin-wall dome after eighty years of operation in a marine climate environment. *Struct. Concr.* **2016**, *17*, 710–717. [CrossRef]
21. Schabowicz, K.; Gorzelańczyk, T.; Szymków, M. Identification of the degree of fibre-cement boards degradation under the influence of high temperature. *Autom. Constr.* **2019**, *101*, 190–198. [CrossRef]
22. Argiz, C.; Moragues, A.; Menéndez, E. Use of ground coal bottom ash as cement constituent in concretes exposed to chloride environments. *J. Clean. Prod.* **2018**, *170*, 25–33. [CrossRef]
23. Rajamane, N.P.; Ambily, P.S.; Nataraja, M.C.; Das, L. Discussion: Modified Bolomey equation for strength of lightweight concretes containing fly ash aggregates. *Mag. Concr. Res.* **2014**, *66*, 1286–1288. [CrossRef]
24. Zhang, X.; Deng, S.; Deng, X.; Qin, Y. Experimental research on regression coefficients in recycled concrete Bolomey formula. *J. Cent. South Univ. Technol.* **2007**, *14*, 314–317. [CrossRef]
25. Abdelgader, H.S.; Saud, A.F.; Othman, A.M.; Fahema, A.H.; El-Baden, A.S. Concrete mix design using the double-coating method. *Betonw. Fert. Plant Precast Technol.* **2014**, *80*, 66–74.
26. Rumman, R.; Bose, B.; Emon, M.A.B.; Manzur, T.; Rahman, M.M. An experimental study: Strength prediction model and statistical analysis of concrete mix design. In Proceedings of the International Conference on Advances in Civil Engineering, Istanbul, Turkey, 21–23 September 2016; pp. 21–23.
27. Pereira, F.C.; Borysov, S.S. Machine Learning Fundamentals. In *Pereira Big Data and Transport Analytics*; Antoniou, C., Dimitriou, L., Pereira, F., Eds.; Elsevier: Amsterdam, The Netherlands, 2019; Chapter 2; pp. 9–29, ISBN 978-0-12-812970-8.
28. Shobha, G.; Rangaswamy, S. Machine Learning. In *Computational Analysis and Understanding of Natural Languages: Principles, Methods and Applications*; Gudivada, V.N., Rao, C.R., Eds.; Elsevier: Amsterdam, The Netherlands, 2018; Volume 38, Chapter 8; pp. 197–228. ISBN 0169-7161.
29. Yang, Z.R.; Yang, Z. Artificial Neural Networks. In *Comprehensive Biomedical Physics*; Araghinejad, S., Ed.; Springer: Dordrecht, The Netherlands, 2014; Volume 6, pp. 1–17, ISBN 9780444536327.
30. Silva, I.N.; Hernane Spatti, D.; Andrade Flauzino, R.; Liboni, L.H.B.; dos Reis Alves, S.F. *Artificial Neural Networks*; Springer: Berlin, Germany, 2017; ISBN 978-3-319-43161-1.
31. Brown, C.J. *Design of Reinforced Concrete Structures*; John Wiley & Sons: New York, NY, USA, 2003; Volume 4, ISBN 9780470279274.
32. Yeh, I.-C. Modeling of Strength of High-Performance Concrete Using Artificial Neural Networks. *Cem. Concr. Res.* **1998**, *28*, 1797–1808. [CrossRef]

33. Lee, S. Prediction of concrete strength using artificial neural networks. *Eng. Struct.* **2003**, *25*, 849–857. [CrossRef]
34. Gupta, R.; Kewalramani, M.A.; Goel, A. Prediction of Concrete Strength Using Neural-Expert System. *J. Mater. Civ. Eng.* **2006**, *18*, 462–466. [CrossRef]
35. Bui, D.K.; Nguyen, T.; Chou, J.S.; Nguyen-Xuan, H.; Ngo, T.D. A modified firefly algorithm-artificial neural network expert system for predicting compressive and tensile strength of high-performance concrete. *Constr. Build. Mater.* **2018**, *180*, 320–333. [CrossRef]
36. Deng, F.; He, Y.; Zhou, S.; Yu, Y.; Cheng, H.; Wu, X. Compressive strength prediction of recycled concrete based on deep learning. *Constr. Build. Mater.* **2018**, *175*, 562–569. [CrossRef]
37. Naderpour, H.; Rafiean, A.H.; Fakharian, P. Compressive strength prediction of environmentally friendly concrete using artificial neural networks. *J. Build. Eng.* **2018**, *16*, 213–219. [CrossRef]
38. Toniolo, G.; di Prisco, M. *Reinforced Concrete Design to Eurocode 2*; Springer: Berlin, Germany, 2017; ISBN 978-3-319-52032-2.
39. Abdi, F.; Shakeri, F. A globally convergent BFGS method for pseudo-monotone variational inequality problems. *Optim. Methods Softw.* **2017**, *34*, 25–36. [CrossRef]
40. Andrei, N. An adaptive scaled BFGS method for unconstrained optimization. *Numer. Algorithms* **2018**, *77*, 413–432. [CrossRef]
41. Battiti, R.; Masulli, F. BFGS optimization for faster and automated supervised learning. In *International Neural Network Conference*; Springer: Berlin, Germany, 1990; pp. 757–760.
42. Berahas, A.S.; Nocedal, J.; Takác, M. A multi-batch l-bfgs method for machine learning. In *Advances in Neural Information Processing Systems;* Curran Associates Inc.: New York, NY, USA, 2016; pp. 1055–1063.
43. Hagan, M.T.; Menhaj, M.B. Training feedforward networks with the Marquardt algorithm. *IEEE Trans. Neural Netw.* **1994**, *5*, 989–993. [CrossRef] [PubMed]
44. Li, D.-H.; Fukushima, M. A modified BFGS method and its global convergence in nonconvex minimization. *J. Comput. Appl. Math.* **2001**, *129*, 15–35. [CrossRef]
45. Grabowska, K.; Szczuko, P. Ship resistance prediction with Artificial Neural Networks. In Proceedings of the Signal Processing: Algorithms, Architectures, Arrangements, and Applications (SPA), Poznan, Poland, 23–25 September 2015; IEEE: Piscataway, NJ, USA, 2015; pp. 168–173.
46. Le, D.; Huang, W.; Johnson, E. Neural network modeling of monthly salinity variations in oyster reef in Apalachicola Bay in response to freshwater inflow and winds. *Neural Comput. Appl.* **2018**, 1–11. [CrossRef]
47. Luenberger, D.G. *Introduction to Linear and Nonlinear Programming*; Pearson Education, Inc.: London, UK, 1973.
48. Yildizel, S.A.; Arslan, Y. Flexural strength estimation of basalt fiber reinforced fly-ash added gypsum based composites. *J. Eng. Res. Appl. Sci.* **2018**, *7*, 829–834.
49. Argiz, C.; Menéndez, E.; Sanjuán, M.A. Effect of mixes made of coal bottom ash and fly ash on the mechanical strength and porosity of Portland cement. *Mater. Construcción* **2013**, *63*, 49–64. [CrossRef]
50. Sanjuán, M.Á.; Estévez, E.; Argiz, C.; del Barrio, D. Effect of curing time on granulated blast-furnace slag cement mortars carbonation. *Cem. Concr. Compos.* **2018**, *90*, 257–265. [CrossRef]
51. Kumar Poloju, K. *Properties of Concrete as Influenced by Shape and Texture of Fine Aggregate*; American Journal of Applied Scientific Research: New York, NY, USA, 2017; Volume 3.
52. Chinchillas-Chinchillas, M.J.; Corral-Higuera, R.; Gómez-Soberón, J.M.; Arredondo-Rea, S.P.; Jorge, L.; Acuña-Aguero, O.H.; Rosas-Casarez, C.A. *Influence of the Shape of the Natural Aggregates, Recycled and Silica Fume on the Mechanical Properties of Pervious Concrete*; The Institute of Research Engineers and Doctors (IRED): New York, NY, USA, 2014; Volume 4, ISBN 9781632480064.
53. Mehnert, A.; Jackway, P. An improved seeded region growing algorithm. *Pattern Recognit. Lett.* **1997**, *18*, 1065–1071. [CrossRef]
54. Huang, G.-B.; Saratchandran, P.; Sundararajan, N. An efficient sequential learning algorithm for growing and pruning RBF (GAP-RBF) networks. *IEEE Trans. Syst. Man Cybern. Part B* **2004**, *34*, 2284–2292. [CrossRef]
55. Arora, P.; Varshney, S. Analysis of k-means and k-medoids algorithm for big data. *Procedia Comput. Sci.* **2016**, *78*, 507–512. [CrossRef]
56. Dariane, A.B.; Azimi, S. Forecasting streamflow by combination of a genetic input selection algorithm and wavelet transforms using ANFIS models. *Hydrol. Sci. J.* **2016**, *61*, 585–600. [CrossRef]
57. Tabakhi, S.; Moradi, P.; Akhlaghian, F. An unsupervised feature selection algorithm based on ant colony optimization. *Eng. Appl. Artif. Intell.* **2014**, *32*, 112–123. [CrossRef]

58. Song, Y.; Liang, J.; Lu, J.; Zhao, X. An efficient instance selection algorithm for k nearest neighbor regression. *Neurocomputing* **2017**, *251*, 26–34. [CrossRef]
59. Yu, H.; Reiner, P.D.; Xie, T.; Bartczak, T.; Wilamowski, B.M. An incremental design of radial basis function networks. *IEEE Trans. Neural Netw. Learn. Syst.* **2014**, *25*, 1793–1803. [CrossRef]
60. Iqbal, S.Z.; Gull, H.; Ahmed, J. Incremental Sorting Algorithm. In Proceedings of the Second International Conference on Computer and Electrical Engineering, Dubai, United Arab Emirates, 28–30 December 2009; pp. 378–381.
61. Gurbuzbalaban, M.; Ozdaglar, A.; Parrilo, P.A. On the convergence rate of incremental aggregated gradient algorithms. *SIAM J. Optim.* **2017**, *27*, 1035–1048. [CrossRef]

4

Residual Magnetic Field Non-Destructive Testing of Gantry Cranes

Janusz Juraszek

Faculty of Materials, Civil and Environmental Engineering, University of Bielsko-Biala; 43-309 Bielsko-Biala, Poland; jjuraszek@ath.bielsko.pl

Abstract: Non-destructive tests of gantry cranes by means of the residual magnetic field (RMF) method were carried out for a duration of 7 years. Distributions of the residual magnetic field tangential and the normal components of their gradients were determined. A database of magnetograms was created. The results show that the gradients of tangential components can be used to identify and localize stress concentration zones in gantry crane beams. Special attention was given to the unsymmetrical distribution of the tangential component gradient on the surface of the crane beam No. 5 (which was the most loaded one). The anomaly was the effect of a slight torsional deflection of the beam as it was loaded. Numerical simulations with the finite element method (FEM) were used to explain this phenomenon. The displacement boundary conditions introduced into the simulations were established experimentally. Validation was carried out using the X-ray diffraction method, which confirmed the location of strain concentration zones (SCZs) identified by means of RMF testing.

Keywords: gantry crane; RMF technique; civil engineering

---

1. Introduction

The analysis of classical standard-related crane tests makes it possible to offer a general assessment of the gantry crane technical condition. This is usually a *post-factum* activity, i.e., an activity performed after the occurrence of a crack or damage. In the case of overhead cranes operated for over 30 years, a different, more thorough assessment is necessary. Standard tests are unable to determine whether stress concentration zones that cause cracks have appeared in the crane structure. This paper presents a new approach that enables a priori diagnostics, allowing the identification of potentially hazardous areas in advance.

The residual magnetic field (RMF) has been used as an inspection tool in transportation, power engineering and in the metal industry [1]. For a ferromagnetic structure, such as a crane, the RMF technique is expected to be one of the solutions that may enable early damage evaluation. It is proved in [2] that the size of the sample does not change the magnetization curve profile but it affects the RMF value. Four sets of Q345 steel samples with different widths and thicknesses were tested in the laboratory. The experimental results are explained by the theory of the interaction between dislocation and the domain wall, as well as the theory of the demagnetizing field. It is proved in [3] that the size of the sample does not change the magnetization curve profile but it affects the RMF value. Four sets of Q345 steel samples with different widths and thicknesses were tested in the laboratory. The experimental results are explained by the theory of the interaction between dislocation and the domain wall, as well as the theory of the demagnetizing field.

Significant static and fatigue experiments were performed with the use of the RMF technique [4–14]. There are numerous publications on the analysis of damage identified in flat specimens using the RMF technique. For example, damage identification in flat 12-mm-thick specimens subjected to tensile stresses was analyzed in [15] by Shui et al. The places in which the normal component Hn is zero indicate locations of damage to the specimen. The RMF also enables the identification of stress concentration zones. An interesting work on this subject is [16], which discussed the impact of the occurrence of stress concentration zones in three-point and four-point bending tests on the distribution of the Hp and Hn components, as well as the reduced stress according to the Huber–Misses–Hencky hypothesis. The RMF technique is used in the diagnostics of ferromagnetic structural elements. The RMF enables the detection of cracks, micro-cracks and closed cracks, which are difficult to detect using traditional methods. Knowing the previous magnetic image of the examined component, it is possible to detect stress concentration zones before actual cracks or micro-cracks appear. It is one of few methods that indicates, a priori, hazardous areas in the structure [17].

2. Materials and Methods

The analyzed gantry crane works in an open area and is used to transport coal from a storage site to a furnace. It was built in 1980. The structure consists of a truss bridge rolling on rails installed on two parallel 13-span assemblies consisting of crane beams. One span is less than 12 m long. The width of the crane is 32.01 m, and the trestle bridge reaches a height of 8.32 m (Figure 1). The cross-section of the crane beam is designed as an I-bar with a height of 824 mm. The web was made of sheet metal with a width of 800 mm and a thickness of 10 mm, while the feet were made of sheet metal with a width of 250 mm and a sheet thickness of 12 mm.

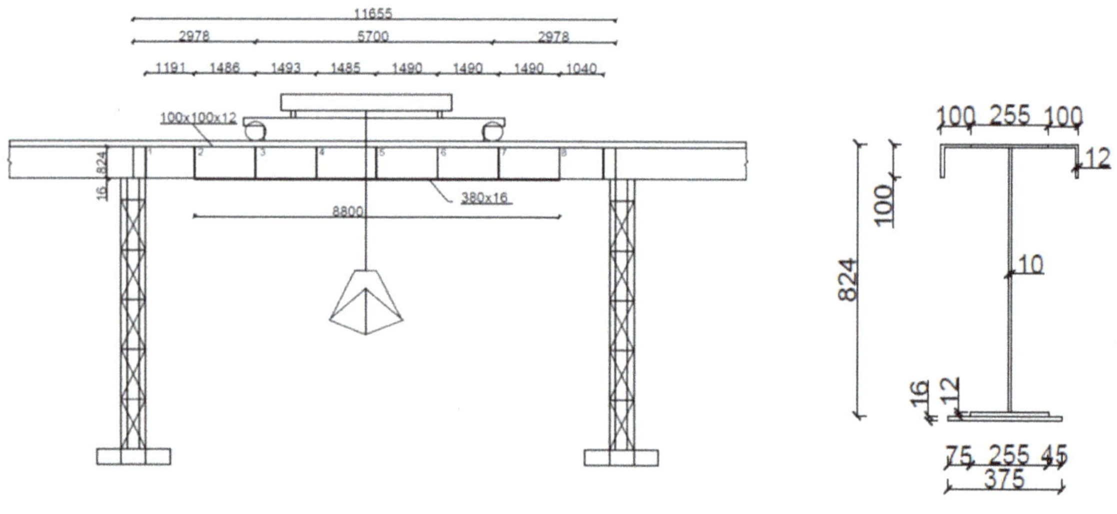

(a) (b)

Figure 1. Diagram: (**a**) crane beam, (**b**) cross-section of the beam.

The approximate weight of the crane carried by the beam and the maximum load weight that can be transported by the structure were taken into account. The load was divided into two spot forces acting in the place where the gantry crane wheel rests on the beam. The total operating load was Fo = 40 kN, while the crane structure weight load was 150 kN distributed over two road wheels, 75 kN each. At this load, the highest stress total was 40.5 MPa.

RMF Methods

In general, local irregularities in the material homogeneity resulting from stress, plastic strain or fatigue-induced changes in the structure generate corresponding, as well as local, changes in the degree of the material magnetization, and thus also in induction. In this situation, an induction component perpendicular to the surface of element B_p appears, which results in the creation of a vertical magnetic field component H_p outside the element, the distributions of which are recorded by the probes used in the RMF method. According to Equation (1):

$$H_p = \frac{1}{\mu_o} B_p \tag{1}$$

where:

B_p is the induction component perpendicular to the metal surface;

$\mu_0 = 2\pi * 10^{-7}$ is the air magnetic permeability.

Since B_p and thus H_p are proportional to the rate of changes in induction B along the tested element, it can be assumed that:

$$H_p \sim \frac{1}{\mu_o} \frac{\Delta B}{\Delta z} \tag{2}$$

where: Δz is the length at which induction B changes.

Juraszek [7] claims that when entering the material area with a permeability μ lower than in the rest of the material, e.g., in the area of a strong plastic strain, the induction flux is scattered and the RMF vertical component above the surface has a positive sign. Moving to an area of higher magnetic permeability (e.g., in the area of increased elastic stresses), the lines of the magnetic field inside the material are focused and the vertical component H_p has a negative sign. Away from areas of different magnetic properties, when the material is homogeneous, the residual magnetic field disappears. In this method, the RMF component H_p is measured by determining the following components:

- tangential component $H_{p(x)}$;
- normal component $H_{p(y)}$.

Based on [11,14], Equation (3) and Equation (4) respectively present simplified formulae describing the values of components $H_{p(x)}$ and $H_{p(y)}$. The values have been confirmed experimentally.

$$H_x = \int_0^b dH_x = \frac{1}{2\pi\mu_0} \int_0^b \rho(l) * \frac{(x-l)}{||r||^2} dl \tag{3}$$

$$H_y = \int_0^b dH_y = \frac{1}{2\pi\mu_0} \int_0^b \rho(l) * \frac{(y)}{||r||^2} dl \tag{4}$$

where:

$r = [x - l, y]$ is a vector, a variable related to the element surface;

ρ is a function defining the boundary condition presenting the distribution of the magnetic charge density on the specimen surface, which depends directly on the dislocation density and indirectly on the stress;

x, y are the coordinates of the point on the tested element surface;

l is the dislocation length;

b is the dislocation width.

In the RMF method, the so-called magnetic stress intensity factor, or the gradient of the magnetic field normal component, was adopted as a measure of the quantitative assessment of the stress concentration level. The gradient of the residual magnetic field is defined by Equation (5):

$$K_{in} = \frac{|\Delta H_p|}{2l_k} \tag{5}$$

where:

K_{in} is the RMF gradient;

$|\Delta H_p|$ is the absolute value of the difference in H_p between two control points located at equal distances l_k on both sides of the line $H_p = 0$.

Due to the magnetoelastic effect, the defective area is very easily detectable under the influence of the load. It is characterized by a different modulus of elasticity, magnetic susceptibility or magnetoelastic properties. Therefore, under the influence of the load, its deformation value as well as the magnetoelastic increase in magnetization will be different from that of the matrix.

The difference in magnetization between the defective area and the matrix can be described using the following formula:

$$\Delta M = \Delta M(H_0) + \Delta M^{\sigma}[H_0; \sigma(x, \alpha)] \tag{6}$$

where:

$\Delta M(H_0)$ is the rise in magnetization under the influence of a weak magnetic field;
$\Delta M^{\sigma}[H_0; \sigma(x, \alpha)]$ is the rise in magnetization under the influence of elastic stresses in the presence of a weak magnetic field. %endenumerate

The result is a clear increase in the residual magnetic field within the defect area, since its value is directly proportional to ΔM.

3. Results of the RMF Scanning

Span No. 5 from the set of 13 spans was selected for analysis. The span is located directly above the coal chute to the conveyor that transports fuel to the turbine boiler (cf. Figure 1, showing a diagram of a single span). Each coal load is moved to span No. 5 using the crane. The crane with the load must stop on span No. 5, causing additional dynamic loads. Then, coal is put into the chute, which in turn causes vibrations of the crane beam. Due to this, the load of span No. 5 is the highest in the entire crane. The measurements involved scanning the bottom surface of the crane beam and determining the distributions of the normal and tangential components of the residual magnetic field. Gradients of both RMF components were also determined.

Next, based on the analysis of the gradient distribution, the highest gradient value of 10 A/m/mm was adopted as the limit criterion based on numerous previous works by the author [7,11]. Areas where the limit gradient value was exceeded were selected. They were then subjected to a further detailed analysis. The beam measurements were carried out in two stages.

The first stage of the analysis consisted in scanning the entire underside of span No. 5 along the previously determined three uniformly distributed measuring lines L1, L2, L3, as shown in Figure 2. It should be noted that line 3 is placed on the chute side.

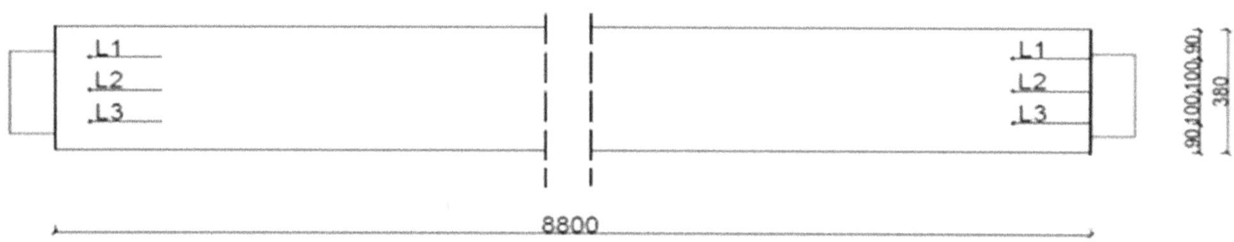

Figure 2. Diagram of measuring lines L1, L2, L3.

Example results of the beam surface scanning in the form of distributions of the tangential and the normal component (Hpx and Hpy, respectively) are shown in Figures 3 and 4 below. The value of the tangential component along the length of measuring line 3 ranges from −100 A/m to 69 A/m. The value of the orthogonal component is in the range of 340 A/m to −80 A/m. The gradient of the tangential and normal components is many times higher than the limit value of 10 A/m/mm.

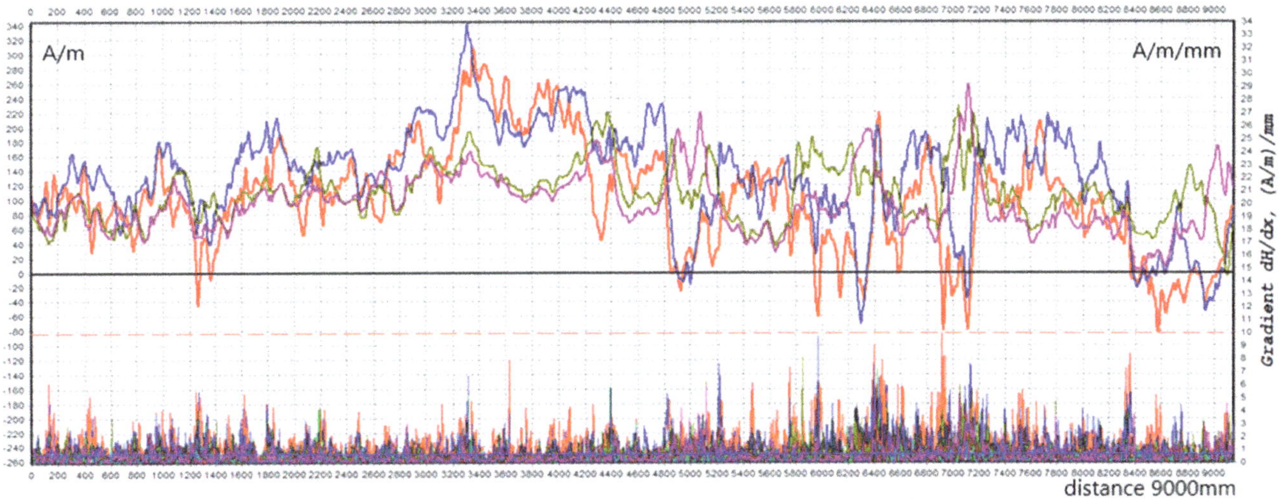

Figure 3. Distribution of tangential component Hpx along L3.

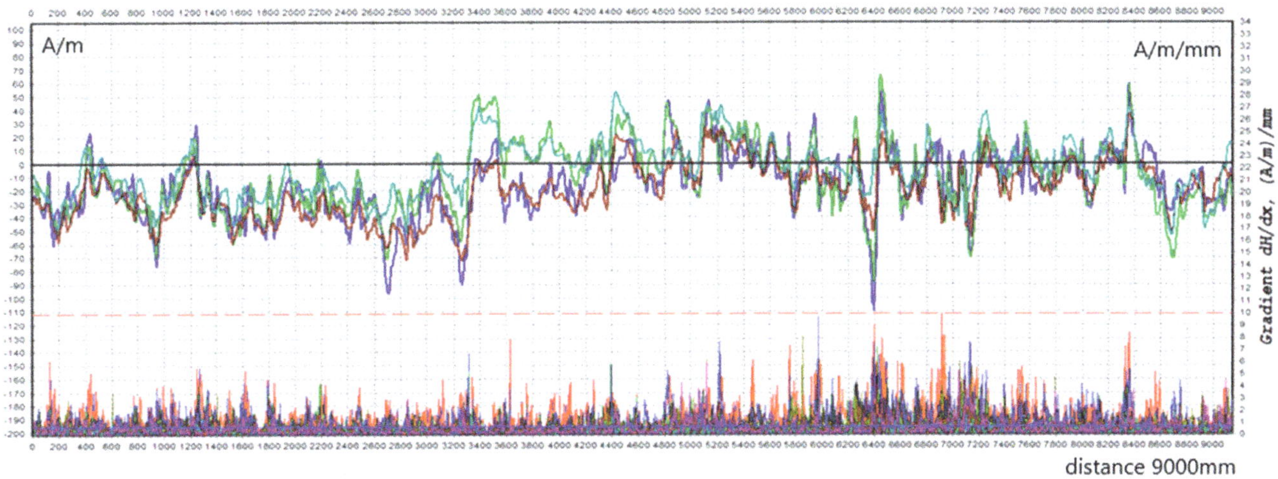

Figure 4. Distribution of normal component Hpy along L3.

This section provides a concise and precise description of the experimental results, their interpretation as well as the conclusions that can be drawn from the experiments. In the second stage, the area with the highest gradient value was selected. For this purpose, the measurements were first carried out with no coal load on the crane, and then the operation was repeated after the crane bucket was filled with coal. In both cases, based on the analysis of the obtained magnetograms and comparing the gradient values of the RMF component Hp with the limit value of 10 A/m/mm, the measurements demonstrated the occurrence of stress concentration zones between 5655 and 7145 mm from the crane left support. Then, the most loaded part of the beam, indicated in red in Figure 5, was tested again. An additional 22 measuring lines with numbers from L27 to L49 were introduced in this area.

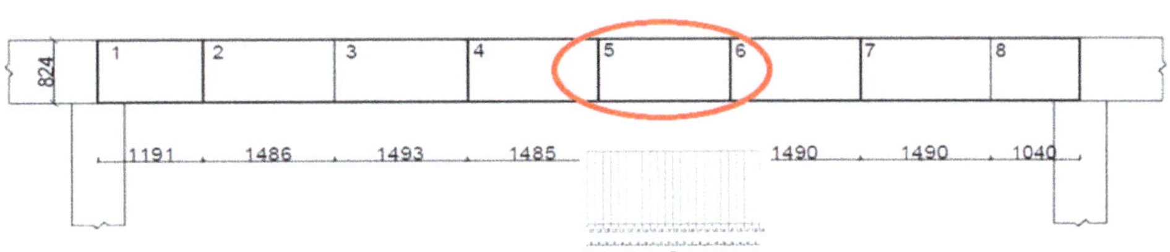

Figure 5. The beam's most loaded part and the additional 22 measuring lines.

The distribution of the changes in the residual magnetic field tangential component gradient for measuring line 3, for the selected sub-area, is presented in Figure 6.

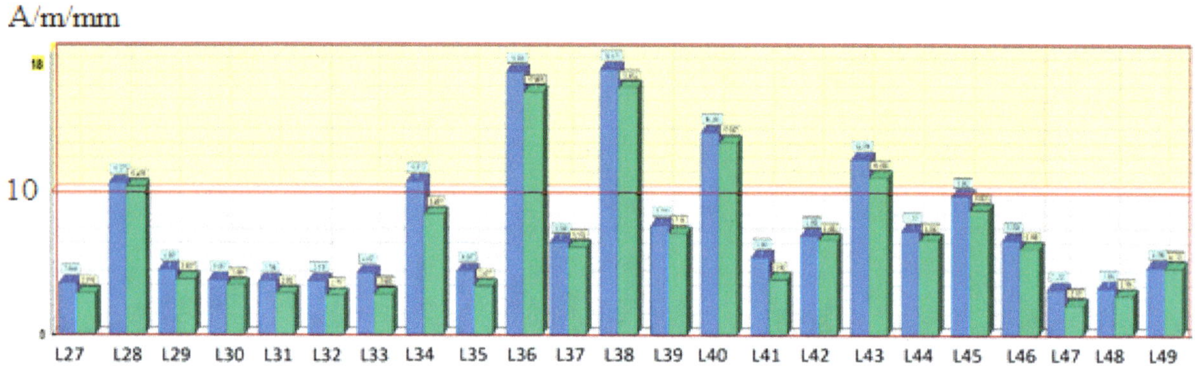

Figure 6. Changes in the tangential component gradient for the selected sub-area under analysis.

The adopted procedure was a gradual refinement of the observation area. Based on the analysis of the tangential component gradient distribution, stress concentration zones were found to occur in places where the limit of 10 A/m/mm was exceeded, as indicated in earlier studies by the author [7]. These are areas covering measuring lines 36–38, while the second area covered lines 40–45. In the next stage of the analysis, 2-D maps of the distribution of the magnetic field component gradients were prepared. The data for the residual magnetic field intensity maps were imported from J&J System software, while the gradients were calculated as a derivative of the RMF component H_p over the measurement line length using the finite difference method according to Equation (7):

$$G = H_p(x)' \approx \frac{H_p(x+a) - Hp(x)}{a} \tag{7}$$

where:

$Hp(x)$ is the RMF intensity value in the point where the gradient was determined;
$H_p(x+a)$ is the RFM intensity value for the next measuring point;
a is the distance between consecutive points on the measuring line.
It was decided that for the magnetic field intensities, the colors would change every 10 units.

The intensities of 0–10 A/m are presented as white. The values of positive intensities depending on the value are represented by warm colors from yellow to brown, while negative values are represented by cold colors from green to purple. For the gradient maps, colors changed every 1 unit. Because the gradient is non-negative, the colors range from white to dark brown. The distribution of the RFM tangential component on the 2-D map for the selected area is shown in Figure 7. It can be noticed that the highest values of 130 A/m were in the range of 607 to 1290 mm (x) and 152 to 380 mm (y). A relatively small area was found for the coordinates of 480 (x) and 16 mm (y).

Further analyses in the selected area concern the distribution of the gradient of the tangential and normal components. Figure 8 presents the distribution of the tangential component gradient for the full load on the crane beam.

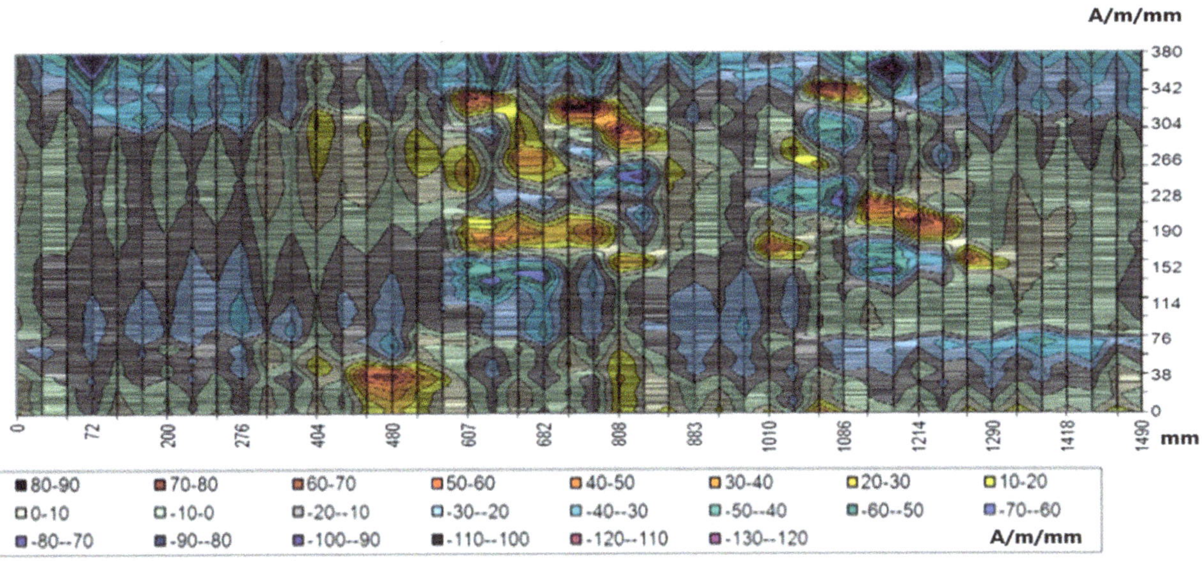

Figure 7. Distribution of tangential component Hpy for the selected area.

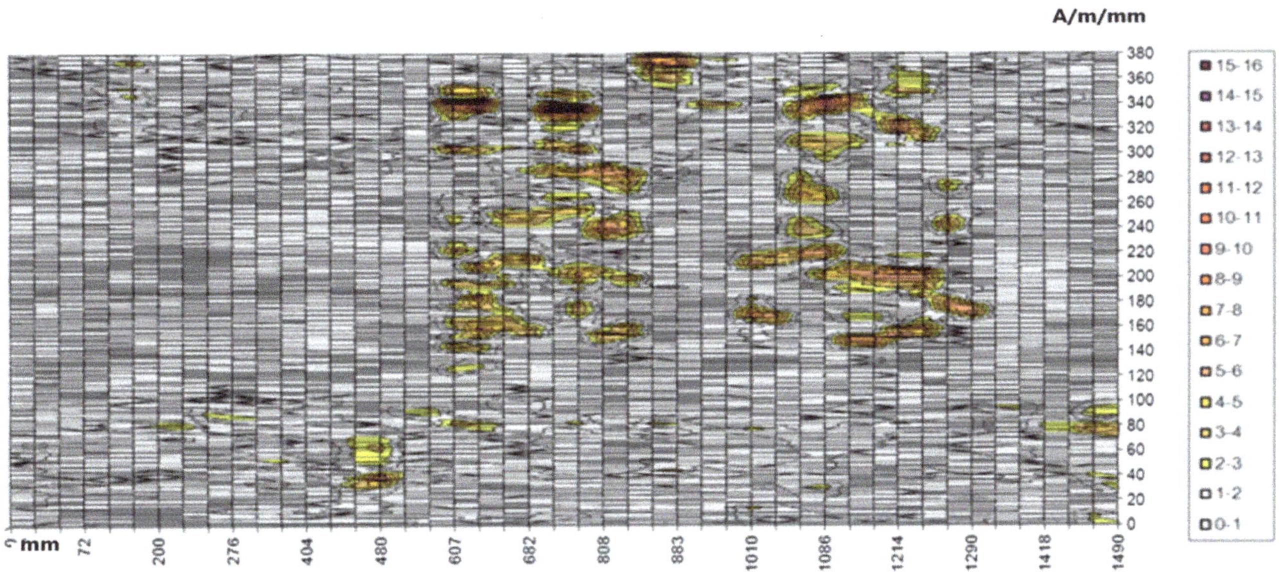

Figure 8. Distribution of the gradient of the tangential component dHpx/dx for the selected area.

If the crane is loaded with a truck only, the gradient values are lower by approximately 2 A/m/mm. Areas with values exceeding 10 A/m/mm also coincide with the areas of the highest values of the tangential component. They are located in the upper left quadrant of the map in the range of 607–1290 mm (x) and 140–380 mm (y). This area is located on the coal chute side—the analyzed part of the beam is located exactly above the coal chute. It is also the most heavily loaded beam, because during each cycle coal is transported at various points of the coal storage, always above the chute, where the load of a full skip hoist is applied first, and then rapid unloading takes place by opening the skip hoist and pouring out the coal. The estimated number of cycles in the heating season is about 60,000. A similar distribution of the normal component gradient confirmed the occurrence of the identified stress concentration zones. The highest values of the gradient of the RMF normal component reached 14 A/m/mm. Figure 9 presents the normal component gradient distribution. The area with the highest gradient value appears in the top left quarter of the 2-D map.

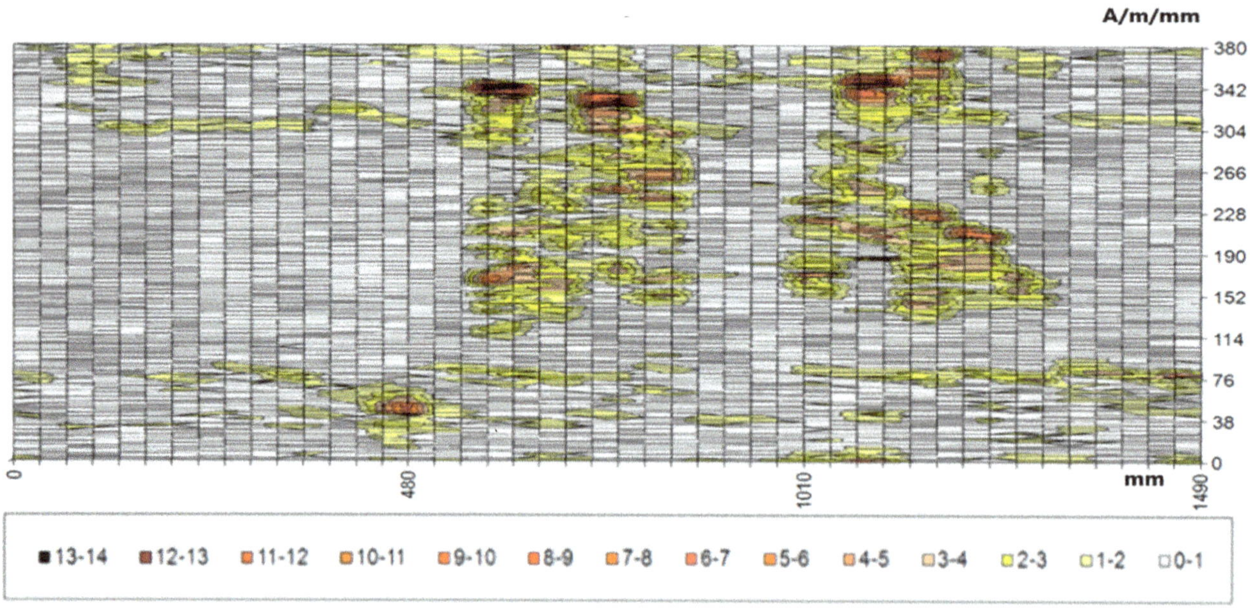

Figure 9. Distribution of the gradient of the normal component for the selected area.

4. Validation and Numerical Simulation

During the crane beam overhaul, samples were taken to determine the distribution of stress components. The samples were collected from places indicated as SCZs. The tests were carried out by means of the X-ray angle diffraction using an AvanceD-8 diffraction instrument. The testing results confirmed stress concentration zones determined using the RMF method. Stresses in the SCZs reached a value of 186 MPa, and outside the zone, of about 60 MPa. The diffraction pattern is shown in Figure 10.

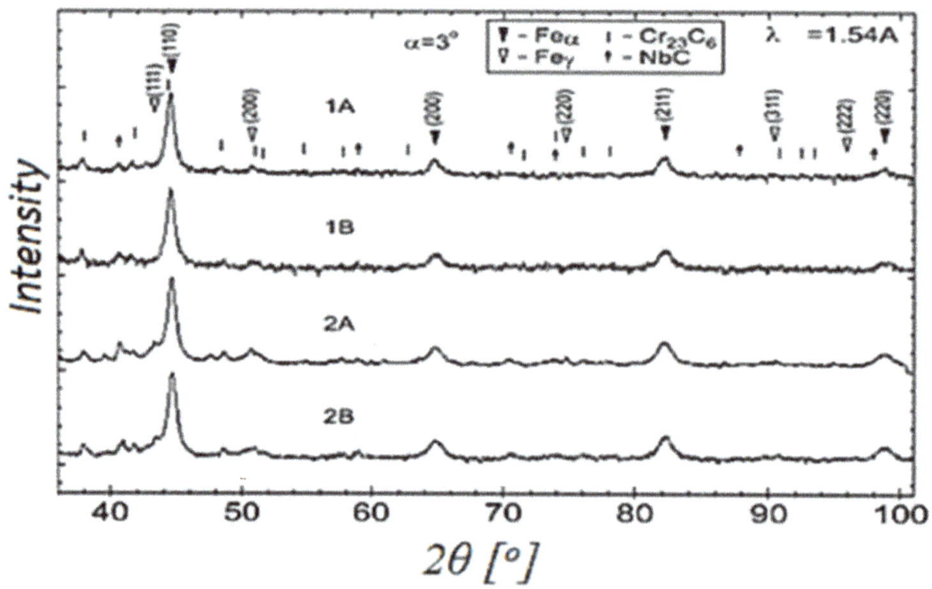

Figure 10. Diffraction pattern of a sample taken from the crane beam.

The detailed and comprehensive experimental and numerical verification consisted in the determination of deflection values halfway through the span length. The values were determined in three locations (points) marked in the figure presenting the crane beam cross-section (cf. Figure 11).

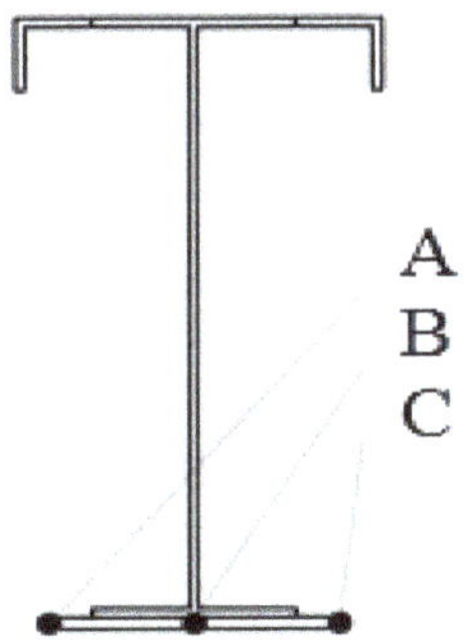

Figure 11. Arrangement of measuring points in the cross-section halfway through the span length.

Figure 12 presents influence lines for span No. 5 determined experimentally (values measured during the gantry crane passage). The measuring points located on the side of the coal chute (inner points) were characterized by the biggest value of deflection of 7.9 mm, whereas the smallest deflection value of 5.2 mm was observed for the opposite (outer) points. The deflection value of the measuring point above the I-beam web (the central point) was included between the values found for the outermost points. It totaled 6.35 mm. The deflection values determined in the crane beam measuring points gave evidence of beam warp. It turns out that this is the effect of the eccentric load of the analyzed crane beam I-section (which does not comply with the gantry crane technical documentation).

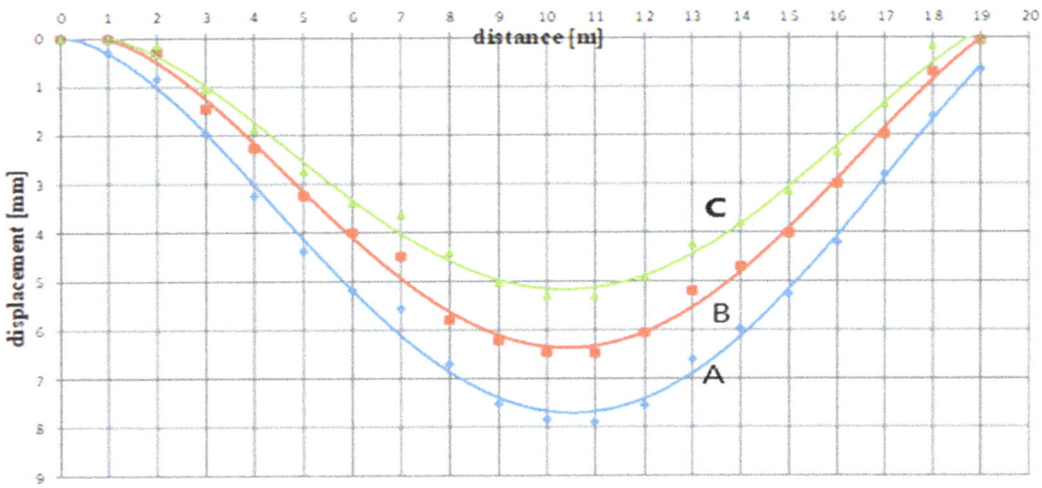

Figure 12. Crane beam deflection during the crane crab passage.

The experimentally determined values were entered into the FEM numerical model. The element side characteristic length was 2–10 mm. Shell 63 element was used. As the aim of the analysis was to determine the deformation of the crane beam and establish the cause of the unsymmetrical distribution of the tangential component gradients at set boundary conditions, a linear elastic material model was used. The constitutive model was linear. The total number of elements was 1.4 million. The numerical testing results confirm the twisting of the beam, which causes a higher level of stress. The experimental results were introduced into the FEM numerical model in the form of boundary conditions for displacement. A 3-D numerical model of the analyzed crane girder was built based on the finite element method ANSYS code, taking into account the upgrades and the span axis displacement detected in relation to the half-distance between the truss poles. The results of the numerical simulations (Figure 13) confirmed the previously indicated two areas, i.e., the area in the middle of the span length and in the places of welding the reinforcing cover. The stress levels in the analyzed areas were also determined. In the middle of the span length, the stresses totaled ~40 MPa.

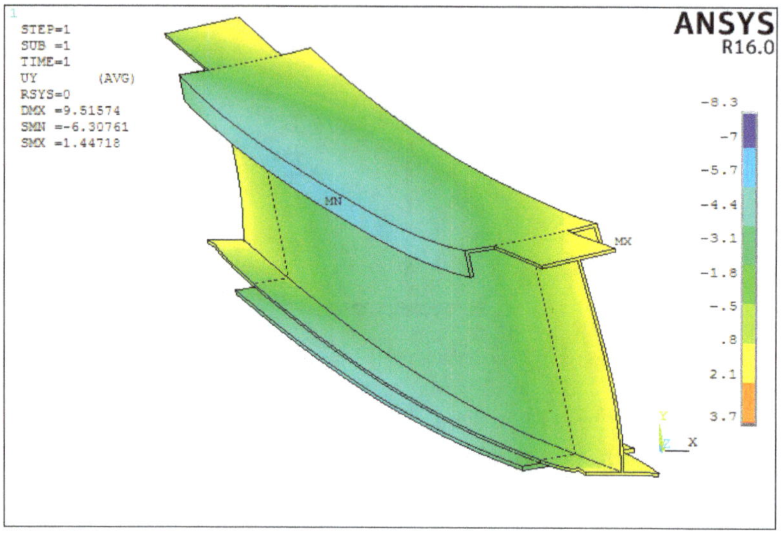

Figure 13. Distribution of the UY displacement.

The reason for this is the slight displacement of the crane travelling wheel towards the coal chute. The determined numerical distribution of the crane beam displacement is shown in Figure 11.

5. Conclusions

1. The RMF technique enabled the effective location of stress concentration zones in the analyzed crane beam by determining the gradient of the RMF tangential component.
2. The experimental verification using the X-ray diffraction method confirmed the stress concentration zones in the beam.
3. Creating a relational database of magnetograms of the ferromagnetic structure from the beginning of its operation seems to be an interesting addition to standard magnetic diagnostic methods.
4. Combining different methods and measurement techniques, for example a fiber optic beam deflection measurement system using the FEM method which implements boundary conditions from optical fiber measurements, with residual magnetic field techniques can significantly contribute to the identification of stress concentration zones.

References

1. Shannon, R.W.F.; Braithwaite, J.C.; Morgan, L.L. Flux-leakage vehicles pass tests for pipeline inspection. *Oil Gas J.* **1988**, *32*. [CrossRef]
2. Fu, M.L.; Bao, S.; Zhao, Z.Y.; Gu, Y.B. Effect of sample size on the residual magnetic field of ferromagnetic steel subjected to tensile stress. *Non-Destr. Test. Cond. Monit.* **2018**, *60*, 90–94. [CrossRef]
3. Bao, S.; Fu, M.L.; Lou, H.J.; Bai, S.Z. Defect identification in ferromagnetic steel based on residual magnetic field measurements. *J. Magn. Magn. Mater.* **2017**, *441*, 590–597. [CrossRef]
4. Yao, L.K.; Wang, Z.D.; Deng, B. Experimental research on metal magnetic memory method. *Exp. Mech.* **2012**, *3*, 305–314. [CrossRef]
5. Wang, Z.D.; Yao, K.; Ding, K.Q. Quantitative study of metal magnetic memory signal versus local stress concentration. *NDT E Int.* **2010**, *6*, 513–518. [CrossRef]
6. Wang, Z.D.; Yao, K.; Ding, K.Q. Theoretical studies of metal magnetic memory technique on magnetic flux leakage signals. *NDT E Int.* **2010**, *43*, 354–359. [CrossRef]
7. Juraszek, J. *Innovative Non–destructive Testing Methods Monography*; ATH, Univ. of Bielsko-Biala: Bielsko-Biała, Poland, 2013.

8. Ren, S.K.; Song, K.; Ren, J.L. Influence of environmental magnetic field on stress magnetism effect for 20 steel ferromagnetic specimen. *Insight* **2009**, *51*, 672–675. [CrossRef]
9. Wilson, J.W.; Tian, G.Y.; Barrans, S. Residual magnetic field sensing for stress measurement. *Sens. Actuators A* **2007**, *135*, 381–387. [CrossRef]
10. Yao, K.; Deng, B.; Wang, Z.D. Numerical studies to signal characteristics with the metal magnetic memory-effect in plastically deformed samples. *NDT E Int.* **2012**, *47*, 7–17. [CrossRef]
11. Juraszek, J. Hoisting machine brake linkage strain analysis. *Arch. Min. Sci.* **2018**, *63*, 583–597.
12. Shi, C.L.; Dong, S.Y.; Xu, B.S.; Peng, H. Stress concentration degree affects spontaneous magnetic signals of ferromagnetic steel under dynamic tension load. *NDT E Int.* **2010**, *43*, 8–12.
13. Dong, L.H.; Xu, B.S.; Dong, S.Y.; Chen, Q.Z.; Wang, D. Stress dependence of the spontaneous stray field signals of ferromagnetic steel. *NDT E Int.* **2009**, *42*, 323–327. [CrossRef]
14. Guo, P.J.; Chen, X.D.; Guan, W.H.; Cheng, H.Y.; Jiang, H. Effect of tensile stress on the variation of magnetic field of low-alloy steel. *J. Magn. Magn. Mater.* **2011**, *23*, 2474–2477.
15. Shui, G.S.; Li, C.W.; Yao, K. Non-destructive evaluation of the damage of ferromagnetic steel using metal magnetic memory and nonlinear ultrasonic method. *Int. J. Appl. Electromagn. Mech.* **2015**, *47*, 1023–1038. [CrossRef]
16. Yi, S.C.; Wei, W.; Su, S.Q. Bending experimental study on metal magnetic memory signal based on von Mises yield criterion. *Int. J. Appl. Electromagn. Mech.* **2015**, *49*, 547–556. [CrossRef]
17. Chen, H.L.; Wang, C.L.; Zuo, X.Z. Research on methods of defect classification based on metal magnetic memory. *NDT E Int.* **2017**, *92*, 82–87. [CrossRef]

5

A Non-Destructive Indirect Approach to Long-Term Wood Moisture Monitoring based on Electrical Methods

Richard Slávik [1], Miroslav Čekon [2,3,*] and Jan Štefaňák [3]

[1] Department of Wood Science and Technology, Faculty of Forestry and Wood Technology, 613 00 Brno, Czech Republic

[2] Department of Physics, Faculty of Civil Engineering, Slovak University of Technology, 810 05 Bratislava, Slovakia

[3] AdMaS Centre, Faculty of Civil Engineering, Brno University of Technology, 602 00 Brno, Czech Republic

[*] Correspondence: miroslav.cekon@stuba.sk or mcekon@gmail.com

Abstract: Wood has a long tradition of use as a building material due its properties and availability. However, it is very sensitive to moisture. Wood components of building structures basically require a certain level of moisture protection, and thus moisture monitoring to ensure the serviceability of such components during their whole lifespan while integrated within buildings is relevant to this area. The aim of this study is to investigate two moisture monitoring techniques promoting moisture safety in wood-based buildings (i.e., new structures, as well as renovated and protected buildings). The study is focused on the comparison of two electrical methods that can be employed for the nondestructive moisture monitoring of wood components integrated in the structures of buildings. The main principle of the two presented methods of the moisture measurement by electric resistance is based on a simple resistor–capacitor (RC) circuit system improved with ICM7555 chip and integrator circuit using TLC71 amplifier. The RC-circuit is easier to implement thanks to the digital signals of the used chip, whilst the newly presented integration method allows faster measurement at lower moisture contents. A comparative experimental campaign utilizing spruce wood samples is conducted in this relation. Based on the results obtained, both methods can be successfully applied to wood components in buildings for moisture contents above 8%.

Keywords: resistance measurement; wood moisture sensing; non-destructive testing; moisture safety

---

1. Introduction

The properties and lifespan of wood strongly depend on many aspects, and therefore the permanent moisture monitoring of wood has a specific application in a wide range of technical fields. Primarily, wood-based wall technologies are widely considered to be suitable building materials for low-environmental impact composites in the building engineering field. Wood framed and composite wood wall technologies that utilize advanced insulation techniques have been well-integrated for several decades in the building sector, and specifically meet thermal and environmental requirements [1]. Thermal properties, such as thermal conductivity and specific heat, are given significant consideration when designing buildings, especially low-energy or passive ones. Building components that are exposed to the weather outdoors are mainly affected by moisture and temperature-related factors. It has been two decades since exterior building surfaces began to be designed to be monitored by methods for the continuous monitoring of temperature and moisture in the micro-environment of structures, and within the wood itself. A system was introduced that maps the climate index for the decay of wood at various geographical levels via the use of existing climatic data, standards, and moisture content measurements [2]. Moisture and temperature can also play an important role indoors and within the structure of a building, i.e., throughout the whole building envelope. The existence of high moisture content can initiate decay or the growth of fungi. Particularly for bio-based building materials such as wood, the use of biological agents should be considered in order to predict service life, particularly with regard to fungal decay and mold growth risks. The control and reduction of wood moisture content is therefore a key instrument for wood protection [3,4]. However, the permanent monitoring of wood components may also have a relevant role to play in this field, especially when such components are incorporated in the building structure and their continuous physical monitoring is impossible. The correct estimation of timber moisture content and the subsequent initiation of potentially necessary measures are therefore essential tasks during the planning, execution, and maintenance of buildings built with wood or wood-based products. This fact has contributed to a recent and considerable rise in interest concerning the in-situ monitoring of the moisture content of structural timber elements [5]. In this connection, the modeling of the outdoor performance of wood products is also attracting specific attention [6]. Furthermore, moisture content measurement has a lot of potential for use in testing the durability of timber products [7].

A technique for the nondestructive evaluation of moisture content distribution during drying using a newly developed soft X-ray digital microscope and absortimetry was investigated by Tanaka et al. [8] and Tanaka and Kawai [9]. X-ray-based methods and diagnostics have already been successfully developed and used in many applications to identify aspects of wood decay [10,11]. In particular, wood temperature and moisture content have a direct impact on fungus and its ability to metabolize and degrade wood cell wall material over time [12]. Moisture requirements for the growth and decay of different fungi and wood species have already been determined, though relationships between wood moisture content, wood temperature, and fungal decay play an important part when applying the method in specific climates [13]. In addition, the relationship between microclimate, material environment, and decay is being studied in order to achieve a better understanding of issues concerned with the service life prediction of wood and wood-based products. Dietsch et al. [5] describes common methods of determining wood moisture content and evaluates them with respect to their applicability in monitoring concepts. Continuous moisture measurements using calibrated load cells and a data logger coupled with a weather station are an efficient way to record moisture in all kinds of material [14]. Unfortunately, the most accurate direct methods, which use oven drying and distillation, are time-consuming and cause the destruction of specimens.

Many indirect methods have been developed based on electric conductance or dielectric properties which allow results to be obtained fast and with satisfactory accuracy. In this relation, wood moisture measurement has a long history. Indirect methods are often applied for in-situ measurements and monitoring. They are based on determining a different property, which is correlated with the water content. Thus, resistivity measurements require material-specific characteristics and a temperature compensation, since both parameters have a significant effect on electrical conductivity [15]. Dunlap [16] discusses twelve commercial electrical moisture meters. Most of them are based on resistance (conductance) measurement. The Wood Handbook [17] states that the resistance of wood ranges from a few petaohms for oven-dry wood to a few kiloohms for wood with fiber saturation. In the range between fiber saturation (appr. 30%) to complete saturation, the change of resistance is not so significant. James [18] mentioned that the conductance and dielectric properties of wood vary consistently with moisture content when it is less than 30%, with a roughly linear relationship between the logarithm of conductance and the logarithm of moisture content. Thanks to this relationship, the moisture content can be determined. The measurement principle is based on the application of direct or alternating current, but higher resistances than a few hundred megaohms are not so easy to measure. Electric-current through such huge resistance is often very small, and direct current measurement in a simple electrical circuit with an appropriate error is not possible. Measuring very high resistance values is a difficult task, since low voltage or currents are present and thus, noise and amplification must be carefully done, especially when low resistance values are required to be measured using the same circuit, too [19]. Moisture content is not the only phenomenon which has an effect on the conductance of wood material; another significant factor is temperature dependency. James [18] mentioned that an increase of 10 °C causes an approximately two-fold increase in conductance in regions with more than 10% moisture content. Another problem is the anisotropy of wood in the direction of the grain. Conductance measured parallel to the grain is approximately double that of perpendicular conductance.

Many methods, complex electrical circuits, and devices have been developed in past decades. The most common and most easily applicable methods are electrical resistance measurements [15]. Typical devices are equipped with probes and a display for showing calculated values. They are portable, and very useful for taking technical measurements or conducting on-site inspections of materials. A long-term moisture measuring and data logging method for wood exposed to weather conditions was developed by Brischke et al. [20]. The method involves measuring the electrical resistance with glued electrodes for a sustainable connection. The measuring points at the tips of the electrodes are glued conductively into the wood while the remaining outer parts of the electrodes are glued with insulating adhesive. For this purpose, special conductive and insulating glues and electrodes were developed and comparatively evaluated in laboratory tests. In a recent study, the comparison of accuracy and ease of operation between voltammetry and digital bridge method for electrical resistance measurement in wood specimens and the factors influencing voltammetry were examined in detail [21]. Another useful approach is based on interdigital capacitance sensing of moisture content in rubber wood where the electrode contact is non-penetrating. It has linear sensitivity and better accuracy at very low moisture content. This represents faster measurement and is more convenient to be used in the industry for the production line quality control monitoring of moisture content of rubber wood [22].

Nowadays, with innovative approaches in many fields, we often use technologies which allow us to monitor the service life of all building elements permanently. Portable independent devices often cannot be integrated into structures for a long time or connected to networks. For that reason, there are potential future applications for sensors which could be connected to a network and placed into a composite structure containing wood for a long period of time (Figure 1). The sophisticated circuits used in commercial portable devices are often trade secrets of their producers and are not so easy to adapt for use in automatized systems. If the resistance measurement approach is used, there are two methods which are applicable: They are subjected to further analysis here. Both of the analyzed methods use the capacitor charge principle in a different way.

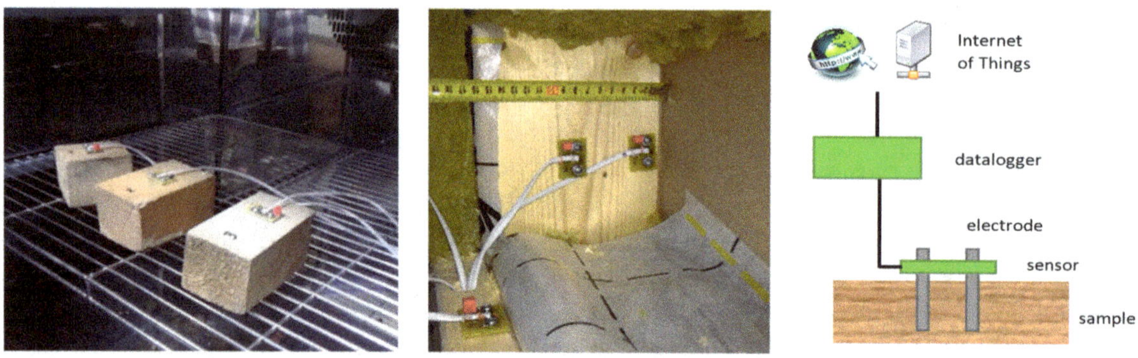

(a) (b) (c)

Figure 1. Implementation of small sensor module at real-world scale; (**a**) sensor testing in experimental chamber; (**b**) real wall structure integration; (**c**) functional scheme.

This study describes experimental work which aimed to obtain wood moisture monitoring data indirectly using two electrical methods. The main aims of this research work are first, to introduce the theoretical principles of both measurement methods used, and second, to evaluate their applicability and finally to verify them based on the experimental results obtained. Both methods are applied to samples made of spruce wood. The experimental results were evaluated via a statistical approach based on Bland-Altman plots. Originally, the Bland-Altman procedure [23] was used in medicine to compare two clinical measurements that each produce some error in the data they measure [24]. Bland-Altman plots are now extensively used to evaluate the agreement between two different instruments or two measurement techniques. They allow the identification of any systematic difference between measurements, or possible outliers. The results of this statistical analysis method determine the applicability of both, monitoring methods and the overall consistency between them. The statistical analysis also identifies the moisture content range for which the methods should obtain acceptable results if properly applied.

2. Description of the Circuits, Theoretical Principles, and Applications

As mentioned above, there are basically two methods which can be used for long-term wood moisture monitoring using electrical resistance and direct current. The first method uses a resistor–capacitor (RC) circuit which has been improved by transforming it into a digital device using a 555 timer chip [25,26]. The second applicable method uses capacitor charging via an operational amplifier connected as an integrator circuit. This method was originally applied to other problems requiring the measurement of high resistance by Aguilar et al. [27]. Both methods can be usefully implemented in small sensor packages (Figure 1). It shows small sensor prototypes which were assembled based on this research work and directly attached to wood samples in a climatic chamber during a test involving long-term monitoring.

The application of circuits was tested on spruce wood samples. Several wood samples with electrodes attached in various orientations relative to the fibers of the wood were measured with both methods. The samples were fitted with electrodes and placed into a desiccator where humidity and temperature were controlled. The measurements were obtained via a digital oscilloscope at intervals during conditioning, and the weights of the samples were measured. The results from both circuits were compared to each other using the obtained data.

2.1. *The RC Circuit Method*

This type of circuit is also known as an RC network or RC filter. It combines a capacitor with a resistor in series and is driven by voltage or a source of current. The main principle of RC circuits is based on the relationship between capacitor and resistor values and the voltage level in the capacitor over time. In a simple example with direct current and a rise in voltage from zero to a particular voltage level at the input rail of the resistor, the time of capacitor charging is proportional to the time constant of the circuit. The time constant is defined by the value of the resistor and capacitor.

At zero time the capacitor *C* is discharged. If the input voltage is applied at resistor *R*, the voltage charge of capacitor *C* starts rising over time until it reaches the maximum value (Figure 2). The time constant (1) represents the time taken by the capacitor to charge to approximately 63.2% of its final voltage value. At a known and stable capacitance and a known time of charging the resistance can be estimated, and vice versa. The relation between the voltage of capacitor V_C and the stable input voltage V_{IN} on the left resistor rail can be described by Equation (2).

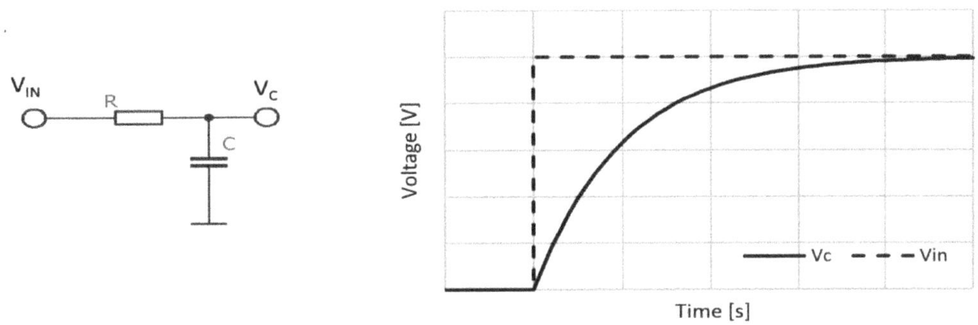

Figure 2. Resistor–capacitor (RC) circuit diagram and working principle.

The described idea is applied in the first method. The charging and discharging of the circuit are managed by a timer-integrated circuit. The most well-known timer, the 555 timer chip, has many applications, from circuits with a blinking LED (Light-Emitting Diode) to circuits which use it to generate an AC (Alternating Current) signal. The use of this chip for wood moisture measurement was already mentioned by Vodicka [25]. It is a logical application in which the chip is used as a monostable timer. This simply means that if a trigger input is pulled to a logical zero, the capacitor is discharged, and then charging starts with the output of the chip set to a high logical level. If the voltage of the capacitor reaches 63% of the input voltage level, the output is set to the zero logical level.

The time of the output pulse is proportional to the RC constant and can be calculated with Equation (3). Testing was performed with an oscilloscope; the circuit (Figure 3) was triggered by a push button and time of charging was captured by oscilloscope (Figure 4). Triggering and time counting could be handled by a small microcontroller in a real world sensor. The problem standard timer chips have with leakage current, which is often higher than the current through a wood sample with a moisture content of under 10%, has been solved by the use of the ICM7555 chip. Its leakage current is within the range of a few picoamperes according the producer's specifications.

$$\tau = RC\ [-] \tag{1}$$

$$V_C = V_{in}\left(1 - e^{\frac{-t}{\tau}}\right)\ [\mathrm{V}] \tag{2}$$

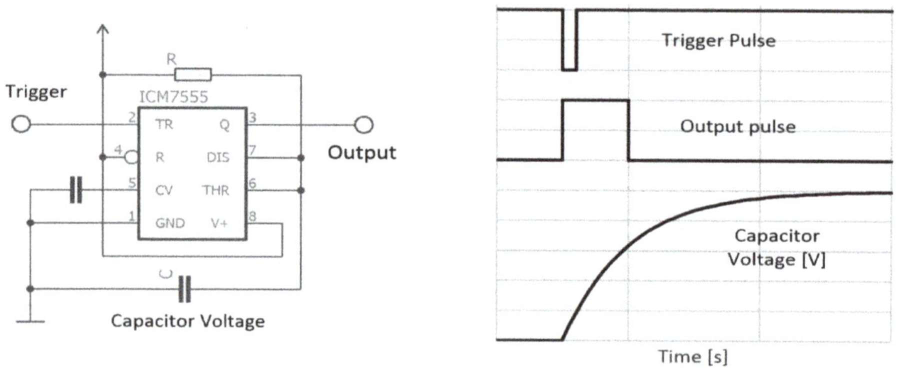

Figure 3. RC circuit improved by a 555 timer chip.

$$t = 1.1RC\ [\mathrm{s}] \tag{3}$$

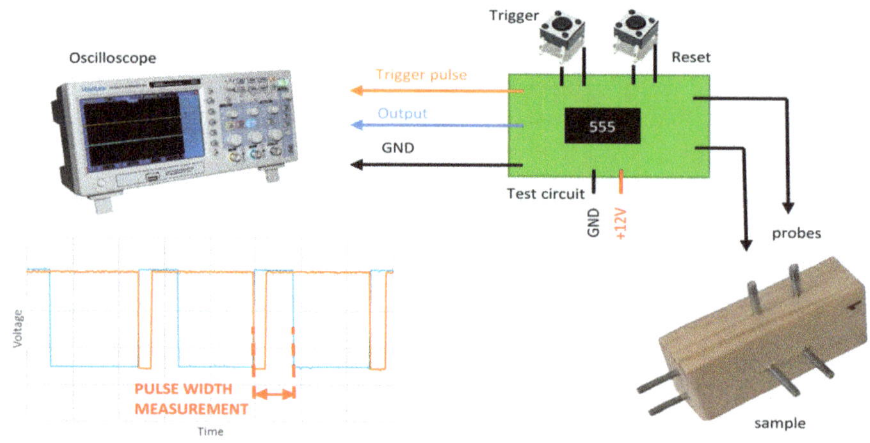

Figure 4. The complete principle of the RC circuit for the moisture content measurement.

2.2. The Integration Method

Operational amplifiers have various uses in electronics. The most common functions are as comparators, amplifiers, differentiators, and integrators. The last of these is the most relevant to high resistance measurement, and the second method implements operational amplifiers in that mode. A TLC71 amplifier in integrator mode has already been used to measure high resistance at the level of hundreds of megaohms [27]. Most operational amplifiers need a symmetric power supply, which is the main disadvantage of these devices in digital electronics.

The integrator works on the principle of capacitor charging. The voltage at the capacitor changes over time according to the voltage value applied to the negative rail of the amplifier (Figure 5).

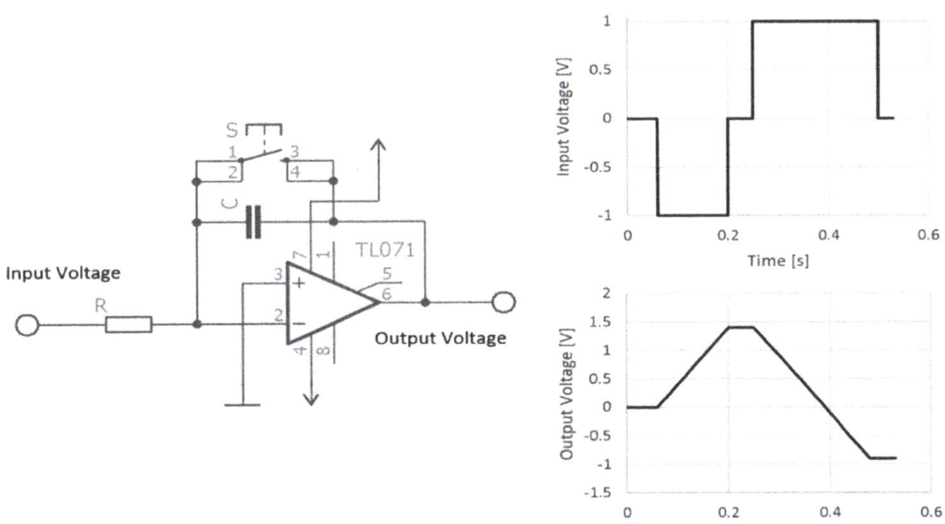

Figure 5. Integration method circuit diagram and working principle.

The positive rail of the amplifier is grounded. If a negative voltage level is reached at input through the resistor, which represents measured resistance, the capacitor voltage grows to a positive supply voltage and vice versa. The slope of voltage change is proportional to the resistance and capacity. The output voltage is measured via an oscilloscope and the resetting of the integrator is triggered by a switch (Figure 6). In the case of a real-world sensor, this could be handled by a transistor managed by a microcontroller which is equipped with an analog to digital convertor for output voltage measurement.

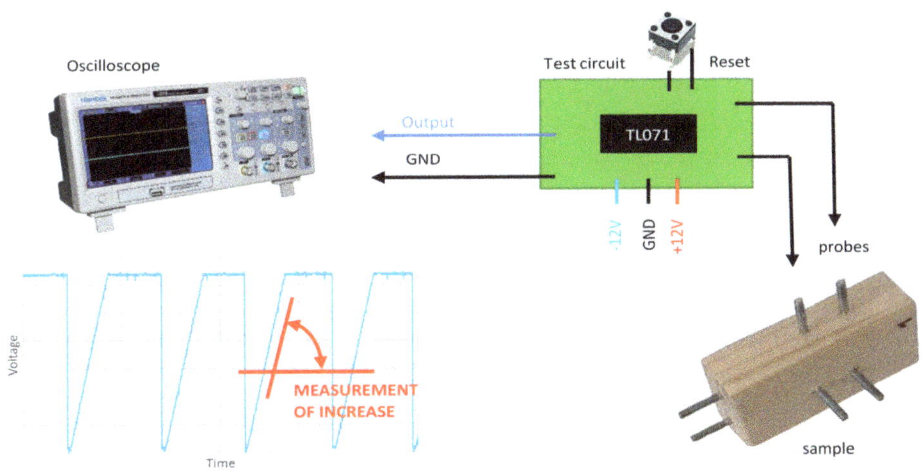

Figure 6. The complete principle of integration method for the moisture content measurement.

3. Materials and Methods

An experimental investigation was carried out using the two abovementioned electrical resistance-based methods, and the measurement results were statistically evaluated. In the initial part of this research, the spruce wood samples were prepared, and test circuits were developed for their calibration using resistor fields. Then, the samples were conditioned, measured, and weighed on a continuous basis over three months. Finally, the obtained data were analyzed and statistically evaluated. Through this approach, a complex characterization of resistance could be comprehensively obtained from the measurements to verify the applicability of both methods used. All of the information on time-dependent moisture parameters measured from the samples was expected to show a clear picture of the evolution of moisture content can be provided if these test circuits are integrated within wood-based material or a real-world structure.

3.1. Sample Preparation Procedure

Three samples were prepared in the form of blocks with a cross section of 47.0 × 27.5 mm and a length of 90.0 mm (Figure 7). They were made of planed spruce wood, dried by electric oven at 105 °C and drilled with diameter 3 mm with deep 20 mm (Figure 7c).

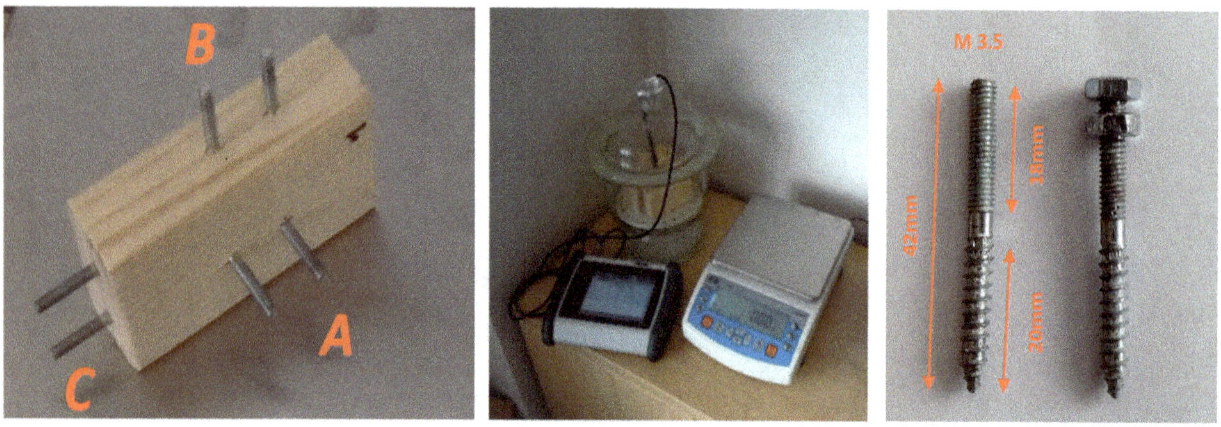

(a) (b) (c)

Figure 7. (**a**) Prepared test sample; (**b**) test conditioning; (**c**) detail of probes.

3.2. Circuit Setup and Calibration Procedure

Test circuits were built on test printed circuit boards (PCBs) with through hole technology (THT) and surface-mount device (SMD) electronic parts. The trigger pulses for measurements were initialized by push buttons. The voltage response of the trigger and the output of the circuits were measured by a digital oscilloscope with a 25 MHz bandwidth. The RC circuit with a 555 timer chip used a 12 V single power supply, while the integrating circuit had a dual power supply with −12 V and +12 V rails. A polypropylene capacitor was chosen with the value 1 nF.

Both circuits were calibrated with two resistor fields. The first test field consisted of 15 resistors with 10 MΩ resistance, and the second test field had 50 resistors with 20 MΩ resistance. Each resistor was measured separately with two digital Ohmmeters and the results were averaged. The total resistance of the first field was set to 148.87 MΩ with a maximum error of 1.5 MΩ, and 993.4 MΩ with a maximum error of 9.9 MΩ. Several combinations of resistors were tested by both experimental circuits. The proportional constants of the circuits were identified according to the obtained data using statistical analysis (Figure 8).

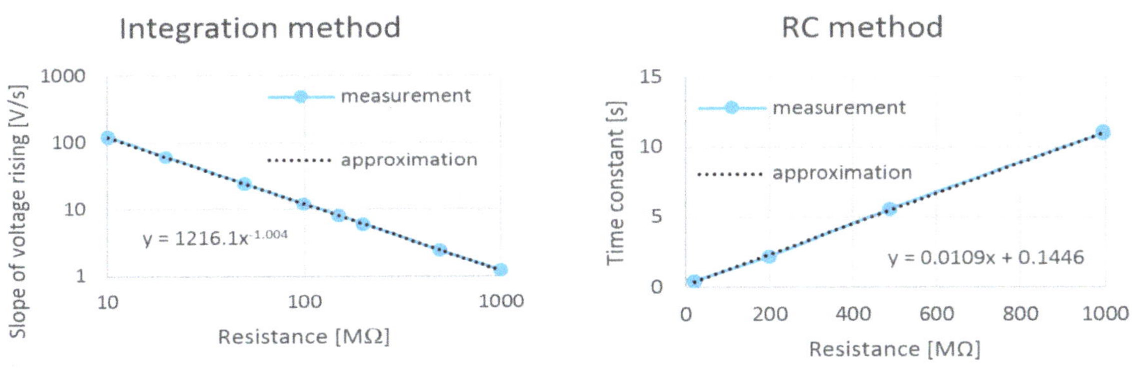

(a) (b)

Figure 8. Calibration of (**a**) integrator and (**b**) RC circuit.

3.3. Sample Conditioning and Measurement Procedure

The samples were conditioned in a desiccator above a solution of salt in accordance with technical standard EN 12751. Sodium chloride and potassium chloride were used in desiccator. Relative humidity at laboratory temperature (21 °C) reached approximately 76% with the first solution and about 86% with the second. The environment in the desiccator was monitored by a temperature and relative humidity probe (Figure 7b). The resistance and weight of the samples were measured at specified intervals by the circuits and a laboratory balance with a resolution of 0.01 g and an accuracy of 0.05 g. Each measurement was taken three times for each electrode location and the results were averaged. The whole conditioning period lasted for roughly three months.

The evolution of sample moisture content was identified by weighing, and the relative humidity in the desiccator was measured (Figure 9). Dry samples were placed into desiccator with solution of sodium chloride for first 1300 h. A logarithmic fast increase of moisture content at the beginning of conditioning can be identified, later continuing more slowly until finally resulting in its equilibrium. In the next step, after 200 h of equilibrium conditions, the samples were placed into another desiccator with higher partial pressure of water vapor. The vapor pressure slowly rises again to reach equilibrium, approaching 2500 h of measurement.

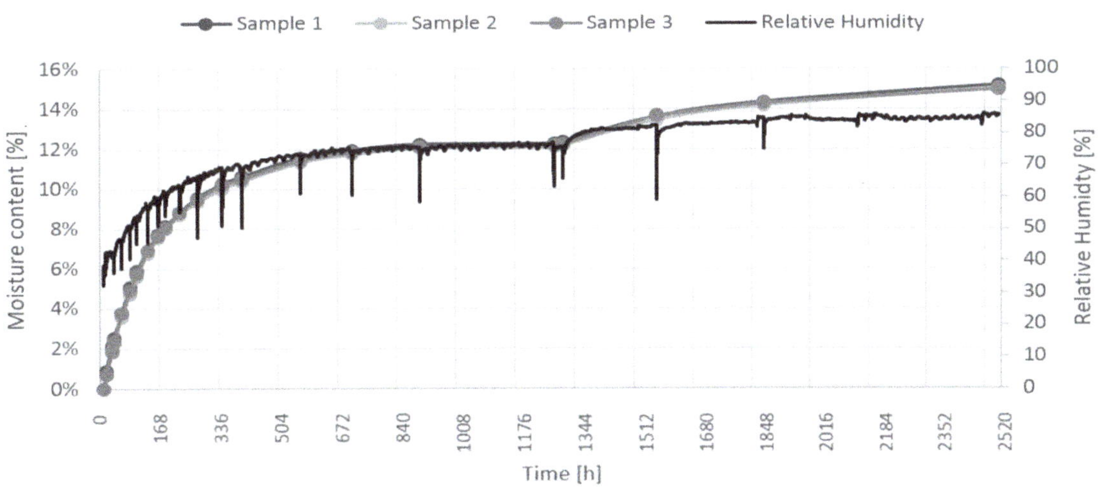

Figure 9. Evolution of moisture content over time.

Resistance value of samples could be distorted by nonlinear distribution of moisture in depth form surface of a sample. The conditioning process takes a long time, however the expectation of quasi linear distribution of moisture in the sample is more realistic closer to states where equilibrium states are reached in desiccator.

3.4. Statistical Analysis

In the real world, every value measured at a given assembly or sensor often differs slightly, thanks to variations in the repeatability and accuracy of methods. The methods used in this research are no exception, and they can be expected to provide slightly varied results, sometimes with higher and sometimes with lower differences. The comparison of both methods requires statistical analysis, which was attempted via the application of the Bland-Altman procedure, which allows the identification of any systematic difference between measurements, as well as possible outliers. The mean difference is the estimated bias, and the standard deviation (SD) of the differences measures the random fluctuations around this mean. The 95% limits of agreement, i.e., ±1.96 SD of the difference, are computed to determine the most likely difference between two measurements conducted using two methods. If the differences within the ±1.96 SD are not physically important, the two methods may be used interchangeably. The 95% limits of agreement are often unreliable estimates of population parameters, especially for small sample sizes. For small sets of data, like those in the presented study, it is appropriate to use a two-sided $1 - (\alpha/2)$ value of Student's t-distribution with $(n - 1)$ degree of freedom as a constant, which multiplies the standard deviation when calculating the limits of agreement.

4. Results

The resistance of the wood samples was obtained from the integration method circuit based on the slope of the rise in voltage. The resistance in megaohms was calculated based on data obtained from the calibration procedure. The obtained results for each electrode position differ slightly. Electrodes A and B obtained very similar values, but the values for electrode C were roughly double those measured by A and B (Table 1). This is in accordance with James [18]. Based on the proportional constant identified during the calibration procedure performed for the RC circuit, the resistance of samples was calculated. Positions A and B provided quite similar values, whilst those gained by position C were nearly two times higher (Table 2).

The relation between moisture content and calculated resistance has approximately linear behavior with resistance in logarithm scale (Figures 10–12). Power regression lines have been added to the plots of individual measurements. The values obtained for each probe orientation have quite similar tendencies and sizes. Both methods reach very close values to each other in whole scale range of measurement. The influence of the probe orientation and the direction of the wood grain is about 2% (Figure 13). The curves obtained by the A and B probes are close to each other and lie below the curves obtained by the C probes, which are shifted horizontally by roughly around 2%. The measurement in perpendicular direction of fibers reached approximately two times higher value than those in parallel directions. It practically confirms general assumptions that are well covered in the literature. Relative conductivity values in the longitudinal, radial, and tangential directions are related by the approximate ratio of 1.0:0.55:0.50 [18].

Table 1. Calculated resistances from the RC method in megaohms (MOhm).

							Sample 1						
	7.73%	**8.14%**	**8.85%**	**9.57%**	**10.26%**	**10.61%**	**11.56%**	**11.96%**	**12.25%**	**12.27%**	**13.63%**	**14.34%**	**15.19%**
A	26,550	14,936	6835	2775	1326	612	282	186	170	130	89	49	27
B	27,101	17,174	7220	3083	1461	665	313	201	191	147	102	58	29
C	46,009	28,257	12,046	5502	2601	1253	599	379	355	268	188	101	61
							Sample 2						
	7.56%	**7.98%**	**8.71%**	**9.33%**	**9.98%**	**10.36%**	**11.29%**	**11.71%**	**12.02%**	**12.10%**	**13.45%**	**14.15%**	**15.00%**
A	27,009	17,358	8752	3372	1735	777	345	224	209	162	105	61	32
B	23,725	14,569	6716	2989	1462	672	299	191	185	140	95	54	29
C	30,183	24,550	11,514	5754	3009	1454	650	428	415	282	183	102	52
							Sample 3						
	7.66%	**8.09%**	**8.83%**	**9.50%**	**10.20%**	**10.51%**	**11.43%**	**11.86%**	**12.13%**	**12.19%**	**13.66%**	**14.28%**	**14.95%**
A	24,349	16,239	6550	3453	1473	697	331	222	207	147	95	55	36
B	27,853	17,780	7321	3428	1427	669	302	193	183	140	94	55	33
C	42,294	24,936	10,945	6396	2599	1286	561	369	346	261	173	102	61

Table 2. Calculated resistances from the integration method in MOhm.

	Sample 1												
	7.73%	8.14%	8.85%	9.57%	10.26%	10.61%	11.56%	11.96%	12.25%	12.27%	13.63%	14.34%	15.19%
A	21,565	14,593	6723	2622	1371	621	281	183	174	135	86	45	26
B	22,452	15,499	7151	2918	1496	673	312	198	196	148	101	56	29
C	33,257	24,589	11,950	5109	2687	1299	575	385	367	279	184	99	60
	Sample 2												
	7.56%	7.98%	8.71%	9.33%	9.98%	10.36%	11.29%	11.71%	12.02%	12.10%	13.45%	14.15%	15.00%
A	24,570	17,251	7513	3756	1675	728	353	225	203	149	103	58	30
B	21,639	14,227	6246	3183	1419	645	308	186	180	137	94	58	27
C	31,784	23,127	10,383	6034	2853	1339	654	430	366	276	184	100	50
	Sample 3												
	7.66%	8.09%	8.83%	9.50%	10.20%	10.51%	11.43%	11.86%	12.13%	12.19%	13.66%	14.28%	14.95%
A	19,681	13,700	6558	2833	1496	747	340	213	209	156	97	57	30
B	23,737	15,293	6945	3019	1437	695	304	194	184	156	92	54	31
C	30,981	21,213	11,066	5170	2612	1328	559	379	359	271	171	99	58

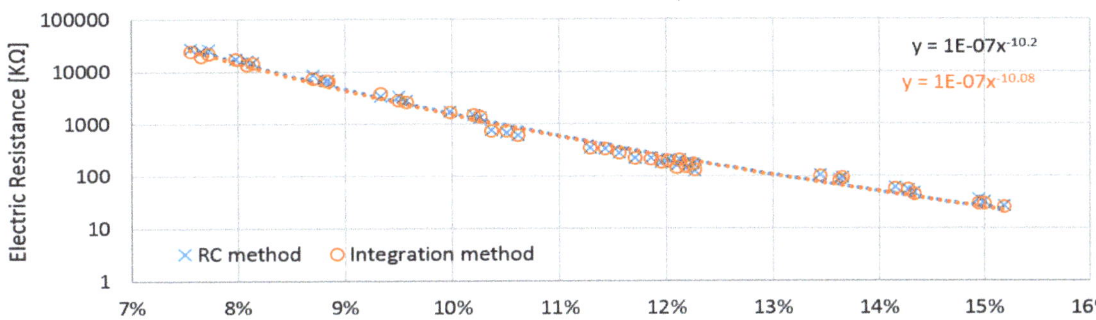

Figure 10. Obtained resistances from the integration and RC methods, the results calculated for the tops of the samples at the "A" electrodes.

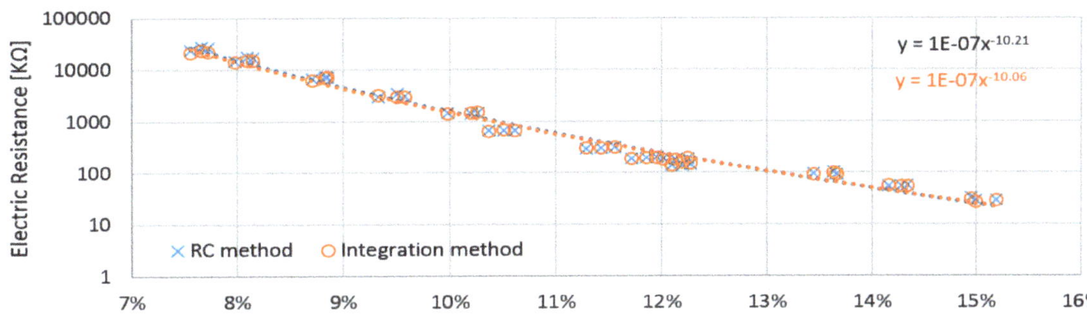

Figure 11. Obtained resistances from the integration and RC methods for the "B" probes placed on the right sides of the samples.

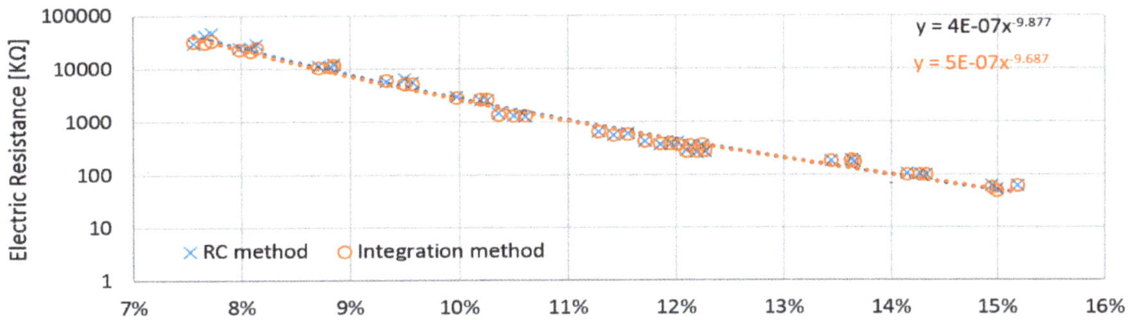

Figure 12. Obtained resistances from the integration and RC methods, the values calculated with the "C" probes placed in the cross sections.

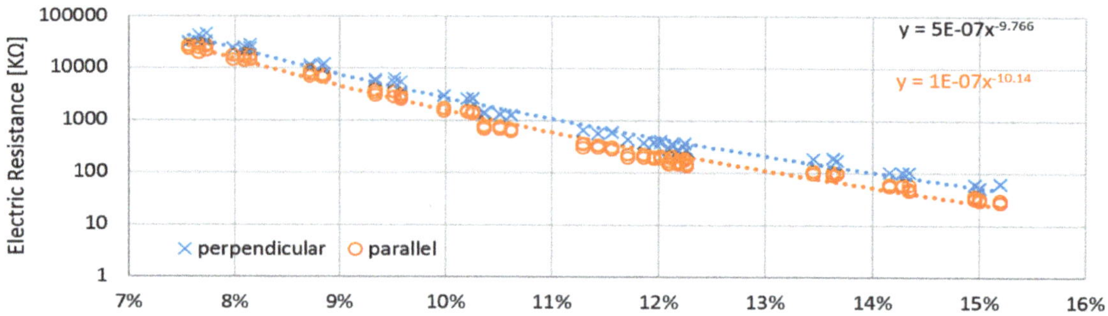

Figure 13. Obtained resistances from the integration and RC methods, the results calculated for the samples at the "A", "B" and "C" electrodes together for parallel and perpendicular direction.

5. Statistical Evaluation of Results

A study by the authors Bland and Altman [28] revealed that any two methods designed to measure the same parameter should show good correlation when a set of samples are chosen for which the parameter to be determined varies considerably. Therefore, a high correlation coefficient obtained for any two methods designed to measure the same property is just a sign that the sample chosen for measurement has a parameter which varies widely. It does not necessarily imply that there is a good agreement between the two methods. Hence, the analysis below was conducted to provide deeper insight into the differences between the two sets of measurements obtained by the two presented methods.

5.1. Construction of Bland-Altman Plots

The x axis represents the mean of the values measured by the RC method and the integration method. The y axis represents the differences between the values obtained by the two described methods. If there is agreement between two methods, the values in Figure 12 are expected to cluster around the mean of the differences (called the bias), and certainly within the limits of agreement. The dashed lines in the plots represent the lower and upper limit of agreement. In the presented case, only two points lie outside the limits of agreement.

5.2. Proportional Measurement Bias

Proportional bias can also be investigated via Bland-Altman plots, which indicate that the methods do not agree equally through the range of measurements. The limits of agreement are then dependent on the actual measurements. When the relationship between differences is identified, e.g., via regression analysis, the regression-based 95% limits of agreement should be provided, or proper transformation of the differences should be conducted. There is dependence in analyzed data between the mean values and differences. The Pearson´s correlation coefficient $r = 0.85$ is high. The found regression curve with regression coefficient R2 (Figure 14). In other words, the limits of agreement are underestimated (too wide) for low values and overestimated (too narrow) for high values. In such cases, logarithmic data transformation can be used. The aim of transformation is to determine the limits of agreement that are valid for the entire range of values. Logarithmically transformed data (resistances measured at position of electrodes A) are derived from the results (Figure 15). The x axis represents the logarithm of mean of the values measured by the RC method and the integration method. The y axis represents the logarithm of differences between the values obtained by both methods.

The regression coefficient decreased ($R^2 = 0.34$) for the transformed data, and the points are more evenly distributed between the new limits of agreement. Nevertheless, three points still lie outside the limits of agreement. Logarithmic transformation was performed also for the two other data sets (resistances measured at position of electrodes B (Figure 16) and at position C (Figure 17), respectively).

Two points corresponding to the moisture 7.33% and 7.66% are outside the limits of agreement in both these cases.

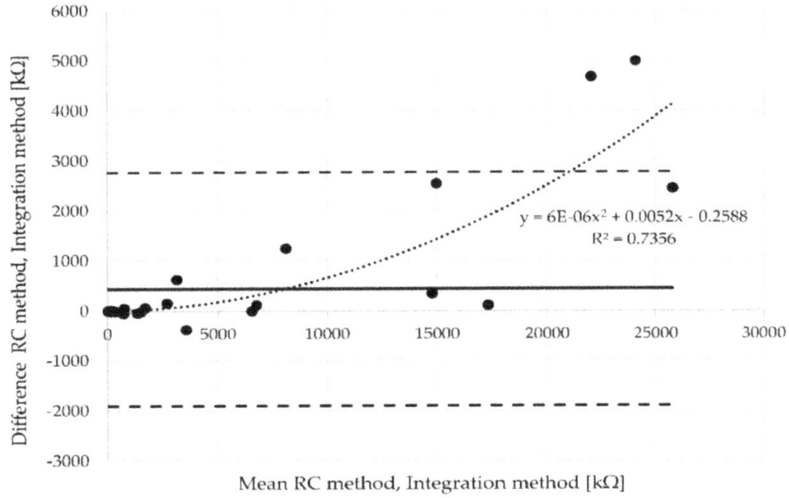

Figure 14. Bland-Altman plot for all samples measured at probe "A".

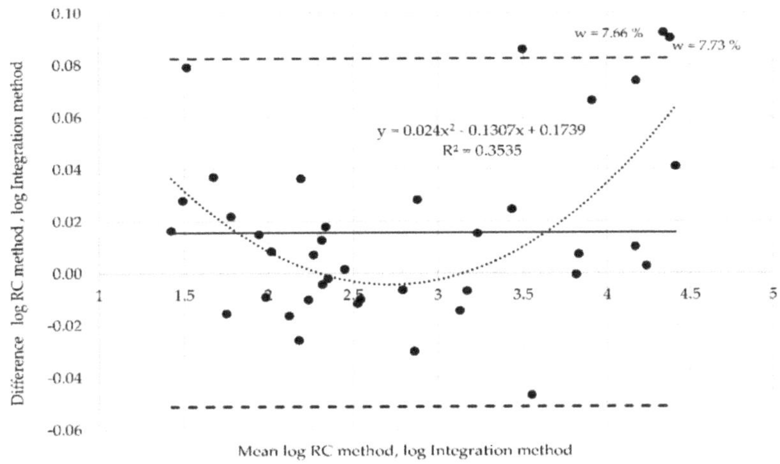

Figure 15. Altman plot for all samples measured at probe "A"—logarithmic transformation.

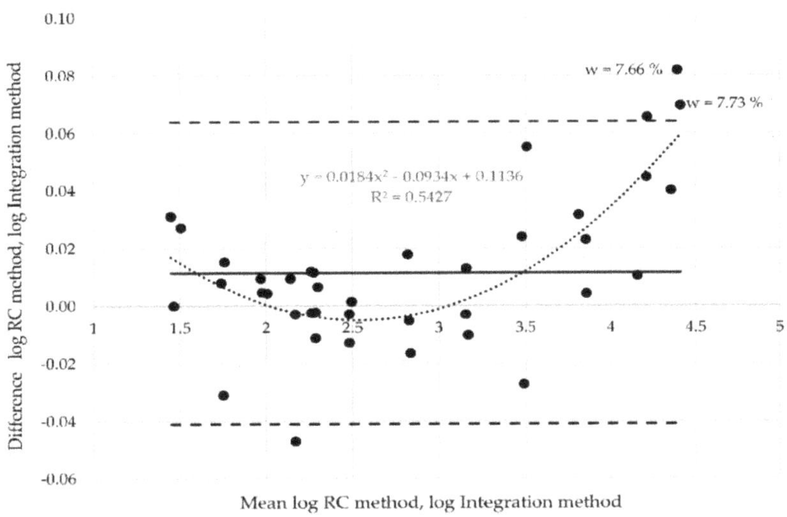

Figure 16. Altman plot for all samples measured at probe "B"—logarithmic transformation.

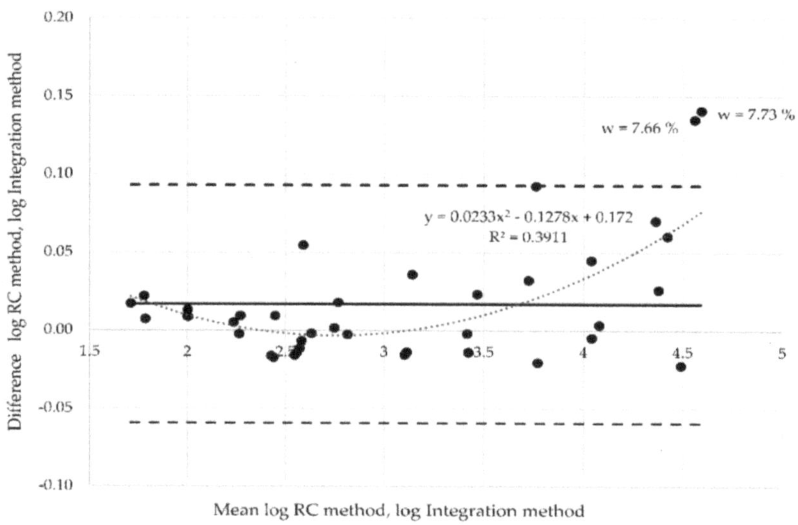

Figure 17. Altman plot for all samples measured at probe "C"—logarithmic transformation.

When the position of the electrode is ignored, there are still a few points out of the limits of agreements. Two of them are of importance, as they are even out of the outlier detector limits (dotted lines) calculated as mean ± 3 standard deviation (SD) (Figure 18).

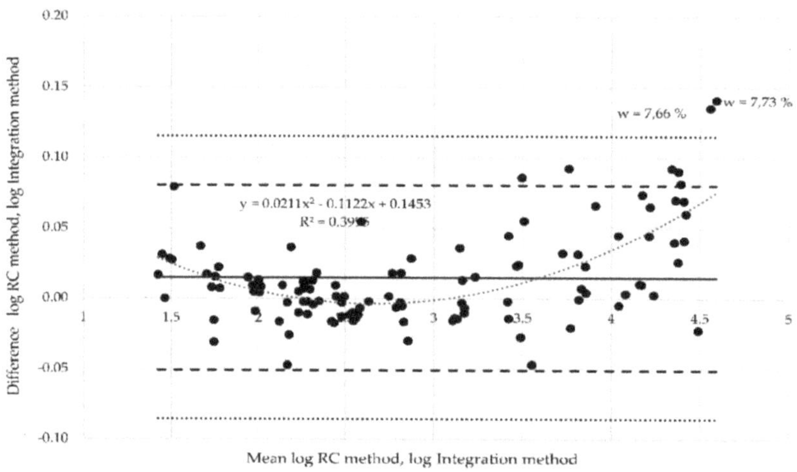

Figure 18. Bland-Altman plot for all samples and all probes—logarithmic transformation.

6. Discussion

The main advantage of RC circuits is the simplicity of adapting timer circuits to emit digital signals. It is easier to integrate this kind of circuit in the sensors utilized by today's digital electronic devices. The key limitation of such circuits is their long capacitor charge period at lower moisture contents. Sometimes, it may be necessary for a capacitor to charge for a relatively huge time; this would not be recorded by the controller, which evaluates the signal from the 555 timer chip at low moisture contents. The second problem is the resolution of charging time, which could be very small at higher moisture contents and, thus, could cause higher repeatability errors to affect measurements.

The integration method shows voltage changes, which allows the identification of lower moisture content much faster than an RC circuit. The main disadvantage of the integration method is the need for a symmetric power supply, as well as other circuits for the post-processing of analog signals. Post-processing circuits for analog signals are expensive and sensitive. The sensitivity of these circuits could lead to worse measurement repeatability and accuracy at lower moisture contents.

The range of differences between the two proposed methods is statistically important for small moisture contents of approximately below 8% (which correspond to high resistances), regardless of the fibers' direction. At moisture contents under 8%, wood has very high resistance and these methods of measurement are affected by significant errors. Based on the results of the statistical analysis, it is recommended that the threshold moisture content for the use of the proposed methods be set at 8%. Both of the proposed methods show good agreement for a range of moisture contents above this threshold.

The threshold value of moisture content 8%, revealed by statistical analysis, is in line with findings of other authors. Papez et al. [29] performed moisture measurement by local resistivity sensors with a measurement range of 7–30%. The calibration curves were developed by Fernandez-Golfin et al. [30] for the estimation of ten hardwoods by means of electrical resistance measurement. All resistance measurements by Fernandez-Golfin et al. were taken by advanced laboratory tool AGILENT 4339B high resistance meter. They reported 8.0% as the lowest value of measurable moisture content.

The direct comparison with laboratory methods was not performed in this study. However, the uncertainty of the presented method can be deduced from comparison with other authors. Some of them, e.g., Fernandez-Golfin et al. [30] and Moron et al. [31], developed similar circuits and calibrated them with laboratory methods. Fernandez-Golfin et al. [30] predicted the moisture content of wood with an error ±1.0%. They also revealed that the measuring direction has negligible influence (error of estimation <±0.5%) on moisture content estimates, below the moisture content of about 15%. Moron et al. [31] compared results of moisture content measurement conducted by their capacitance meters with those based on the variation of electromagnetic transmittance of timber. They reported similar accuracy (below 1%) between both methods in measurement range 1 MΩ to 100 GΩ. Their results show how low power and low-cost circuits can be similar to high precision, cost, and size instruments.

7. Conclusions

Based on the data obtained, both tested methods—RC and integration method—can successfully identify moisture changes within the range of 8–15%, with a resolution and high accuracy of about 0.5% of moisture content. Furthermore, the advantages and disadvantages of the methods have been analyzed. The RC circuit is easier to implement thanks to the digital signals of the 555 chip, whilst the integration method allows faster measurement at lower moisture contents. The RC method seems to be more suitable for intelligent sensors integrated within the structure of a building to perform its long-term monitoring. Low moisture content significantly limits the effectiveness of the two methods due to the high resistance of the wood, and both methods can fail in this situation. At moisture contents higher than 8%, both methods seem to be adequately suitable. Based on an analysis of the results of both methods, moisture values ranging from 8–15% also reach similar levels from a statistical point of view (the differences between both methods are within the limits of agreement for this moisture content range). When two methods are compared to determine whether they can be used interchangeably, the 'true' value of the measured quantity is unknown. Hence, the comparability differs from calibration, which is the case when a true measure is compared with measurement by a new method. Linear regression is sometimes used inaccurately for comparison of two methods. A high correlation coefficient is just a sign that a widespread sample has been chosen to measure; however it does not imply that there is a good agreement between the two methods. The Bland-Altman technique was adopted for assessment of the magnitude of disagreement here, and based on the statistical analysis, the conformity of the measurement with the integrator and the RC circuit was provided. The accuracy of measurements with RC resistance is then supported by comparison with RC resistance that is already introduced by other authors [25,26]. Statistically, the relation between moisture content and calculated resistance has approximately linear behavior with resistance in logarithm scale. Furthermore, it was clearly demonstrated that the influence of the probe orientation and the direction of the wood grain is about 2%.

It is planned that further research will focus on the improvement of circuits and signal processing from an electrical point of view. Further measurements with higher moisture contents and various kinds of wood need to be analyzed using the two methods to expand the options for their application in buildings.

Author Contributions: R.S. conceived and designed the main concept of the presented research. J.Š. analyzed and described the statistical analysis of the presented study. M.Č. compiled the texts and prepared the draft of the manuscript. R.S. proposed and performed the experimental analysis and described it in the manuscript.

References

1. Kosny, J.; Asiz, A.; Smith, I.; Shrestha, S.; Fallahi, A. A review of high-value wood framed and composite wood wall technologies using advanced insulation techniques. *Energy Build.* **2014**, *72*, 441–456. [CrossRef]
2. Norberg, P. Monitoring wood moisture content using the WETCORR method Part 1: Background and theoretical considerations. *Holz Als Roh Werkstoff* **1999**, *57*, 448–453. [CrossRef]
3. Isaksson, T.; Thelandersson, S. Experimental investigation on the effect of detail design on wood moisture content in outdoor above ground applications. *Build. Environ.* **2013**, *59*, 239–249. [CrossRef]
4. Németh, R.; Tsalagkas, D.; Bak, M. Effect of soil contact on the modulus of elasticity of beeswax-impregnated wood. *BioResources* **2015**, *10*, 1574–1586. [CrossRef]
5. Dietsch, P.; Franke, S.; Franke, B.; Gamper, A.; Winter, S. Methods to determine wood moisture content and their applicability in monitoring concepts. *J. Civ. Struct. Health Monit.* **2015**, *5*, 115–127. [CrossRef]
6. Brischke, C.; Thelandersson, S. Modelling the outdoor performance of wood products—A review on existing approaches. *Constr. Build. Mater.* **2014**, *66*, 384–397. [CrossRef]
7. Brischke, C.; Meyer, L.; Bornemann, T. The potential of moisture content measurements for testing the durability of timber products. *Wood Sci. Technol.* **2013**, *43*, 869–886. [CrossRef]
8. Tanaka, T.; Avramidis, S.; Shida, S. Evaluation of moisture content distribution in wood by soft X-ray imaging. J Wood Sci. Modelling the outdoor performance of wood products—A review on existing approaches. *Constr. Build. Mater.* **2009**, *55*, 69–73.
9. Tanaka, T.; Kawai, Y. A new method for nondestructive evaluation of solid wood moisture content based on dual-energy X-ray absorptiometry. *Wood Sci. Technol.* **2013**, *47*, 1213–1229. [CrossRef]
10. Maeda, K.; Ohta, M.; Momohara, I. Relationship between the mass profile and the strength property profile of decayed wood. *Wood Sci. Technol.* **2015**, *49*, 331–344. [CrossRef]
11. Hultnas, M.; Fernandez-Cano, V. Determination of the moisture content in wood chips of Scots pine and Norway spruce using Mantex desktop scanner based on dual energy X-ray absorptiometry. *J. Wood Sci.* **2012**, *58*, 309–314. [CrossRef]
12. Meyer, L.; Brischke, C. Fungal decay at different moisture levels of selected European-grown wood species. *Int. Biodeter. Biodeger.* **2015**, *103*, 23–29. [CrossRef]
13. Brischke, C.; Rapp, O.A. Influence of wood moisture content and wood temperature on fungal decay in the field: Observations in different micro-climates. *Wood Sci. Technol.* **2008**, *42*, 663–677. [CrossRef]
14. Van den Bulcke, J.; Van Acker, J.; De Smet, J. An experimental set-up for real-time continuous moisture measurements of plywood exposed to outdoor climate. *Build. Environ.* **2009**, *44*, 2368–2377. [CrossRef]
15. Otten, K.A.; Brischke, C.; Meyer, C. Material moisture content of wood and cement mortars—Electrical resistance-based measurements in the high ohmic range. *Constr. Build. Mater.* **2017**, *153*, 640–646. [CrossRef]
16. Dunlap, M.E. *Electrical Moisture Meters for Wood*; United States Department of Agriculture, Forest Service, Forest Product Laboratory: Madison, WI, USA, 1944.
17. *Wood Handbook—Wood as an Engineering Material*; General Technical Report FPL-GTR-190; Department of Agriculture, Forest Service, Forest Products Laboratory: Madison, WI, USA, 2010; p. 508.
18. James, W.L. *Electric Moisture Meters for Wood*; General Technical Report FPL-GTR-6; United States Department of Agriculture, Forest Service, Forest Products Laboratory: Madison, WI, USA, 1988; Available online: http://www.fpl.fs.fed.us/documnts/fplgtr/fplgtr06.pdf (accessed on 1 March 2019).
19. Casans, S.; Iakymchuk, T.; Rosado-Muñoz, A. High resistance measurement circuit for fiber materials: Application to moisture content estimation. *Measurement* **2018**, *119*, 167–174. [CrossRef]

20. Brischke, C.; Rapp, O.A.; Bayerbach, R. Measurement system for long-term recording of wood moisture content with internal conductively glued electrodes. *Build. Environ.* **2008**, *43*, 1566–1574. [CrossRef]
21. Gao, S.; Bao, Z.; Wang, L.; Yue, X. Comparison of voltammetry and digital bridge methods for electrical resistance measurements in wood. *Comput. Electron. Agr.* **2018**, *145*, 161–168. [CrossRef]
22. Chetpattananondh, P.; Thongpull, K.; Chetpattananondh, K. Interdigital capacitance sensing of moisture content in rubber wood. *Comput. Electron. Agric.* **2017**, *142*, 545–551. [CrossRef]
23. Bland, J.M.; Altman, D.G. Statistical methods for assessing agreement between two methods of clinical measurement. *Lancet* **1986**, *327*, 307–310. [CrossRef]
24. Hanneman, S.K. Design, analysis and interpretation of method-comparison studies. *AACN Adv. Critic. Care* **2009**, *19*, 19–234.
25. Vodicka, A. Increased Moisture Detection and Early Warning System for Buildings. Ph.D. Thesis, Czech Technical University, Praha, Czech Republic, 2015. Available online: https://dspace.cvut.cz/ (accessed on 1 April 2019).
26. Mlejnek, P.; Vodička, A.; Včelák, J. Zařízení pro detekci vlhkosti dřeva. Available online: https://stavba.tzb-info.cz/drevene-konstrukce/12437-zarizeni-pro-detekci-vlhkosti-dreva (accessed on 19 September 2018).
27. Aguilar, H.M.; Landín, R.O. A simple technique for high resistance measurement. *Phys. Edu.* **2012**, *47*, 599–602. [CrossRef]
28. Bland, J.M.; Altman, D.G. Measuring agreement in method comparison studies. *Stat. Methods Med. Res.* **1999**, *8*, 135–160. [CrossRef] [PubMed]
29. Papez, J.; Kic, P. Wood moisture of rural timber construction. *Agron. Res.* **2013**, *11*, 505–512.
30. Fernandez-Golfin, J.I.; Conde, M.; Calvo, R.; Baonza, M.V.; De Palacios, A.P.; Palacios, Y. Curves for the estimation of the moisture content of ten hardwoods by means of electrical resistance measurements. *Forest Syst.* **2012**, *21*, 121–127. [CrossRef]
31. Moron, C.; Garcia-Fuentevilla, L.; Garcia, A.; Moron, A. Measurement of moisture in wood for application in the restoration of old buildings. *Sensors* **2016**, *16*, 697. [CrossRef]

6

Identification of the Degree of Degradation of Fibre-Cement Boards Exposed to Fire by Means of the Acoustic Emission Method and Artificial Neural Networks

Krzysztof Schabowicz [1], Tomasz Gorzelańczyk [1,*] and Mateusz Szymków [1]

Faculty of Civil Engineering, Wrocław University of Science and Technology, Wybrzeże Wyspiańskiego 27, 50-370 Wrocław, Poland; krzysztof.schabowicz@pwr.edu.pl (K.S.); mat.szymkow@gmail.com (M.S.)

* Correspondence: tomasz.gorzelanczyk@pwr.edu.pl

Abstract: This paper presents the results of research aimed at identifying the degree of degradation of fibre-cement boards exposed to fire. The fibre-cement board samples were initially exposed to fire at various durations in the range of 1–15 min. The samples were then subjected to three-point bending and were investigated using the acoustic emission method. Artificial neural networks (ANNs) were employed to analyse the results yielded by the acoustic emission method. Fire was found to have a degrading effect on the fibres contained in the boards. As the length of exposure to fire increased, the fibres underwent gradual degradation, which was reflected in a decrease in the number of acoustic emission (AE) events recognised by the artificial neural networks as accompanying the breaking of the fibres during the three-point bending of the sample. It was shown that it is not sufficient to determine the degree of degradation of fibre-cement boards solely on the basis of bending strength (*MOR*).

Keywords: fibre-cement boards; non-destructive testing; acoustic emission; degree of degradation

1. Introduction

The fibre-cement board is a building material that has been used in construction since the beginning of the last century. The Czech engineer Ludwig Hatschek developed and patented the technology for the production of this composite material. The first boards contained asbestos fibres. After asbestos had been found to be carcinogenic, they were replaced with cellulose fibres and synthetic fibres [1]. Being composed of cement, cellulose fibres, synthetic fibres and various innovative additives and admixtures, the currently produced fibre-cement boards are a completely different building product. The added components and fillers of fibre-cement boards are: limestone powder, mica, perlite, kaolin, microspheres and recycled materials [2,3]. As a result, fibre-cement boards continue to be innovative products consistent with the principles of sustainable development [4]. Fibre-cement boards are used in construction mainly for ventilated façade cladding [5], as shown in Figure 1. In the course of their service life, fibre-cement boards are exposed to variable environmental factors, chemical aggressivity (acid rains), and physical aggressivity (ultraviolet radiation). Moreover, fibre-cement boards are exposed to accidental operational factors, mainly the high temperature that a building fire generates.

(a)

(b)

Figure 1. Exemplary uses of fibre-cement boards as ventilated façade cladding: (**a**) University building in Łódz, Poland, (**b**) University building in Wrocław, Poland.

After a building fire occurs, it is necessary to determine the impact of the high temperatures on the façade cladding of the building and the neighbouring buildings with regard to their further service life. This means that identifying the degree of degradation of fibre-cement boards caused by the impact of fire is vital from both the scientific and practical point of view. Most of the research on fibre-cement boards to date has been limited to determining their standard physicomechanical properties, the effect of operational factors, such as soak–dry cycles, freeze–thaw cycles, heating, raining and high temperatures, and the effect of the use of various types of fibres and production processes, solely through bending (*MOR*) tests [6]. In the literature on the subject, one can find only a few nondestructive tests carried out on fibre-cement boards, limited to the imperfections arising during production [7–10]. The effect of fire is one of the most destructive accidental factors for many building products, especially the composite ones containing reinforcement in the form of fibres of various kinds, particularly cellulose fibres, considering that at a temperature above 200 °C cellulose fibres undergo pyrolysis. Thus, the action of fire critically affects the durability of the whole composite [6]. In order to verify this hypothesis, experiments were conducted during which fibre-cement board samples were exposed to fire at a temperature of about 400 °C at various durations in the range of 1–15 min. This time range and the temperature of 400 °C had been experimentally determined through many preliminary trials. The aim of the experimental research is to observe the advance of the degradation of the fibres due to fire and the associated high temperature. Preliminary tests conducted at the temperature of 500 °C very quickly, i.e., after 1–2 min, resulted in the complete destruction of the board, which no longer could be subjected to the bending test. A reduction in the temperature to 250–300 °C would considerably lengthen the duration of the tests. Preliminary investigations showed that, at a temperature of 250 °C, the fibres in the boards would be completely destroyed after 2–3 h. Therefore, in the authors' opinion, the choice of the temperature of 400 °C was optimal for conducting experiments aimed at identifying the degree of degradation of the fibres contained in fibre-cement boards under the high temperature generated by fire.

After they had been exposed to fire, the board samples were subjected to three-point bending, during which the acoustic emission (AE) was recorded, and the results were then analysed by means of artificial neural networks. Research carried out by the authors has shown that the identification of the degree of degradation of fibre-cement boards exposed to fire, based on only bending strength (*MOR*), is inadequate [11,12]. Thanks to the use of the acoustic emission method, the degree of degradation can be determined on the basis of not only the mechanical parameters, but also the acoustic phenomena that can occur in fibre-cement boards. From the recorded AE signals, model (reference) characteristics of the acoustic spectra that accompany the breaking of the cement matrix and the fibres during bending were derived. Under the continuing exposure to fire, the fibres in the boards would undergo gradual degradation, which was reflected in a decrease in the counts of AE events identified as accompanying fibre breaking. This phenomenon is described in more detail in this paper.

2. Survey of the Literature

Most of the research on fibre-cement boards to date has been devoted to the effect of operational factors [13–15] and the effect of high temperatures, investigated by testing the physicomechanical parameters of the boards, mainly their bending strength (*MOR*). The paper by Ardanuy et al. [6] presents the results of research on, inter alia, the effect of high temperature on fibre-cement boards, but only with reference to bending strength *MOR*. Li et al. [16] studied the effect of high temperatures on composites produced using the extrusion method, but also solely on the basis of their mechanical properties. The nondestructive investigations of fibre-cement boards have been mainly limited to the detection of imperfections arising at the production stage. Papers by Drelich et al. [8] and Schabowicz and Gorzelańczyk [17] presented the possibility of using Lamb waves in a noncontact ultrasonic scanner to detect defects in fibre-cement boards at the production stage. Paper [18] by Stark describes a method of detecting delaminations in composite elements by means of a moving ultrasonic probe. An ultrasonic device and a way of detecting delaminations are described in work [7] by Dębowski, Lewandowski, Mackiewicz, Ranachowski and Schabowicz. In works by Berkowski et al. [19], Hoła and Schabowicz [20] and Davis et al. [21] it was proposed to use the impact-echo method jointly with the impulse response method to identify delaminations in concrete members. Since the impulse response method is used to test elements thicker than 100 mm, it is not suitable for testing fibre-cement boards. Preliminary research had shown that striking such a board with a hammer could damage it, which is another reason why the impulse response method is not suitable for fibre-cement boards. Also, the impact echo method is not used to test fibre-cement boards. The drawback of this method is that multiple reflections of waves cause disturbances, making the interpretation of the obtained image difficult [19]. Therefore, it is not recommended to use this method for fibre-cement boards thicker than about 8 mm.

In the literature, there is little information on the use of other nondestructive methods for testing fibre-cement boards. The research described in the works of Chady et al. [9] and Chady and Schabowicz [22] showed the terahertz (T-Ray) method to be suitable for testing fibre-cement boards. Terahertz signals have a very similar character to that of ultrasonic signals, but their interpretation is more complicated. In work [23], the microtomography method was used to identify delaminations and low-density regions in fibre-cement boards. The test results indicate that this method clearly reveals differences in the microstructure of the boards. Therefore, the microtomography method can be a useful tool for testing the structure of fibre-cement boards in which defects can arise due to production errors. However, this method can be used only for small-sized boards. It should be noted that, so far, few cases of testing fibre-cement boards by means of an acoustic emission have been reported in the literature. Ranachowski and Schabowicz et al. [23] carried out pilot studies of fibre-cement boards produced using the extrusion method, including boards subjected to the temperature of 230 °C for 2 h, in which the acoustic emission method was used to determine the effect of cellulose fibres on the strength of the boards and where attempts were made to distinguish between the AE events emitted by the fibres and the cement matrix. The results of this research confirmed the suitability of this method for testing fibre-cement boards. In paper [11] by Gorzelańczyk et al., it was proposed to use the acoustic emission method to investigate the impact of high temperature on fibre-cement boards. It should be noted that the effects of high temperatures on concrete, and the interdependences involved in this process, have been widely described using the acoustic emission method; examples are the works [24,25] by Ranachowski. One should also note that the acoustic emission method is often used to test thin materials, for example steel and polymeric composites, and even to test brittle food products [26,27]. During acoustic emission measurements a lot of data are recorded, which need to be properly analysed and interpreted. For this purpose, it can be highly effective to combine the acoustic emission method with artificial intelligence, including artificial neural networks (ANNs). In the literature, ANNs are used to analyse and recognise the signals registered during the failure of various materials [28]. In paper [29], Schabowicz used ANNs to analyse the results of nondestructive tests carried out on concrete. Łazarska et al. [30] and Woźniak et al. [31] successfully used the acoustic emission method and artificial neural networks to analyse its results. Rucka and Wilde [32,33] successfully used the ultrasonic method to investigate damage to concrete structures.

Considering the above information, it was decided that the acoustic emission method combined with artificial neural networks would be proper for identifying the degree of degradation of fibre-cement boards exposed to fire.

3. Strength Tests

The effect of fire was investigated for two series of fibre-cement boards, designated respectively A and B. The basic parameters of the two series are specified in Table 1.

Table 1. The tested fibre-cement boards series A and B and their basic parameters.

Series Name	Board Thickness e (mm)	Board Colour	Application	Board Bulk Density ρ (g/cm^3)	Bending Strength MOR (MPa)
A	8.0	natural	exterior	1.60	25
B	8.0	pigmented	exterior	1.60	30

The effect of fire on the fibre-cement boards was investigated by applying a flame generating a temperature of about 400 °C to the surface of the board, for about 1–15 min, every 2.5 min. The reference samples under an air-dry condition were designated as A and B. The designations of the other series of boards are listed in Table 2.

Table 2. A list of fibre-cement board series and test cases with the adopted sample designations.

Series Name/Test Case	SERIES A	SERIES B
Air-dry condition (reference board)	A_1	B_1
Exposure to the temperature of 400 °C: 1 min	A_2	B_2
Exposure to the temperature of 400 °C: 2.5 min	A_3	B_3
Exposure to the temperature of 400 °C: 5 min	A_4	B_4
Exposure to the temperature of 400 °C: 7.5 min	A_5	B_5
Exposure to the temperature of 400 °C: 10 min	A_6	B_6
Exposure to the temperature of 400 °C: 12.5 min	A_7	B_7
Exposure to the temperature of 400 °C: 15 min	A_8	B_8

Since not the whole fire separation area, but the fibre-cement board alone was the object of the test, on the basis of experiments the duration of exposure to fire was limited to 15 min. A stand for exposing the samples to fire at the temperature of 400 °C is shown in Figure 2.

Figure 2. The stand for exposing samples to fire at the temperature of 400 °C.

In order to identify the effect of the high temperature generated by fire on the fibre-cement boards, acoustic emission investigations were carried out during three-point bending. Before the investigations, the samples were exposed to fire, then they were placed in a strength testing machine and subjected to three-point bending. In the course of the three-point bending, the breaking force F, the strain ε and acoustic emission signals were registered. The tested samples are shown in Figure 3, while the stand for three-point bending tests and the set of equipment for acoustic emission measurements are shown in Figure 4.

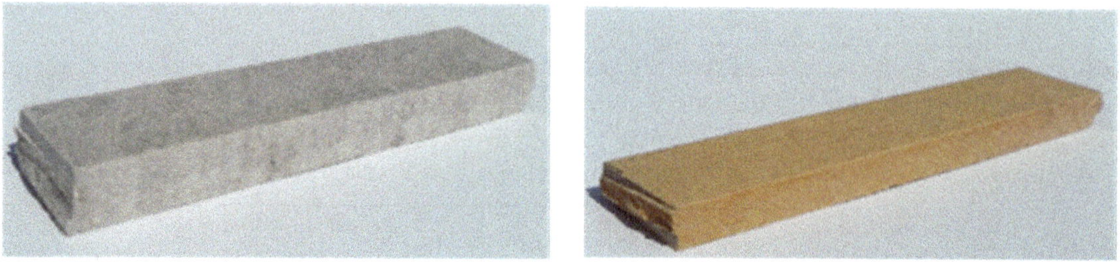

(a) (b)

Figure 3. The tested fibre-cement board samples: (**a**) board A, (**b**) board B.

Figure 4. The test stand for the acoustic emission measurements, and a fibre-cement board during a test.

The trace of flexural strength σ_m, bending strength *MOR*, notch toughness W_f, limit of proportionality (*LOP*) and strain ε were analysed. Bending strength *MOR* was calculated from the standard formula [34]:

$$MOR = \frac{3Fl_s}{2b\,e^2} \tag{1}$$

where:

F is the loading force (N);
l_s is the length of the support span (mm);
B is the specimen width (mm); and
e is the specimen thickness (mm).

Notch toughness was calculated from the following formula given in [35]:

$$W_f = \frac{1}{S} \int_{F0}^{F0,4max} F\,da \tag{2}$$

where:

F is the loading force (N);

S is the specimen cross-section (m2); and
a is the specimen deflection under the loading roller (m).

Figure 5 shows diagrams of the $\sigma-\varepsilon$ dependence under bending for the samples of all the tested fibre-cement boards.

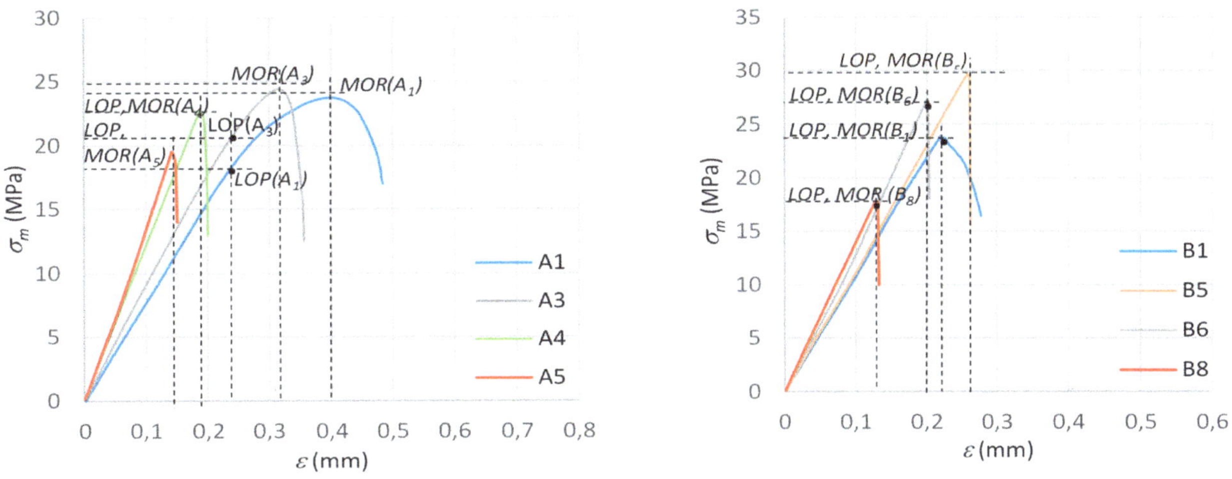

(**a**) (**b**)

Figure 5. $\sigma-\varepsilon$ dependence under bending for fibre-cement boards: (**a**) series A, (**b**) series B. *LOP*, limit of proportionality.

The results presented in Figure 5 show that, under high temperature, the bending strength *MOR* initially increases by 3–15% and then sharply decreases in comparison with that of the reference fibre-cement boards. In the initial phase of exposure to fire, especially in the time interval of 1–2.5 min, the sample would dry, but no fibres would be destroyed yet. It should be noted that the reference samples were under an air-dry condition and their bulk moisture content amounted to 6–8%. One can suppose that the dampness of the fibre-cement boards affects the bending strength *MOR* mainly as a result of the weakening of the bonds between the crystals in the structural lattice of the cement matrix. The weakening is due to the fact that, as the material's dampness increases, the bonds partially dissolve, whereby the bending strength slightly decreases. The decrease in bending strength *MOR* caused by dampness is partially reversible; i.e., after it is dried, the material regains most of its lost strength and so ultimately its strength is close to the initial one. In the case of the B_1–B_8 series samples, no significant changes in the $\sigma-\varepsilon$ dependence diagrams for the reference boards and the ones exposed to fire were observed, except for the changes in bending strength values. In the case of the A series boards, the $\sigma-\varepsilon$ dependence changed markedly with the time of their exposure to fire. On this basis, for the boards of series A, one can determine the effect of high temperature, reflected not only in the initial increase in bending strength *MOR*, followed by its decrease, but also in changes in the plot of $\sigma-\varepsilon$ dependence. For the series A boards, it was also observed that as the time of their exposure to the temperature of 400 °C is extended, the structure of the fibre-cement board becomes stiffer and more brittle. For this series, as the time of exposure to the temperature generated by fire is extended, the extent of the nonlinear increase in bending stress is reduced until the bending strength *MOR* comes to be level with the proportionality limit *LOP*. In the case of the reference boards, the *MOR* and *LOP* values were clearly separated.

Figure 6 shows the obtained values of notch toughness W_f of the tested fibre-cement boards.

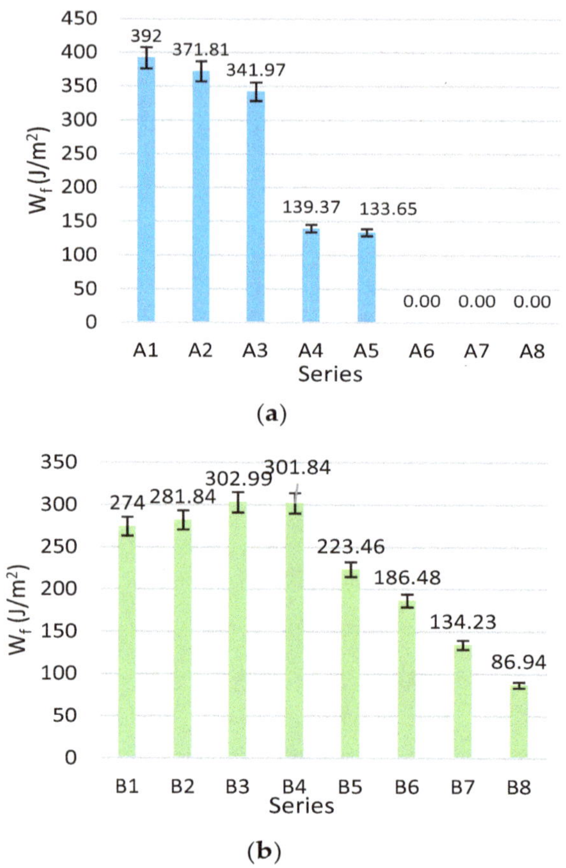

Figure 6. The values of notch toughness W_f for air-dry samples (case 1) and samples exposed to the temperature of 400 °C (cases 2–8): (**a**) series A, (**b**) series B.

An analysis of the results presented in Figure 6 showed that under the impact of high temperature, the notch toughness W_f of the boards of series B initially increases by 2–10% and then markedly decreases. For the boards of series A, the notch toughness W_f decreases with the time of exposure to fire. Hence, one can conclude that, as a result of the high temperature, the fibre-cement board matrix stiffens and becomes more brittle. The investigations carried out for the boards of series B showed that the identification of the degree of degradation on the basis of bending strength *MOR*, notch toughness W_f, proportionality limit *LOP* and $\sigma-\varepsilon$ dependence is inadequate. The results obtained for the fibre-cement boards of series A indicate that destructive changes took place in their structure. However, the fact that the bending strength *MOR* increased in the initial phase of the exposure to the high temperature is worth noting. The above results show that the identification of the degree of degradation solely on the basis of mechanical parameters, such as *MOR* and W_f, is insufficient to determine the destructive effect of high temperature on fibre-cement boards in the initial phase of their exposure to fire.

4. Tests Using the Acoustic Emission Method and Artificial Neural Networks

The next step in the investigations of the degree of degradation caused by fire consists in analysing the AE signals registered during three-point bending, such as the events rate N_{ev}, the events sum $\sum N_{ev}$, the events energy E_{ev} and the frequency distribution of the AE signal. For AE measurements, a broadband sensor with a frequency band of 5–500 kHz and an AE signal analyser (made by IPPT PAN, Warsaw, Poland), were used [35]. In the AE analyser the signal from the sensor is amplified and prefiltered to remove the acoustic background from the surroundings of the monitored element. Then, the signal is converted into a digital form. Further processing of digital records was carried out using audio file analysis software (made by IPPT PAN, Warsaw, Poland). Figure 7 shows exemplary values of events sum $\sum N_{ev}$ for the boards of series A1–A6 and B1–B8. No three-point bending test was carried out for cases A6-A8 because the fibre-cement boards had been completely destroyed by fire.

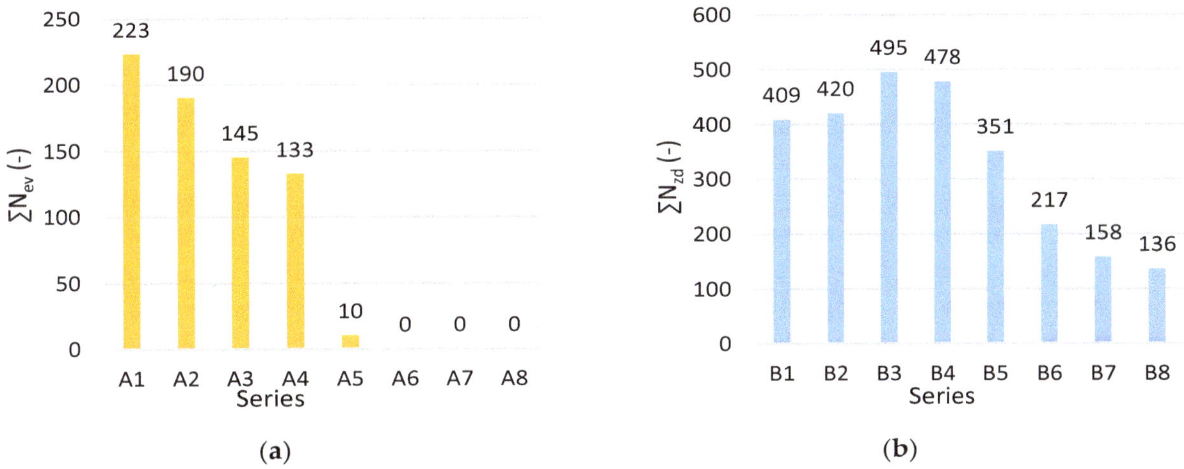

Figure 7. Exemplary registered values of events sum $\sum N_{ev}$ for boards of series A under an air-dry condition (case 1) and exposed to the temperature of 400 °C (cases 2–8): (**a**) series A, (**b**) series B.

In order to more closely analyse the course of the bending test and how it was affected by the degrading factor, in the form of the high temperature of 400 °C lasting from 1 to 15 min, the dependence between events rate $\sum N_{ev}$ and bending stress σ_m over time for selected cases is presented in Figure 8.

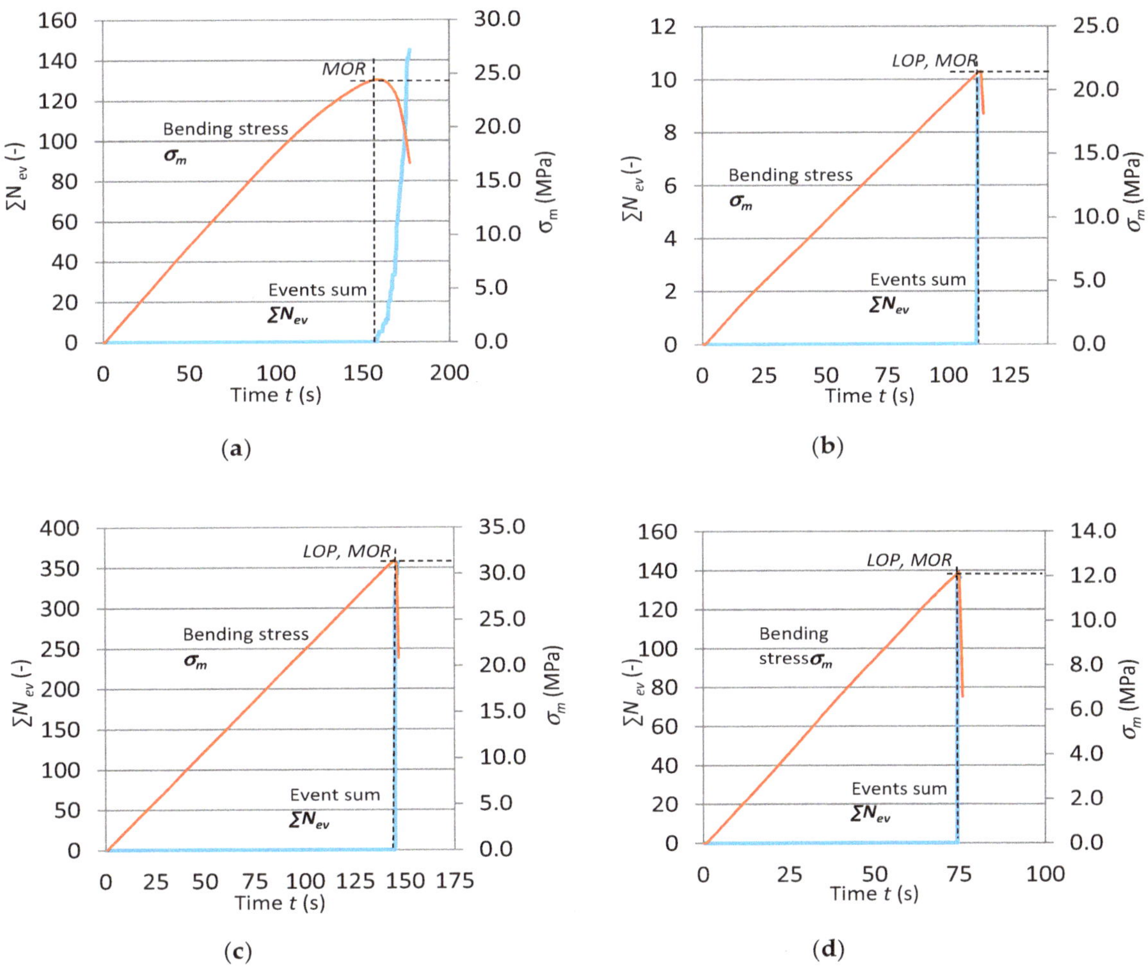

Figure 8. Dependence between σ_m and $\sum N_{ev}$ as function of time t for fibre-cement boards: (**a**) series A_3, (**b**) series A_5, (**c**) series B_5, (**d**) series B_8.

An analysis of the obtained results showed that AE signals were registered only after the stress corresponding to bending strength *MOR* had been exceeded. The analysis of the course of the three-point bending test, based on AE events sum $\sum N_{ev}$, confirmed the changes taking place in the fibre-cement boards. Marked qualitative changes in the registered events in comparison with the reference boards were visible. At the high temperature of about 400 °C, the sample is initially dried, whereby the strength of the board and the acoustic activity of the AE events registered during bending slightly increase. An analysis of the AE descriptors indicates that the brief effect of the high temperature of 400 °C proved to be significant for the fibre-cement boards. The graphs for series and A and B show a noticeable decrease in the registered events and a change in the path of events sum $\sum N_{ev}$. Besides the decrease in the value of events sum $\sum N_{ev}$ due to the exposure to the high temperature of 400 °C for 7.5 min in the case of series A_5 and for 12.5 min in the case of series B_7, it was also noticed that all the registered events were within one 1.5 second time interval. Therefore, it can be concluded that the events originated from a single fracture of the cement matrix. This indicates that the exposure to the high temperature of 400 °C for about 7.5 min was more destructive for the fibre-cement boards of series A. Because of the destructive effect of the high temperature of about 400 °C, the registered AE event counts declined.

By analysing the acoustic activity traces in the time-frequency system during the bending of the boards of series A and B exposed to the temperature of 400 °C for over 7.5 min, model (reference) acoustic spectra for the breaking of the cement matrix devoid of fibres were selected. A model characteristic of the fibre breaking acoustic spectrum was selected from the spectra read for the boards of series A_1 to B_1 under an air-dry condition, with a similar characteristic repeating itself in the frequency range of 10–24 kHz, clearly standing out against the characteristic of the cement matrix. The characteristic of the background acoustic spectrum originating from the strength tester was determined on the basis of the initial bending phase by averaging the characteristics obtained from all of the tested fibre-cement boards of series A and B. The selected spectral characteristics of fibre-breaking are understood to be the signal accompanying the breaking of the cement matrix together with the fibres, whereas the model matrix spectral characteristic is understood to be the signal accompanying the breaking of the matrix alone. The selected model (reference) acoustic spectrum characteristics were recorded in 80 intervals at every 0.5 kHz. Figure 9 shows the record of the model characteristics of the acoustic spectrum accompanying the breaking of, respectively, the cement matrix and the fibres, and of the background acoustic spectrum.

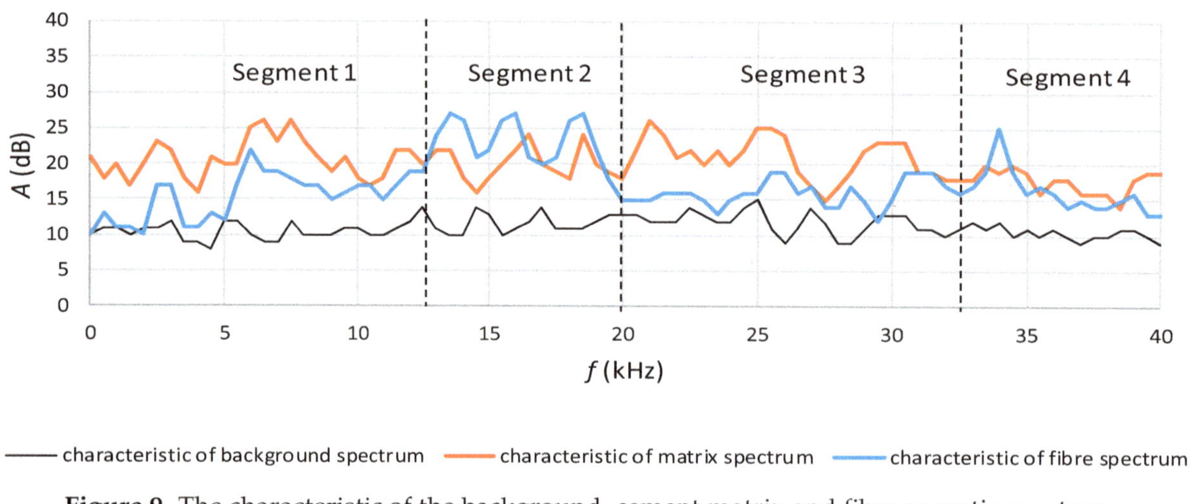

Figure 9. The characteristic of the background, cement matrix and fibre acoustic spectrum.

According to Figure 9, the acoustic activity of the background is in the range of 10–15 dB. The characteristic of the cement matrix acoustic spectrum reached the acoustic activity of 25 dB in the range of 5–10 kHz (segment 1) and 20–32 kHz (segment 3). In the case of the fibres, the activity was above 25 dB in the frequency range of 12–18 dB (segment 2) and 32–38 kHz (segment 4). The above reference spectra for the cement matrix, the fibres and the background were implemented in artificial neural networks (ANNs) for training and testing. A unidirectional multilayered ANN structure with error backpropagation with momentum was adopted. Eight appropriate learning sequences were adopted in an iterative manner to achieve optimal compatibility of the learned ANN with the training pattern. After the ANN had been trained on the input data, its mapping correctness was verified using the training data and the testing data. For this purpose, two pairs of input data were fed, i.e., the data used for training the ANN, to check its ability to reproduce the reference spectra, and the one used for testing the ANN, to check its ability to identify the reference spectral characteristics originating from the fibres and the cement matrix during the bending test. For the eight performed training sequences, ANN compliance was obtained with the training standard at 0.995. Then, a record of the ANN output in the form of recognised acoustic spectra for, respectively, fibre breaking, matrix breaking and the background was obtained.

Figure 10 shows the results of the acoustic spectrum pattern recognition for the cement matrix and the fibres. The results are marked on the record of events rate N_{ev} and flexural stress σ_m versus time. The graphs are for the reference samples of series A_1 and the samples of series A_3–A_5 exposed to fire. In order to better distinguish between the recognised acoustic spectra, the matrix patterns were marked green while the ones for the fibres were marked in violet.

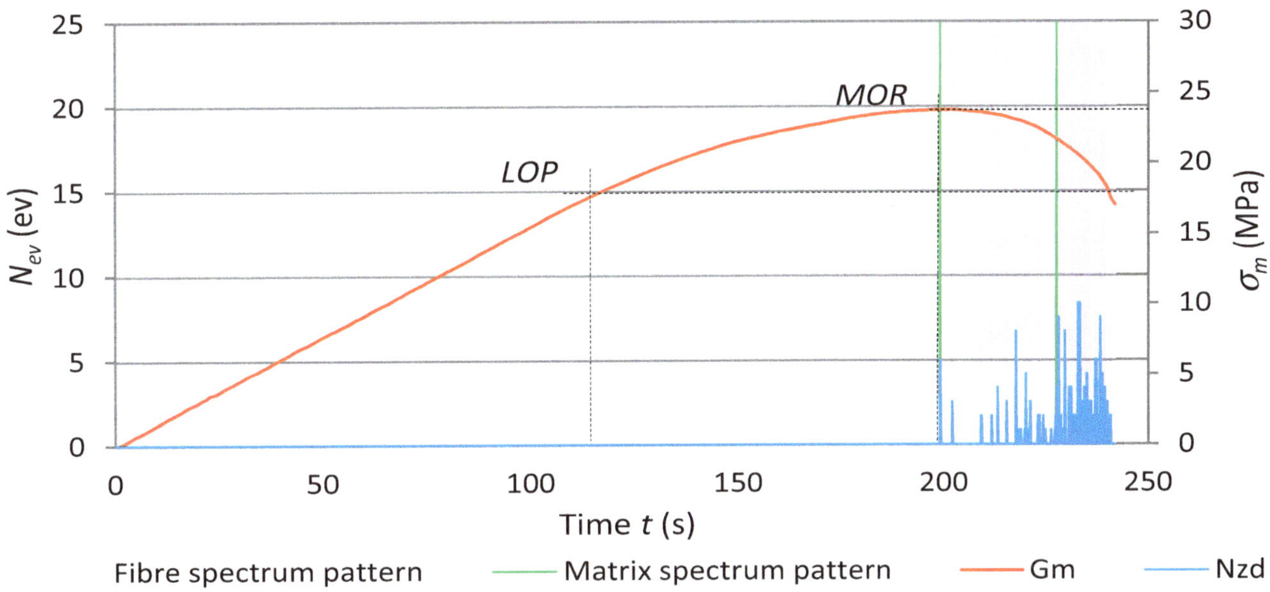

(a)

Figure 10. *Cont.*

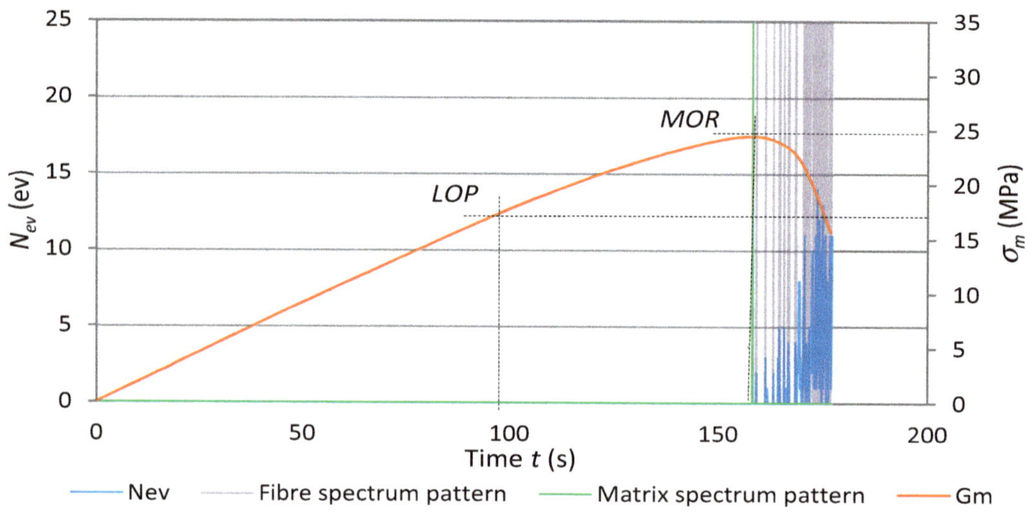

(**b**)

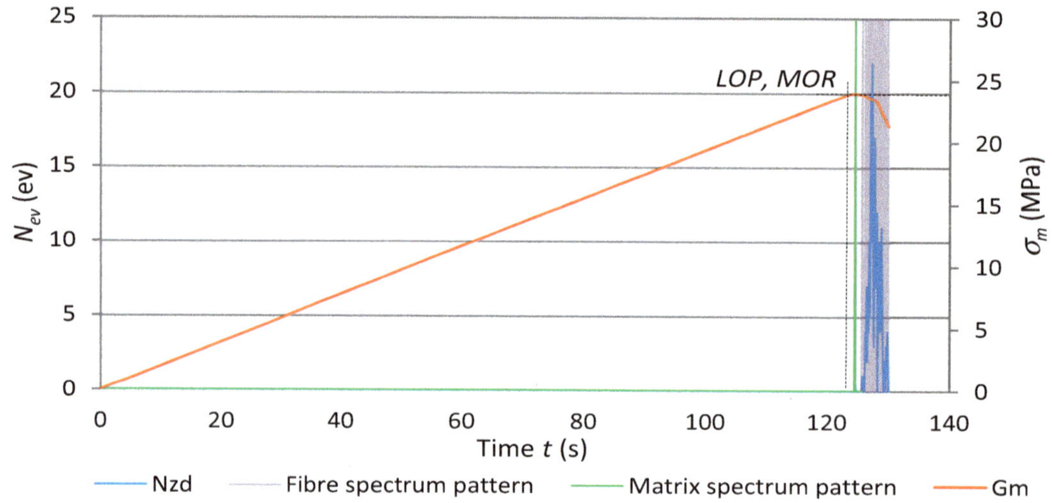

(**c**)

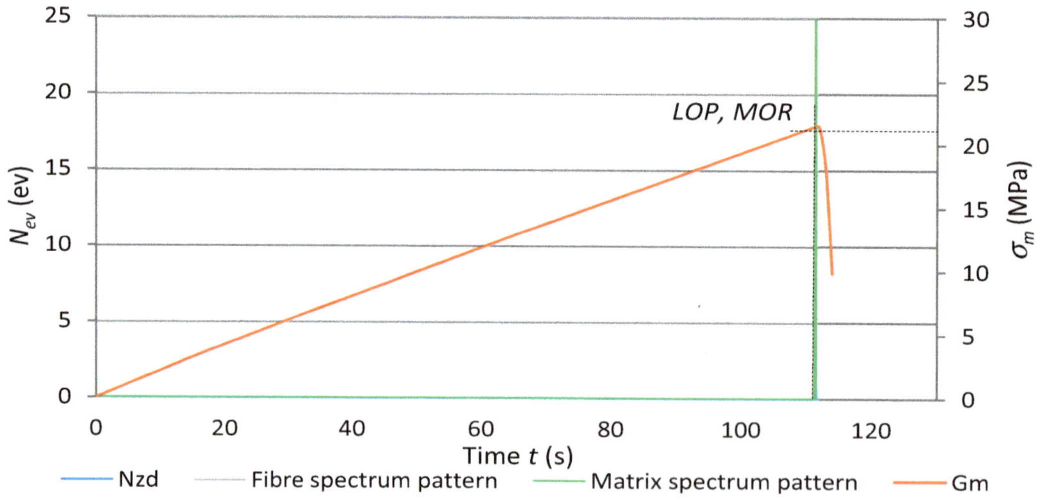

(**d**)

Figure 10. The events rate N_{ev} and flexural stress σ_m versus time for exposure to fire, with marked recognised reference spectral characteristics: (**a**) series A_1, (**b**) series A_3, (**c**) series A_4, (**d**) series A_5.

An analysis of the results presented in Figure 10 showed that the registered acoustic events repeatedly predominate after the bending strength *MOR* is exceeded. The recognised events result from the breaking of the fibres and the cement matrix. An event originating from cement matrix breaking initiates subsequent events originating from fibre breaking. The matrix fractures when the bending strength *MOR* is reached. Subsequently, events originating from the breaking of the fibres build up until the board's cross-section fractures. In Figure 10c, which shows the recognised events for the boards of series A_4 exposed to the high temperature for 5 min, one can clearly see that the drop in the counts of acoustic events originating from the fibres subjected to the high temperature is smaller. Similarly to the case of the boards of series A_3, the fracture of the cement matrix initiates the breaking of the fibres. The graph of flexural stress σ_m has no segment characterised by the linear increase of stress relative to strain. The limit of proportionality coincides with the bending strength *MOR*.

For the series A_5 sample, exposed to the high temperature of 400 °C for 7.5 min, only events recognised as originating from the breaking of the matrix were registered within the time of 0.1 s. All of the fibres in this board were destroyed. The samples of series A_6–A_8 exposed to fire for over 7.5 min underwent complete destruction, whereby no further tests could be carried out. Table 3 lists the events recognised as accompanying fibre breaking and cement matrix breaking for the fibre-cement boards of series A_2–A_5.

Table 3. A list of events recognised as accompanying fibre breaking and cement matrix breaking for boards of series A_1–A_8.

Series	Events Sum $\sum N_{ev}$	Sum of Recognised Events $\sum N_{ev,r}$	Sum of Events Ascribed to Fibre Breaking $\sum N_{ev,f}$	Sum of Events Ascribed to Matrix Breaking $\sum N_{ev,m}$
A_1	223	203	195	8
A_2	190	183	169	14
A_3	145	144	70	70
A_4	133	130	45	85
A_5	92	92	0	92

Figure 11 shows diagrams for samples of series B1, B5 and B8.

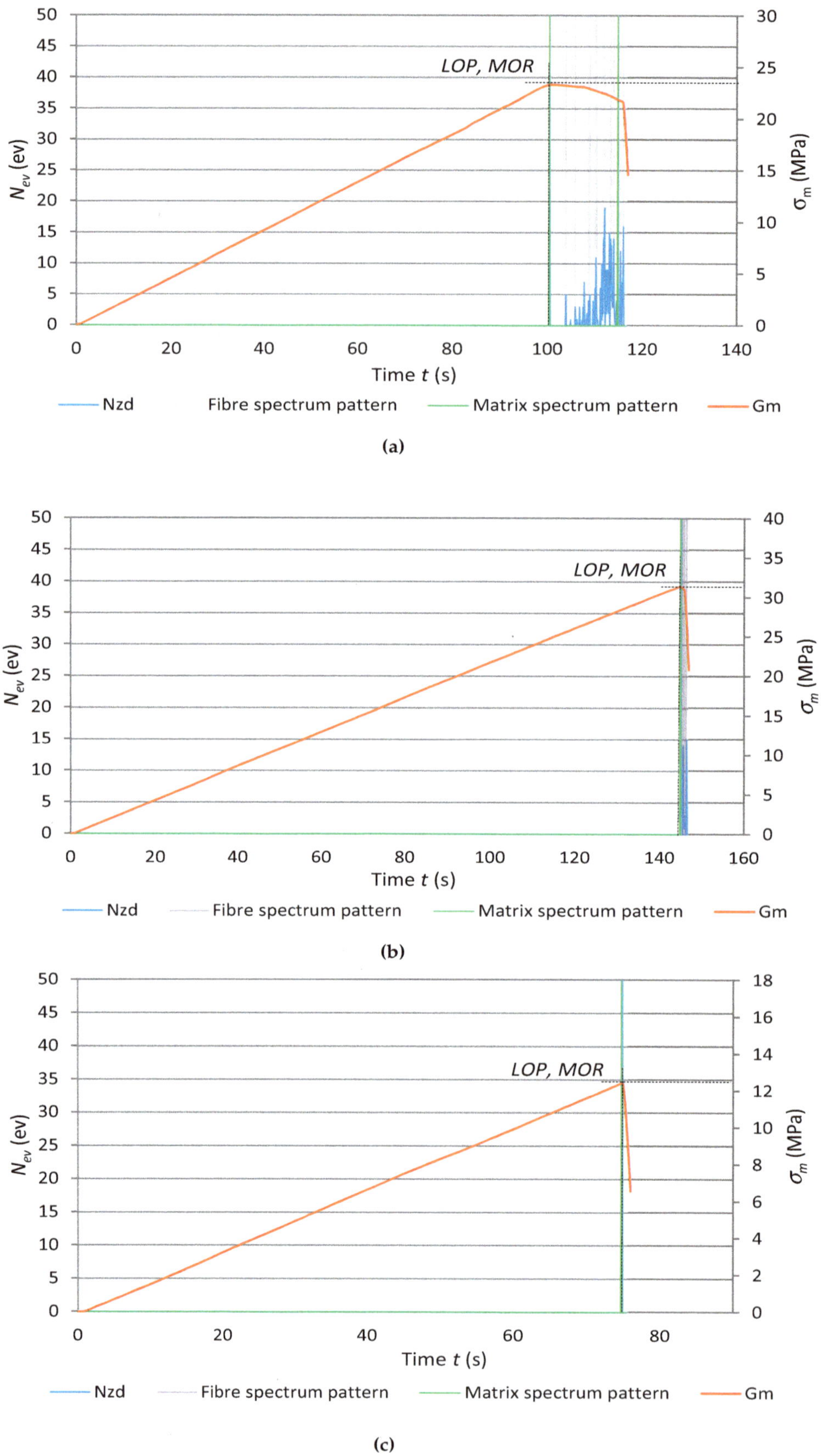

Figure 11. The events rate N_{ev} and flexural stress σ_m versus time for exposure to fire, with marked recognised reference spectral characteristics: (**a**) series B_1, (**b**) series B_5, (**c**) series B_8.

In Figure 11, one can see a decline in the counts of events recognized for the fibres and a considerably shorter time segment in which events were registered. It should be noted that, in the case of the fibre-cement boards of series B, distinct changes in the registered AE descriptors occurred only when the boards were exposed to the high temperature for 7.5 min. Moreover, the proportionality limit of the boards of series B was equal to their bending strength *MOR*. In Figure 11c, which shows the results for series B_8 exposed to the high temperature of 400 °C for 15 min, one can see that the recognised events originate from cement matrix breaking, but there are no recognised events originating from the fibres.

Table 4 lists the events recognised as accompanying fibre breaking and cement matrix breaking for boards of series B_1–B_8.

Table 4. A list of events recognised as accompanying fibre breaking and cement matrix breaking for boards of series B_1–B_8.

Series	Events Sum $\sum N_{ev}$	Sum of Recognised Events $\sum N_{ev,r}$	Sum of Events Ascribed to Fibre Breaking $\sum N_{ev,f}$	Sum of Events Ascribed to Matrix Cracking $\sum N_{ev,m}$
B_1	409	405	325	80
B_2	420	413	319	94
B_3	495	484	336	148
B_4	478	475	315	160
B_5	351	349	198	151
B_6	217	207	54	153
B_7	158	158	0	158
B_8	114	114	0	114

The above investigations have shown that it is insufficient to determine the degree of degradation solely on the basis of the value of bending strength *MOR*. As part of the investigations, the degree of degradation was determined not only on the basis of bending strength *MOR*, but also using the number of the registered AE events recognised (by artificial neural networks) as accompanying fibre breaking, and the dependence between notch toughness W_f and the events recognised as originating from fibre breaking. In this way, the degree of degradation caused by the high temperature could be unequivocally identified. As the number of events recognised as fibre breaking decreases, the notch toughness W_f also commensurately decreases. The sharp drop in notch toughness W_f is connected with the large number of fibres contained in fibre-cement boards and their rapid degradation under high temperature. It was experimentally found that the coefficient U_i determines the significance of the degradation of fibre-cement boards caused by high temperature. The coefficient U_i is expressed by the formula:

$$U_i = \frac{W_{fi}}{W_{f,ref}} \tag{3}$$

where: U_i is a value of the notch toughness coefficient common for samples exposed to high temperature and for the reference samples, W_{fi} is the notch toughness of a fibre-cement board exposed to high temperature, and $W_{f,ref}$ is the notch toughness of the reference board.

For the notch toughness coefficient U_i, its average value U_{av} and the lower and upper limit of the confidence interval were determined. The confidence level was assumed to amount to 95% of the average value expressed by the formulas:

$$U_L = U_{av} - 0.75 \times \overline{s} \tag{4}$$

$$U_H = U_{av} + 0.75 \times \overline{s} \tag{5}$$

where: U_L is the lower limit of the confidence interval (the confidence amounting to 95% of the average value of the notch toughness coefficient U), U_H is the upper limit of the confidence interval (the confidence amounting to 95% of the average value of notch toughness coefficient U), U_{av} is the average value of the notch toughness coefficient, and $\bar{s}$ is the standard deviation.

$U_H = 0.85$ was adopted as the boundary value of the upper limit of the confidence interval.

Hence, insignificant degradation is characterised by an insignificant decrease in registered events recognised as accompanying fibre breaking or its absence, in comparison with the reference fibre-cement boards. An insignificant decrease in events recognised as accompanying fibre breaking is such a decrease for which the notch toughness W_f satisfies the condition: $U_L > 0.85$. At the same time, the bending strength *MOR* condition, defined as $R_L > 0.75$, must be satisfied. Significant degradation is characterised by a significant drop in registered events recognised as accompanying fibre breaking, in comparison with the reference board. A significant drop in registered events recognised as accompanying fibre breaking is such a drop for which the notch toughness W_f satisfies condition $U_L < 0.85$. Simultaneously, the bending strength *MOR* condition, defined as $R_L > 0.75$, must be satisfied. Critical degradation is characterised by the total absence of or a significant drop in events recognised as accompanying fibre breaking, in comparison with the reference board, and when the fibre-cement boards do not satisfy the bending strength *MOR* condition, defined as $R_L < 0.75$.

Tables 5 and 6 show the degrees of degradation identified for the tested samples of the fibre-cement boards of, respectively, series A and B.

Table 5. The identified degrees of degradation of the tested fibre-cement boards of series A.

Series	R_L (-)	U_L (-)	Degree of Degradation
A_2	0.95	0.92	Insignificant degradation
A_3	1.00	0.79	Significant degradation
A_4	0.80	0.35	Significant degradation
A_5	0.7	–	Critical degradation

Table 6. The identified degrees of degradation of the tested fibre-cement boards of series B.

Series	R_L (-)	U_L (-)	Degree of Degradation
B_2	0.96	0.99	Insignificant degradation
B_3	0.94	1.00	Insignificant degradation
B_4	1.00	1.00	Insignificant degradation
B_5	1.00	0.81	Significant degradation
B_6	0.91	0.68	Significant degradation
B_7	0.56	–	Critical degradation
B_8	0.54	–	Critical degradation

5. Conclusions

High temperatures by nature have a degrading effect on most building products. The latter's resistance to high temperatures is measured by the length of time during which they preserve the properties required by the standards. The tests carried out on the fibre-cement boards of series A and B showed that they differed in the length of time during which they were resistant to fire. It should be emphasised that the tests have shown that it not sufficient to determine the resistance to high temperature solely on the basis of bending strength *MOR*. An analysis of the events registered during the bending test and their assignment to the signals accompanying the breaking of the fibres or the cement matrix broaden the range of investigations whereby one can correctly determine the high-temperature resistance of fibre-cement boards. Thanks to the use of the nondestructive methods and artificial neural networks, the degree of degradation of the fibre-cement boards exposed to the high temperature generated by fire was successfully determined. The following three degrees of degradation: insignificant degradation, significant degradation and critical degradation, were defined on the basis of the test results. Owing to the formulation of the dependence between the notch toughness W_f and the events originating from fibre breaking, registered during three-point bending, the degree of degradation of the fibre-cement boards exposed to high temperatures could be properly parameterised. The dependence between the notch toughness W_f and the counts ($\sum N_{ev,f}$) of recognised events accompanying fibre breaking during the bending of the fibre-cement boards was used for this purpose.

Author Contributions: K.S. conceived of and designed the experimental work and completed the experiments; T.G. analysed the test results and performed editing; and M.S. prepared the specimens.

References

1. Schabowicz, K.; Gorzelańczyk, T. Fabrication of fibre cement boards. In *The Fabrication, Testing and Application of Fibre Cement Boards*, 1st ed.; Ranachowski, Z., Schabowicz, K., Eds.; Cambridge Scholars Publishing: Newcastle upon Tyne, UK, 2018; pp. 7–39. ISBN 978-1-5276-6.
2. Bentchikou, M.; Guidoum, A.; Scrivener, K.; Silhadi, K.; Hanini, S. Effect of recycled cellulose fibres on the properties of lightweight cement composite matrix. *Constr. Build. Mater.* **2012**, *34*, 451–456. [CrossRef]
3. Savastano, H.; Warden, P.G.; Coutts, R.S.P. Microstructure and mechanical properties of waste fibre–cement composites. *Cem. Concr. Compos.* **2005**, *27*, 583–592. [CrossRef]
4. Coutts, R.S.P. A Review of Australian Research into Natural Fibre Cement Composites. *Cem. Concr. Compos.* **2005**, *27*, 518–526. [CrossRef]
5. Schabowicz, K.; Szymków, M. Ventilated facades made of fibre-cement boards. *Materiały Budowlane* **2016**, *4*, 112–114. (In Polish) [CrossRef]
6. Ardanuy, M.; Claramunt, J.; Toledo Filho, R.D. Cellulosic Fibre Reinforced Cement-Based Composites: A Review of Recent Research. *Constr. Build. Mater.* **2015**, *79*, 115–128. [CrossRef]
7. Dębowski, T.; Lewandowski, M.; Mackiewicz, S.; Ranachowski, Z.; Schabowicz, K. Ultrasonic tests of fibre-cement boards. *Przegląd Spawalnictwa* **2016**, *10*, 69–71. (In Polish) [CrossRef]
8. Drelich, R.; Gorzelanczyk, T.; Pakuła, M.; Schabowicz, K. Automated control of cellulose fibre cement boards with a non-contact ultrasound scanner. *Autom. Constr.* **2015**, *57*, 55–63. [CrossRef]
9. Chady, T.; Schabowicz, K.; Szymków, M. Automated multisource electromagnetic inspection of fibre-cement boards. *Autom. Constr.* **2018**, *94*, 383–394. [CrossRef]
10. Schabowicz, K.; Jóźwiak-Niedźwiedzka, D.; Ranachowski, Z.; Kudela, S.; Dvorak, T. Microstructural characterization of cellulose fibres in reinforced cement boards. *Arch. Civ. Mech. Eng.* **2018**, *4*, 1068–1078. [CrossRef]
11. Gorzelańczyk, T.; Schabowicz, K.; Szymków, M. Non-destructive testing of fibre-cement boards, using acoustic emission. *Przegląd Spawalnictwa* **2016**, *88*, 35–38. (In Polish) [CrossRef]
12. Adamczak-Bugno, A.; Gorzelańczyk, T.; Krampikowska, A.; Szymków, M. Non-destructive testing of the structure of fibre-cement materials by means of a scanning electron microscope. *Badania Nieniszczące i Diagnostyka* **2017**, *3*, 20–23. (In Polish)
13. Claramunt, J.; Ardanuy, M.; García-Hortal, J.A. Effect of drying and rewetting cycles on the structure and physicochemical characteristics of softwood fibres for reinforcement of cementitious composites. *Carbohydr. Polym.* **2010**, *79*, 200–205. [CrossRef]
14. Mohr, B.J.; Nanko, H.; Kurtis, K.E. Durability of kraft pulp fibre-cement composites to wet/dry cycling. *Cem. Concr. Compos.* **2005**, *27*, 435–448. [CrossRef]
15. Pizzol, V.D.; Mendes, L.M.; Savastano, H.; Frías, M.; Davila, F.J.; Cincotto, M.A.; John, V.M.; Tonoli, G.H.D. Mineralogical and microstructural changes promoted by accelerated carbonation and ageing cycles of hybrid fibre–cement composites. *Constr. Build. Mater.* **2014**, *68*, 750–756. [CrossRef]
16. Li, Z.; Zhou, X.; Bin, S. Fibre-Cement extrudates with perlite subjected to high temperatures. *J. Mater. Civ. Eng.* **2004**, *3*, 221–229. [CrossRef]
17. Schabowicz, K.; Gorzelańczyk, T. A non-destructive methodology for the testing of fibre cement boards by means of a non-contact ultrasound scanner. *Constr. Build. Mater.* **2016**, *102*, 200–207. [CrossRef]
18. Stark, W. Non-destructive evaluation (NDE) of composites: Using ultrasound to monitor the curing of composites. In *Non-Destructive Evaluation (NDE) of Polymer Matrix Composites. Techniques and Applications*, 1st ed.; Karbhari, V.M., Ed.; Woodhead Publishing Limited: Cambridge, UK, 2013; pp. 136–181. ISBN 978-0-85709-344-8.
19. Berkowski, P.; Dmochowski, G.; Grosel, J.; Schabowicz, K.; Wójcicki, Z. Analysis of failure conditions for a dynamically loaded composite floor system of an industrial building. *J. Civ. Eng. Manag.* **2013**, *19*, 529–541. [CrossRef]

20. Hoła, J.; Schabowicz, K. State-of-the-art non-destructive methods for diagnostic testing of building structures—Anticipated development trends. *Arch. Civ. Mech. Eng.* **2010**, *10*, 5–18. [CrossRef]
21. Davis, A.; Hertlein, B.; Lim, K.; Michols, K. Impact-echo and impulse response stress wave methods: Advantages and limitations for the evaluation of highway pavement concrete overlays. In Proceedings of the Conference on Nondestructive Evaluation of Bridges and Highways, Scottsdale, AZ, USA, 4 December 1996; pp. 88–96. [CrossRef]
22. Chady, T.; Schabowicz, K. Non-destructive testing of fibre-cement boards, using terahertz spectroscopy in time domain. *Badania Nieniszczące i Diagnostyka* **2016**, *1–2*, 62–66. (In Polish)
23. Schabowicz, K.; Ranachowski, Z.; Jóźwiak-Niedźwiedzka, D.; Radzik, Ł.; Kudela, S.; Dvorak, T. Application of X-ray microtomography to quality assessment of fibre cement boards. *Constr. Build. Mater.* **2016**, *110*, 182–188. [CrossRef]
24. Ranachowski, Z. The Application of neural networks to classify the acoustic emission waveforms emitted by the concrete under thermal stress. *Arch. Acoust.* **1996**, *21*, 89–98.
25. Ranachowski, Z.; Jóźwiak-Niedźwiedzka, D.; Brandt, A.M.; Dębowski, T. Application of acoustic emission method to determine critical stress in fibre reinforced mortar beams. *Arch. Acoust.* **2012**, *37*, 261–268. [CrossRef]
26. Jawaid, M.; Khalil, H.P.S.A. Cellulosic/synthetic fibre reinforced polymer hybrid composites: A review. *Carbohydr. Polym.* **2011**, *86*, 1–18. [CrossRef]
27. Marzec, A.; Lewicki, P.; Ranachowski, Z.; Debowski, T. The influence of moisture content on spectral characteristic of acoustic signals emitted by flat bread samples. In Proceedings of the AMAS Course on Nondestructive Testing of Materials and Structures, Centre of Excellence for Advanced Materials and Structures, Warszawa, Poland, 20–22 May 2002; Deputat, J., Ranachowski, Z., Eds.; pp. 127–135.
28. Yuki, H.; Homma, K. Estimation of acoustic emission source waveform of fracture using a neural network. *NDT E Int.* **1996**, *29*, 21–25. [CrossRef]
29. Schabowicz, K. Neural networks in the NDT identification of the strength of concrete. *Arch. Civ. Eng.* **2005**, *51*, 371–382.
30. Łazarska, M.; Woźniak, T.; Ranachowski, Z.; Trafarski, A.; Domek, G. Analysis of acoustic emission signals at austempering of steels using neural networks. *Met. Mater. Int.* **2017**, *23*, 426–433. [CrossRef]
31. Woźniak, T.Z.; Ranachowski, Z.; Ranachowski, P.; Ozgowicz, W.; Trafarski, A. The application of neural networks for studying phase transformation by the method of acoustic emission in bearing steel. *Arch. Metal. Mater.* **2014**, *59*, 1705–1712. [CrossRef]
32. Rucka, M.; Wilde, K. Experimental study on ultrasonic monitoring of splitting failure in reinforced concrete. *J. Nondestruct. Eval.* **2013**, *32*, 372–383. [CrossRef]
33. Rucka, M.; Wilde, K. Ultrasound monitoring for evaluation of damage in reinforced concrete. *Bull. Pol. Acad. Sci. Tech. Sci.* **2015**, *63*, 65–75. [CrossRef]
34. EN 12467. *Cellulose Fibre Cement Flat Sheets. PRODUCT Specification and Test Methods*; Standards Australia International Ltd: Strathfield, Australia, 2000.
35. Ranachowski, Z.; Schabowicz, K. The contribution of fibre reinforcement system to the overall toughness of cellulose fibre concrete panels. *Constr. Build. Mater.* **2017**, *156*, 1028–1034. [CrossRef]

7

Waste Brick Dust as Potential Sorbent of Lead and Cesium from Contaminated Water

Barbora Doušová [1,*], David Koloušek [1], Miloslav Lhotka [1], Martin Keppert [2], Martina Urbanová [3], Libor Kobera [3] and Jiří Brus [3]

1 University of Chemistry and Technology Prague, Faculty of Chemical Technology, Technická 5, 166 28 Praha 6, Czech Republic; kouloused@vscht.cz (D.K.); lhotkam@vscht.cz (M.L.)
2 Faculty of Civil Engineering, Czech Technical University in Prague, Thákurova 7, 166 29 Praha 6, Czech Republic; martin.keppert@fsv.cvut.cz
3 Institute of Macromolecular Chemistry AS CR, Heyrovského nám. 2, 162 06 Praha 6, Czech Republic; urbanova@imc.cas.cz (M.U.); kobera@imc.cas.cz (L.K.); brus@imc.cas.cz (J.B.)
* Correspondence: dousovab@vscht.cz

Abstract: Adsorption properties of waste brick dust (WBD) were studied by the removing of Pb^{II} and Cs^{I} from an aqueous system. For adsorption experiments, 0.1 M and 0.5 M aqueous solutions of Cs^+ and Pb^{2+} and two WBD (Libochovice—LB, and Tyn nad Vltavou—TN) in the fraction below 125 μm were used. The structural and surface properties of WBD were characterized by X-ray diffraction (XRD) in combination with solid-state nuclear magnetic resonance (NMR), supplemented by scanning electron microscopy (SEM), specific surface area (S_{BET}), total pore volume and zero point of charge (pH_{ZPC}). LB was a more amorphous material showing a better adsorption condition than that of TN. The adsorption process indicated better results for Pb^{2+}, due to the inner-sphere surface complexation in all Pb^{2+} systems, supported by the formation of insoluble $Pb(OH)_2$ precipitation on the sorbent surface. A weak adsorption of Cs^+ on WBD corresponded to the non-Langmuir adsorption run followed by the outer-sphere surface complexation. The leachability of Pb^{2+} from saturated WBDs varied from 0.001% to 0.3%, while in the case of Cs^+, 4% to 12% of the initial amount was leached. Both LB and TN met the standards for Pb^{II} adsorption, yet completely failed for any Cs^{I} removal from water systems.

Keywords: waste brick dust; adsorption; lead; cesium; surface complexation; precipitation; solid-state NMR spectroscopy

1. Introduction

The contamination of waters and soils by toxic carcinogens and radioactive nuclides is a pressing environmental problem. There is particular concern for accumulative poisonous effects in local environments.

Lead (Pb) is one such element that has well known chronic influence on the central nervous system, where it replaces the residing zinc in neurons. This process can cause mental retardation, behavioral changes, paralysis and anemia. Pb is easily accumulated in body over the lifetime, yet its expulsion is very difficult. It is generally believed that the anthropogenic sources of Pb are limited to the use of tetraethyl lead, however, other Pb origins, such as battery manufacturers, lead smelters and ammunition industries are all major Pb polluters of the environment [1,2]. Cesium as ^{137}Cs is an important source of radioactivity from nuclear waste, as it is the major radionuclide of spent nuclear fuel [3]. Due to nuclear accidents over the last decades (e.g., Fukushima in 2011, Chernobyl in 1986), ^{137}Cs has infiltrated soils and groundwater and even further into the biosphere thanks to its high solubility [4]. A potential risk of both Pb and ^{137}Cs originates from their toxicity and harmful concentrations to organisms in various part of the biosphere [5]. While Pb poisons the soil microorganisms, which limits heterotrophic breakdown of organic matter [6], ^{137}Cs is toxic due to its long radio-isotopic half-life (30 years) [7]. Furthermore, ^{137}Cs is metabolically similar to potassium, causing it to accumulate in plants and therefore it is incorporated into many food chains [8].

Despite a number of modern cleaning technologies (ultrafiltration, reverse osmosis, biological methods, etc.), adsorption remains an effective technique due to its simple application in wastewater treatment. The demand for environmentally and economically friendly technologies has resulted in the development and testing of new natural and synthetic adsorbents, preferably low-cost [5,9–12].

Waste building materials have attained increased professional concern due to their availability and properties analogous to aluminosilicates, such as composition, chemical stability, fine structure and environmentally safe nature. Generally, they are recycled either as concrete admixtures [13] or when fine waste particles have pozzolanic activity, they can be added to cement-based materials to declining the Portland cement consumption [14].

Dousova et al. [15] followed out the pilot study on waste brick dusts (WBD) as potential adsorbents of a series of risk cations, anions and radioactive residues. As has been found, WBD showed to be selective cation-active sorbents. The anionic As form (AsO_4^{3-}) was also adsorbed onto WBD, but with more than four times higher sorbent dosage. In terms of possible technological applications, the long-term stability of toxic ions in saturated WBD, as well as its successful incorporation into cement building materials, are crucial.

The aim of this work was to study the characteristics of WBD as a cation-active adsorbent of Pb^{II} and Cs^{I} particles from aqueous solutions. A deep structural characterization of the materials of interest was conducted for potential technological applications based on the combination of X-ray diffraction (XRD) and solid-state NMR spectroscopy; these complementary methods are especially informative. While XRD analysis can easily describe crystalline fractions and changes in mineralogical composition, solid-state NMR spectroscopy gives information about the structure of both amorphous and crystalline phases, local framework defects and extra framework components. Specifically, ^{133}Cs and ^{207}Pb solid-state NMR spectra can solely provide insight into the chemical nature of adsorbed and solidified contaminants. The information on the structural and surface changes related to the adsorption process may lead to an adsorption mechanism, in which the adsorption rate can be characterized through important parameters such as adsorption capacity, adsorption efficiency, surface complexation and binding energy.

2. Materials and Methods

2.1. *Waste Brick Dust*

Waste brick dusts (WBD) arose as a waste (grinding powder) during the production of vertically perforated ceramic blocks intended for thin joint masonry. The amount of recyclable WBD is limited, as surplus ceramic waste is typically dumped.

The fraction below 125 μm of two WBD samples from different brick factories in the Czech Republic (Libochovice—LB and Tyn nad Vltavou—TN) were tested as selective cation-active adsorbents for Pb/Cs removal from model aqueous systems. The chemical characteristics of both materials (Table 1) indicates more than 10 times alkalinity, slightly less Fe content and almost double S_{BET} of LB compare to TN. Both WBD were mineralogically similar, with dominant quartz content (ca. 25%), and the presence of illite, albite, anorthite and hematite. The presence of calcite was detected in LB only, while sillimanite was found in TN.

Table 1. Chemical composition, pH of the zero point of charge (pH_{ZPC}) and specific surface area (S_{BET}) of waste brick dust (WBD) from Libochovice (LB) and Tyn nad Vltavou (TN).

Composition (wt%)	LB	TN
Na_2O/K_2O	1.1/4.3	2.5/2.8
MgO/CaO	1.5/14.4	1.3/1.6
Al_2O_3	19.7	32.6
SiO_2	46.4	60.3
P_2O_5	0.4	0.3
SO_3	1.5	0.6
Fe_2O_3	3.7	5.4
TiO_2	0.4	0.8
MnO_2	0.03	0.04
S_{BET} (m^2 g^{-1})	4.1	2.5
V_{total} (cm^3 g^{-1})*)	0.01	0.006
pH **)	~9–10	~5–6
pH_{ZPC}	4.3	5.5

*) total pore volume of mesopores; **) water leachate at 20 °C, solid-liquid ratio 1:30

2.2. Model Solutions

Model solutions of Pb^{2+} and Cs^+ were prepared from inorganic salts ($PbCl_2$ and CsCl, respectively) of analytical grade and distilled water, with concentrations of 0.1 and 0.5 mM L^{-1} and their natural pH values (≈3.5). The concentration ranges were selected as appropriate to simulate a slightly increased amount of the contaminant in a water system to a heavily contaminated solution.

2.3. Sorption Experiments

The suspension of model solution (50 mL) and defined dosage (1–15 g L^{-1} for Pb and 1–30 g L^{-1} for Cs) of WBD was agitated in a batch manner at laboratory temperature (20 °C) for 24 h [16]. The product was filtered, and the filtrate was analyzed for residual concentration of cations, while the solid residue (saturated WBD) was kept for subsequent solid-state analyses (NMR, SEM, S_{BET}).

All adsorption data were fitted to the Langmuir isotherm [17,18], according to the equations (1) and (2), which was verified in many papers [9,15–17] as a suitable adsorption model for natural sorbents, including aluminosilicates and soils. The obvious adsorption parameters (q_{max}—maximum equilibrium adsorption capacity; Q_t—theoretical adsorption capacity; R^2—correlation factor; K_L—Langmuir adsorption constant) were then used to evaluate the effectiveness of adsorption systems.

$$q = \frac{Q_t \cdot K_L \cdot c}{1 + K_L \cdot c}, \tag{1}$$

and its linearized form:

$$\frac{1}{q} = \frac{1}{Q_t} + \frac{1}{Q_t \cdot K_L \cdot c}, \tag{2}$$

where q is an equilibrium concentration of adsorbed ion (adsorbate) in solid phase [mol g^{-1}], c is an equilibrium concentration of adsorbate in solution [mol L^{-1}], Q_t indicates the theoretical adsorption capacity [mol g^{-1}], and K_L is a Langmuir adsorption constant [L mol^{-1}].

The equilibrium amount (q) of adsorbate caught in the solid phase was calculated from experimental data by Equation (3):

$$q = \frac{V_0(c_0 - c)}{m}, \tag{3}$$

where V_0 is the volume of solution [L], c_0 is the initial concentration of adsorbate in solution [mol L^{-1}], and m is the mass of solid phase [g].

2.4. Leaching Test

The stability of cations in the saturated WBDs was tested by a standard procedure EN 12457–2 [19]. First, a suspension of dry sorbent and distilled H_2O at a ratio of 1:10 was agitated in a sealed polyethylene bottle at 20 °C for 24 h. The suspension was filtered out and the filtrate was analyzed for residual Pb^{2+}/Cs^{+} concentration.

2.5. Analytical Methods

Powder XRD was measured with a Seifert XRD 3000P diffractometer (Seifert, Ahrensburg, Germany) with CoK_{α} radiation (λ= 0.179026 nm, graphite monochromator, goniometer with Bragg–Brentano geometry) in the 2θ range of 5–60° with a step size of 0.05° 2θ.

The X-ray Fluorescence (XRF) analyses of the solid phase were carried out using an ARL 9400 XP+ spectrometer (ARL, Ecublens, Switzerland) at a voltage of 20–60 kV, probe current of 40–80 mA, and with an effective area of 490.6 mm^2. UniQuant 4 software was used to evaluate the data (Thermo ARL, Ecublens, Switzerland).

The S_{BET} was measured on a Micromeritics ASAP 2020 (accelerated surface area and porosimetry) analyzer using gas sorption. The ASAP 2020 model (Micromeritics, Norcross, GA, USA) assesses single and multipoint BET surface area, Langmuir surface area, Temkin and Freundlich isotherm analysis, pore volume and pore area distributions in the micro- and macro-pore ranges. The macro-pore and micro-pore samples were analyzed by the Horvath-Kavazoe method (BJH method), respectively. The BHJ method used N_2 as the analysis adsorptive and an analysis bath temperature of −195.8 °C. Samples were degassed at 313 K for 1000 min.

Solid-state NMR spectra were measured at 11.7 T using a Bruker AVANCE III HD 500 WB/US NMR spectrometer (Bruker, Karlsruhe, Germany). The ^{27}Al Magic Angle Spinning Nuclear Magnetic Resonance Spectroscopy (MAS NMR)spectra were acquired at a spinning frequency of 11 kHz, Larmor frequency of 130.287 MHz and recycle delay of 2 s. The spectra were referenced to the external standard $Al(NO_3)_3$ (0 ppm). The ^{29}Si MAS NMR spectra were acquired at a spinning frequency of 11 kHz, Larmor frequency of 99.325 MHz and recycle delay of 10 s. The number of scans for the acquisition of a single ^{29}Si MAS NMR spectrum was 6144. The spectra were referenced to the external standard M_8Q_8 (−109.8 ppm). ^{133}Cs MAS NMR spectra were acquired at a spinning frequency of 11 kHz, Larmor frequency of 65.601 MHz, recycle delay of 2 s and the number of scans of 73,000. The spectra were referenced to the external standard $CsOH{\cdot}H_2O$ (124.1 ppm). Due to the large chemical shift anisotropy the ^{207}Pb NMR spectra were measured using the ^{207}Pb MAS NMR and WURST-QCPMG NMR experiments combined with variable offset cumulative spectroscopy (VOCS) technique [20]. The Larmor frequency was 104.640 MHz and the recycle delay was 2 s for all ^{207}Pb NMR spectra. The number of scans was set to 30000 for each sub-spectrum. The applied experimental parameters were optimized using a mixture of PbO and $Pb(OH)_2$ the resonance frequency. The ^{207}Pb NMR spectra were referenced to the external standard $Pb(NO_3)_2$ (−3473.6 ppm). High-power ^{1}H decoupling (SPINAL64 for MAS experiments and Continuous Wave (CW) for static experiments) was used to eliminate heteronuclear dipolar couplings in all measurements. The NMR experiments were performed at a temperature of 303 K, and temperature calibration was done to compensate for the frictional heating of the samples [21].

The structure of samples was determined using scanning electron microscopy (SEM) on the Tescan Vega 3 (Brno-Kohoutovice, Czech Republic). Energy dispersion spectrometry (EDS) was conducted on the Inca 350 spectrometer (Oxford Instruments, Abingdon, UK).

The concentration of Pb and Cs in aqueous solutions was measured by atomic absorption spectrometry (AAS) using a SpectrAA-880 VGA 77 unit (Varian, Palo Alto, CA, USA) in flame mode. An accuracy of AAS analyses was guaranteed by the Laboratory of Atomic Absorption Spectrometry of UCT Prague, CR, with the detection limit of 0.5 $\mu g\ L^{-1}$, with a standard deviation ranging from 5%–10% of the mean).

3. Results and Discussion

3.1. Specification of WBD Adsorbents in Relation to Pb/Cs Chemistry and pH Value

The chemical and mineralogical composition of both WBD are similar apart from the high content of alkalis in LB as compared to TN, which results in the former exhibiting a significant higher pH (see Table 1). Whereas the pH/pH_{ZPC} values (Table 1) are responsible for the selectivity of adsorption, the points of pH_{ZPC} in relation with pH ranges of adsorption solution are shown in Figure 1. At the pH_{ZPC} of LB (the crosses in both phase diagrams) lower than the pH range of adsorption solution (the dashed areas in the phase diagrams), when $pH/pH_{ZPC} > 2$, the sorbent surface primarily deprotonates to balance the pH gradient and the negatively charged surface of LB then attracts cations from the solution [22]. In the case of TN, the negligible pH/pH_{ZPC} gradient ($pH/pH_{ZPC} \approx 1$, spotted areas vs. the empty rings in the phase diagrams) caused a weak attraction at the solid-liquid interface. In terms of WBD with the average pH_{ZPC} of 5, a higher pH of adsorption has been essential for selective removal of Pb^{2+}/Cs^{+}. The optimal pH resulting from adsorption experiments ranged from 7.5 to 10.

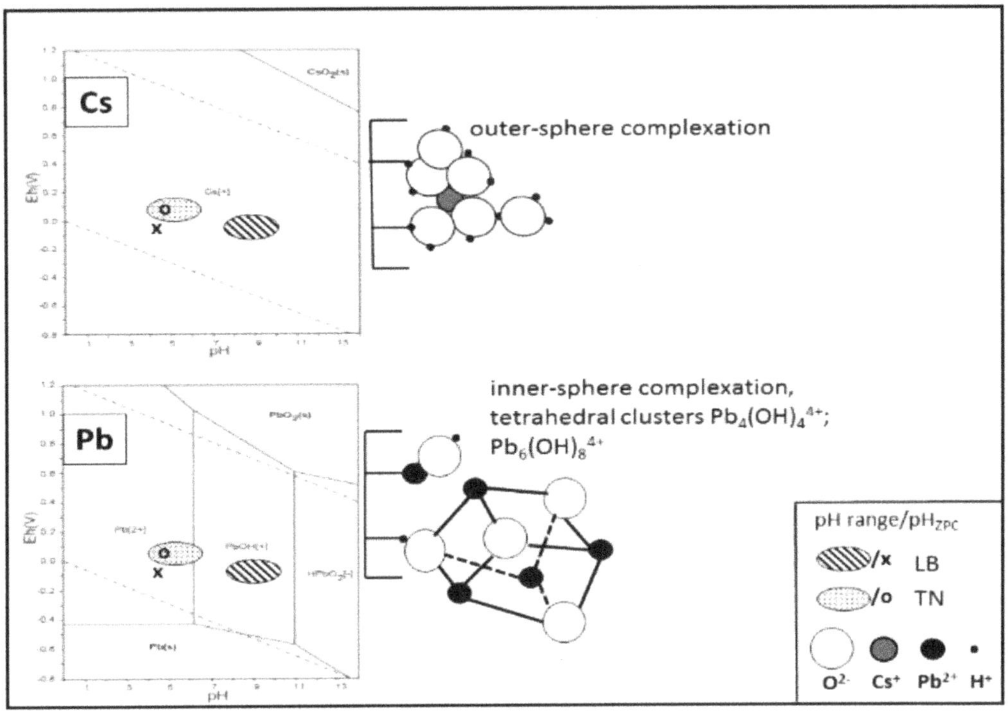

Figure 1. Surface chemistry and expected complexation of Cs^{+}/Pb^{2+} in aqueous system.

As shown in Table 1, the S_{BET} and total porosity values also favored the adsorption selectivity of LB to TN, because a more porous material with a larger surface area would generally provide more sorption sites for surface binding.

The adsorption behavior of the investigated systems was also based on the different aqueous chemistry of Pb and Cs, respectively (Table 2, Figure 1). In an aqueous environment, Cs^{+} was expected to form a small hydrated ion, because of its low charge and large crystallographic radius (1.69 Å). While the adsorption of Cs was minimally pH dependent, Cs^{+} ions were bound primarily to cation exchange sites, forming only outer-sphere complexes. On the other hand, Pb^{2+} (1.32 Å) was subjected to high hydrolysis, easily losing the part of its primary hydration shell, especially at high pH [23]. The hydrolyzed ions initially formed strong inner-sphere complexes with sorbent surfaces and eventually led to the formation of polynuclear complexes or surface precipitates [24].

Table 2. Ionic radius and thermodynamic properties of hydrated Cs^+/Pb^{2+} [25].

Cation	r(Å)	r′(Å)	n	ΔG_{hydr} (kJ mol^{-1})
Cs^+	1.69	3.28	2.1	−250
Pb^{2+}	1.32	4.03	6.1	−1425

r—ionic radius; r′—hydrated radius; n—number of water molecules in hydration shell; ΔG_{hydr}—molar Gibbs energy of ion hydration.

3.2. Structural Changes Following Adsorption

The SEM micro-observation of initial and Cs^+/Pb^{2+} saturated WBD (Figure 2) showed completely different morphology of surface particles, confirming the above described unequal hydrolyses and binding strategies of Cs^+ and Pb^{2+} ions. Separately hydrated Cs^+ ions, that formed outer-sphere complexes with active surface sites, did not change the surface morphology remarkably (Figure 2), while the predicted formation of surface precipitates of $Pb_4(OH)_4{}^{4+}/Pb_6(OH)_8{}^{4+}$ (Figure 1) led to the homogeneous covering of sorbent surface.

Figure 2. SEM images of WBD before and after Cs^+/Pb^{2+}adsorption on LB and TN (magnification of 300×).

The MAS NMR spectra of both materials recorded before and after Cs^+/Pb^{2+}adsorption indicate structural changes, crystallinity and framework defects resulting from the adsorption process. The ^{27}Al MAS NMR spectra (Figure 3) demonstrate that aluminum sites in both systems occupy tetrahedral coordination Al^{IV}, as reflected by the most intense signals at 60 and 58 ppm for LB and TN systems, respectively. The low-intensity signal at approximately 5 ppm reflects a small amount of hexagonally coordinated aluminum species present in the LB system. Additional information relating to the crystallinity of investigated systems can be deduced from the line-widths of the recorded signals. While the LB system is represented by a relatively broad signal with the half-width of about 1.6 kHz, the TN system is reflected by considerably narrower signal (1.3 kHz). This finding, which agrees with the ^{29}Si MAS NMR spectra (discussed later), indicates a more disordered and amorphous character of aluminosilicate framework of the LB system.

The ^{29}Si MAS NMR spectra (Figure 4) provide more detailed insight into the framework of investigated systems and further highlight their distinct structures. The main signal at −91 ppm is clearly attributed to the Q^3(0Al) sites in dehydroxylated illite, whereas the shoulder at −85 ppm is assigned to a mixture of Q^3(1Al) and Q^3(2Al) usually present in natural illite and related systems. The signal resonating at approximately -75 ppm likely reflects the incompletely condensed Q^1 and Q^2

units of mineral defects and the amorphous SiO_2 fraction. The narrow signal at −108 ppm therefore reflects the Q^4 units of crystalline SiO_2 (quartz, approx. 20%).

On the other hand, the ^{29}Si MAS NMR spectrum of neat brick dust from TN reflects a much narrower distribution of aluminosilicate building blocks. The signal is dominated by the narrow resonance at −108 ppm, reflecting the Q^4 units of crystalline SiO_2 (quartz, approx. 70%). Deconvolution of the recorded spectrum further revealed additional much broader resonance centered at −92 ppm, which would reflect the Q^3(0Al) units in dehydroxylated illite (approx. 30%).

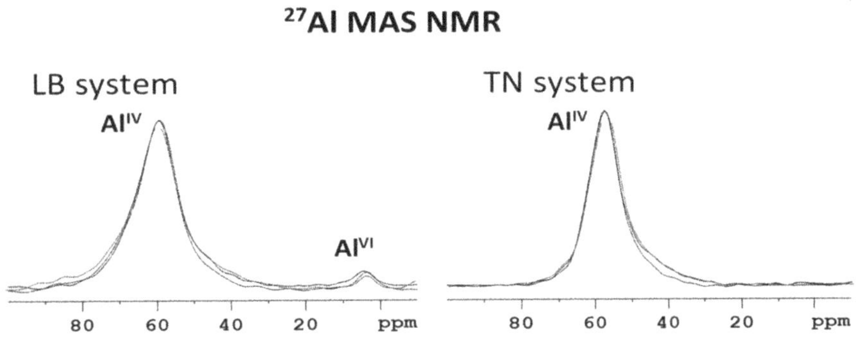

Figure 3. Comparison of ^{27}Al MAS NMR spectra of neat LB and TN WBD systems, before and after Cs^+/Pb^{2+} adsorption.

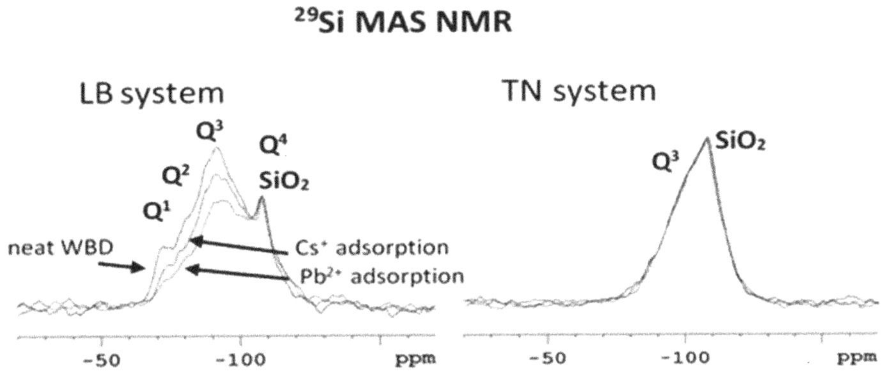

Figure 4. Comparison of ^{29}Si MAS NMR spectra of neat LB and TN WBD systems, before and after Cs^+/Pb^{2+} adsorption.

The structural changes associated with the sorption process were mostly observed at the ^{29}Si MAS NMR spectra of LB system (Figure 4). The intensities of broad signals of Q^1, Q^2 and Q^3 units considerably decreased relative to the signal intensity of Q^4 units of crystalline SiO_2. In the case of Pb^{2+} adsorption, the observed decrease is more intense in comparison with the decrease observed during the sorption of Cs^+ ions, which indicates the partial dissolution of disordered and amorphous fractions rich in the SiO_2 phase. In contrast, there were no considerable changes in the ^{29}Si MAS NMR spectra of the TN system, thus exhibiting structural stability of this material during the adsorption of Pb^{2+} and Cs^+ ions.

3.3. *Adsorption of Pb^{2+}/Cs^+ on WBD*

The adsorption series were performed under the same conditions (described in Section 2.3) in 'order to evaluate the applicability of WBD for the removal of toxic cations from aqueous systems.

The Langmuir adsorption isotherms (Figure 5) illustrate improved adsorption properties of LB for both cations and a considerably higher affinity of Pb^{2+} for both WBD. As shown by the adsorption parameters calculated from the linearized Langmuir equation (Section 2.3, Table 3), only Pb^{2+} adsorption corresponded to the Langmuir model indicating an inner-sphere surface complexation [26,27] for all Pb^{2+} adsorptions. The differences in theoretical adsorption capacities (Q_{theor}) and binding energies (K_L) of LB/Pb^{2+} as compared to the TN/Pb^{2+} systems resulted from the different pH/pH_{ZPC} ratios (Section 3.1), which could be considered the prior adsorption power [15,16]. Accordingly, LB/Pb^{2+} adsorption system showed much more robust process in all aspects (Q_{theor}, K_L, R^2) (Table 3).

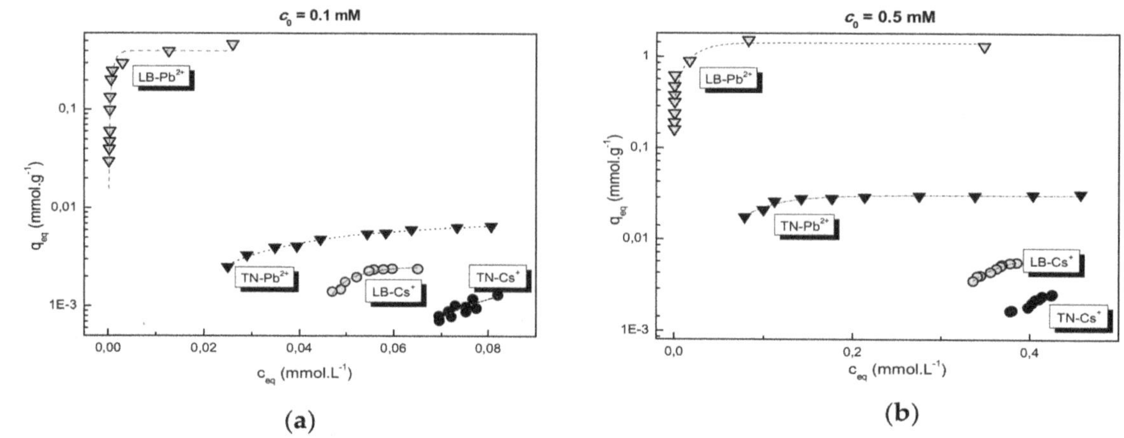

Figure 5. Langmuir adsorption isotherms of Pb^{2+}/Cs^{+} adsorption on LB and TN at different initial concentrations; (**a**) $c_0 = 0.1$ mmol L^{-1}, (**b**) $c_0 = 0.5$ mmol L^{-1}. Sorbent dosage $\approx$ 1–15 g L^{-1} for Pb^{2+} and 1–30 g L^{-1} for Cs^{+}.

Table 3. Adsorption parameters for Pb and Cs adsorption from model solutions on LB and TN.

Sorbent/ c_0 (mM)	Pb^{2+}				Cs^{+}			
	q_{max} (mmol g^{-1})	Q_{theor} (mmol g^{-1})	K_L (L $mmol^{-1}$)	R^2	q_{max} (mmol g^{-1})	Q_{theor} (mmol g^{-1})	K_L (L $mmol^{-1}$)	R^2
LB/0.5	1.32	1.6	776.4	0.973	5.5×10^{-3}	-	-	-
LB/0.1	0.85	1.1	171.5	0.930	2.4×10^{-3}	-	-	-
TN/0.5	3.4×10^{-2}	4.5×10^{-2}	16.7	0.827	2.9×10^{-3}	-	-	-
TN/0.1	5.5×10^{-3}	6.8×10^{-3}	39.8	0.838	1.1×10^{-3}	-	-	-

On the other hand, the adsorption of hydrated Cs^{+} ions did not follow Langmuir run, but typical physical adsorption with a weak electrostatic surface binding (outer-sphere complexation) was observed. Maximum equilibrium sorption capacities (q_{max}) of both LB and TN for Cs^{+} were one to three orders of magnitude lower than for Pb^{2+}.

An almost 100% adsorption efficiency at low sorbent consumption for Pb^{2+} (Figure 6) supported the abovementioned formation of poorly soluble $Pb(OH)_2$ clusters on the sorbent surface, which markedly increased the adsorption yields.

The fate of adsorbed Cs^{+}/Pb^{2+} ions was also observed in the ^{113}Cs and ^{207}Pb solid-state NMR spectra. The spectra were recorded only for the LB-system, which in both cases exhibited considerably higher sorption capacities. The ^{133}Cs MAS NMR spectrum (Figure 7) shows relatively broad signal covering the frequency range from −7 to approx. −20 ppm. According to literature data, these resonances can be assigned to surface species, in which more shielded (more negative) resonances reflect Cs^{+} sites more tightly bound to the surface, whereas the less shielded resonances correspond to less tightly bound species. This phenomenon can be attributed to the formation of multilayer deposits of Cs^{+} ions.

When measuring ^{207}Pb solid-state NMR spectra, there is relatively low natural abundance of ^{207}Pb spins (approx. 22.6 %, I=1/2), which is combined with usually very large chemical shift anisotropy (CSA). Consequently, under the standard spinning speeds (10–30 kHz) manifolds of the central signals and spinning sidebands in ^{207}Pb MAS NMR spectra can cover extremely broad spectral windows 2000–4000 ppm, rendering them nearly undetectable. Therefore, the attempt here to record

^{207}Pb solid-state NMR spectra was nearly unsuccessful. This effect was observed even in the model system, which consisted of a mixture of neat powdered PbO and $Pb(OH)_2$ (Figure 8a). However, the application of an alternative experimental technique, ^{207}Pb Wideband Uniform-Rate Smooth Truncation-Quadrupolar Carr-Purcell Meiboom-Gill (WURST-QCPMG) NMR, seems to be more suitable for the measurement of broad spectral lines. As demonstrated in Figure 8b, the intensities of the recorded ^{207}Pb WURST-QCPMG NMR signals are considerably higher than that recorded by the standard ^{207}Pb MAS NMR experiment.

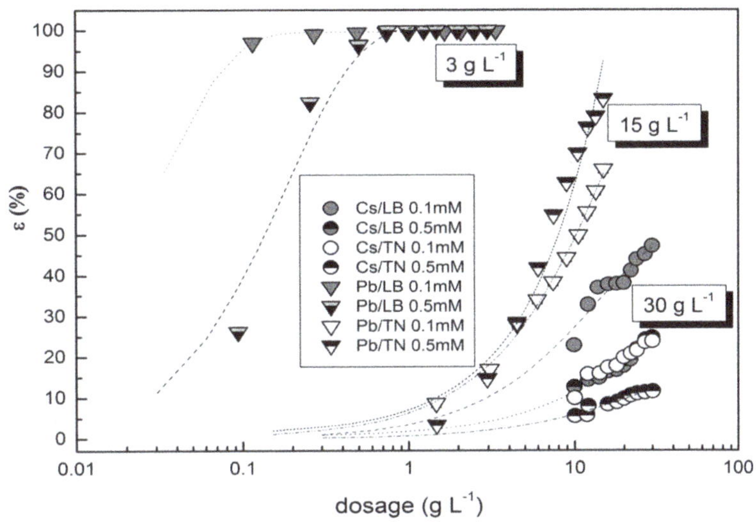

Figure 6. Adsorption efficiency of Cs^+/Pb^{2+} adsorption on LB and TN in relation to sorbent dosage.

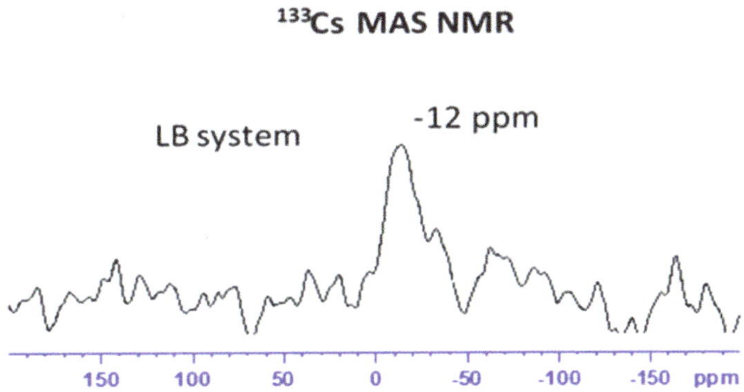

Figure 7. ^{133}Cs MAS NMR spectrum of LB system after Cs^+ adsorption.

In spite of a long accumulation time (48 h), the resulting spectrum is very poor, indicating very low amounts of the surface Pb^{2+} species (Figure 8c). Nevertheless, two spectral regions with ^{207}Pb NMR resonances were identified in the recorded spectrum. One of them covers frequencies from 1200 to 2200 ppm, whereas the second one ranges from −400 to −1000 ppm. These results imply that Pb^{2+} ions on the surface of LB-system form two chemically distinct fractions. One of these fractions is likely structurally related to precipitated hydroxide species, whereas the second one can be rather attributed to Pb^{2+} binding to a silicate tetrahedral [28].

Consistent with the previously mentioned results, the adsorption of Cs^+ on WBDs was worse in all aspects when compared to Pb^{2+}. A low effect of Cs^+ adsorption was mainly given by steric properties. According to Rajec et al. [29], the majority of active sorption sites for Cs^+ binding have been associated with permanent (pH independent) charge cation exchange sites on zeolites and smectite clays rather than with surface hydroxyl sites on quartz and feldspars. Therefore, zeolites have been frequently used to remove radionuclides and metal cations because of their molecular sieve structure and high cation exchange capacity [30]. Contrastingly, WBDs consist mostly of quartz, feldspars and hematite and exhibit rather poor Cs adsorbing behavior [31].

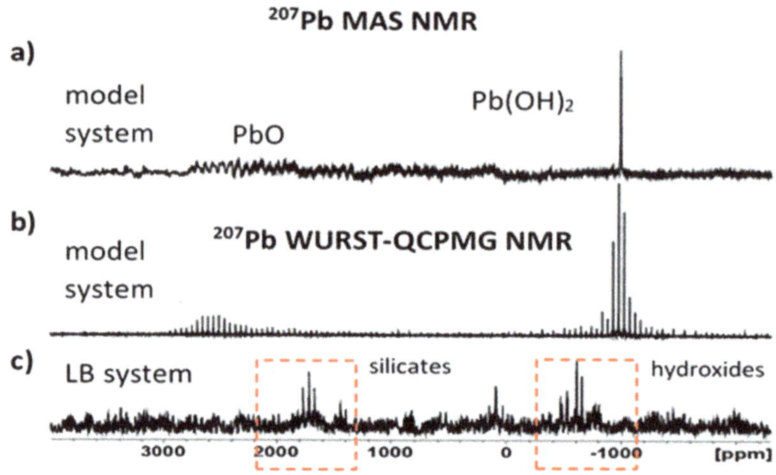

Figure 8. ^{207}Pb solid-state NMR spectra. (**a**) Standard 207 Pb MAS NMR spectrum of a reference mixture of neat powdered PbO and $Pb(OH)_2$ recorded at 12 kHz (NS = 50,000); (**b**) the ^{207}Pb WURST-QCPMG NMR spectrum of the reference system (NS = 8000); and (**c**) the corresponding ^{207}Pb WURST-QCPMG NMR spectrum of LB system after Pb^{2+} adsorption (NS = 50,000).

3.4. Leachability of Pb^{2+}/Cs^{+} from Saturated Sorbents

The stability of cations in saturated WBD was tested by the leaching test (Section 2.4). Not surprisingly, the stability of Pb^{2+}/Cs^{+} in saturated WBD was closely related to the selectivity of a particular adsorption (Table 4); Pb^{2+} adsorption on LB carried out by high sorption capacities and efficiencies, linking the formation of very stable inner-sphere surface complexes at both initial concentrations (only about 0.001% of Pb^{2+} was leached). The favorable stability (less than 0.3% of the original amount leached) was recorded for Pb^{2+} in TN. The non-selective Cs^{+} adsorption resulted in a low binding stability during the leaching test, where 4 to 12% of the initial amount was leached. According to the standards (CSN EN 12457), LB and TN saturated with Pb^{2+} fall into the non-hazardous waste. Although the regulation has not specified the limits for a Cs hazard, a high leachability of Cs^{+} from both saturated WBDs (≈ 10% wt.) indicates its environmental risk. In line with the previous results, LB proved better adsorption selectivity and leaching stability to cations than TN. The stability of testified adsorption systems decreased in the following order: Pb^{2+}/LB >> Pb^{2+}/TN >> Cs^{+}/LB > Cs^{+}/TN.

Table 4. Leachability of Cs/Pb from WBDs.

Element/Initial Concentration	Initial Amount in Saturated WBD ($mg \cdot g^{-1}$)	Leached in H_2O (%)	Class of Leachability*
	Libochovice (LB)		
Cs^{+}/0.1 mM	0.32	3.7	not limited
Cs^{+}/0.5 mM	0.74	7.4	not limited
Pb^{2+}/0.1 mM	146.3	1.2×10^{-3}	IIa—other waste
Pb^{2+}/0.5 mM	212.8	6.6×10^{-4}	I—inert waste
	Tyn nad Vltavou (TN)		
Cs^{+}/0.1 mM	0.15	12.3	not limited
Cs^{+}/0.5 mM	0.33	11.4	not limited
Pb^{2+}/0.1 mM	0.93	0.25	IIa—other waste
Pb^{2+}/0.5 mM	5.98	0.18	IIa—other waste

* Regulation No. 294/2005 (CSN EN 12457).

4. Conclusions

The adsorption selectivity of WDBs to Pb(II) and Cs(I) was initially affected by the structural properties and morphology of WBDs being tested as Pb^{2+}/Cs^{+} adsorbents. The unequal binding forms of both cations presented the next important aspect of adsorption process; while Cs^{+} ions formed separate particles weakly bound by outer-sphere surface complexes, Pb^{2+} appeared in two co-existed phases, namely as the hydrolyzed particles Pb^{2+} bound by strong inner-sphere complexes, and the poorly soluble $Pb(OH)_2$ clusters that catch on the sorbent surface. Therefore, the adsorption of Pb^{2+} was much more effective, strongly supported by the almost insoluble surface precipitation ($Pb(OH)_2$). In connection with that process, the adsorption of Pb^{2+} was pH dependent, with an optimal pH range of 8 to 11. On the other hand, Cs^{+} ions were equally adsorbed within a wide range of pHs.

The stability of Cs^{+}/Pb^{2+} saturated sorbents subjected to the standard leaching tests corresponded well with the quality and effectiveness of adsorption in particular; the better parameters of Pb^{2+} adsorption, following the Langmuir adsorption model, indicated a high binding energy of surface complexation that resulted in a considerably higher stability of WBD that was saturated with Pb^{2+} when compared to Cs^{+}. The stability of Pb^{2+} saturated WBDs was about three orders in magnitude higher than that of Cs^{+}. In all aspects, LB proved to be a much better sorbent than TN. Both WBDs (LB and TN) are promising prospective adsorbents for Pb^{II}.

Author Contributions: B.D. wrote the paper, carried out adsorption experiments, managed the methodology of research steps; D.K. performed XRD analyses; M.L. measured and evaluated the S_{BET} and porosity of WBDs; M.K. provided suitable material and its characterization; M.U. and L.K. performed NMR study, while J.B. supervised and evaluated the MMR experiments.

References

1. Wong, C.S.C.; Li, X.D. Pb contamination and isotopic composition of urban soils in Hong Kong. *Sci. Total Environ.* **2004**, *319*, 185–195. [CrossRef]
2. Jiang, M.; Wang, Q.; Jin, X.; Chen, Z. Removal of Pb(II) from aqueous solution using modified and unmodified kaolinite clay. *J. Hazard. Mater.* **2009**, *170*, 332–339. [CrossRef] [PubMed]
3. Benedicto, A.; Missana, T.; Fernández, A.M. Interlayer Collapse Affects on Cesium Adsorption onto Illite. *Environ. Sci. Technol.* **2014**, *48*, 4909–4915. [CrossRef] [PubMed]
4. Cornell, R.M. Adsorption of cesium on minerals: A review. *J. Radioanal. Nuclear Chem.* **1993**, *171*, 483–500. [CrossRef]
5. Dahiya, S.; Tripathi, R.M.; Hegde, A.G. Biosorption of heavy metals and radionuclide from aqueous solutions by pre-treated arcashell biomass. *J. Hazard. Mater.* **2008**, *150*, 376–386. [CrossRef]
6. Honeyman, B.D.; Santsch, P.H. Metals in aquatic environment. *Environ. Sci. Technol.* **1988**, *22*, 862–871. [CrossRef] [PubMed]
7. Song, K.C.; Lee, H.K.; Moon, H.; Lee, K.J. Simultaneous removal of the radiotoxic nuclides ^{137}Cs & ^{129}I from aqueous solution. *Sep. Purif. Technol.* **1997**, *12*, 215–217.
8. Jalali, R.R.; Ghafourian, H.; Asef, Y.; Dalir, S.T.; Sahafipour, M.H.; Gharanjik, B.M. Biosorption of cesium by native and chemically modified biomass of marine algae: Introduce the new biosorbent for biotechnology applications. *J. Hazard. Mater.* **2004**, *B116*, 125–134. [CrossRef] [PubMed]
9. Han, R.; Lu, Z.; Zou, W.; Daotong, W.; Shi, J.; Jiujun, Y. Removal of copper(II) and lead(II) from aqueous solution by manganese oxide coated sand. II. Equilibrium study and competitive adsorption. *J. Hazard. Mater.* **2006**, *B137*, 480–488. [CrossRef]
10. Mousavi, H.Z.; Seyedi, S.R. Kinetic and equilibrium studies on the removal of Pb(II) from aqueous solution using nettle ash. *J. Chil. Chem. Soc.* **2010**, *55*, 307–311. [CrossRef]
11. Ding, D.; Lei, Z.; Yang, Y.; Feng, C.; Zhang, Z. Selective removal of cesium from aqueous solutions with nickel (II) hexacyanoferrate (III) functionalized agricultural residue–walnut shell. *J. Hazard. Mater.* **2014**, *270*, 187–195. [CrossRef]

12. Badescu, I.S.; Bulgariu, D.; Ahmad, I.; Bulgariu, L. Valorisation possibilities of exhausted biosorbents loaded with metal ions—A review. *J. Environ. Manag.* **2018**, *224*, 288–297. [CrossRef]
13. Silva, R.V.; de Brito, J.; Dhir, R.K. Properties and composition of recycled aggregates from construction and demolition waste suitable for concrete production. *Con. Build. Mat.* **2014**, *65*, 201–217. [CrossRef]
14. Vejmelková, E.; Keppert, M.; Rovnaníková, P.; Ondráček, M.; Keršner, Z.; Černý, R. Properties of high performance concrete containing fine-ground ceramics as supplementary cementitious material. *Cement Concrete Compos.* **2012**, *34*, 55–61. [CrossRef]
15. Dousova, B.; Kolousek, D.; Keppert, M.; Machovic, V.; Lhotka, M.; Urbanova, M.; Brus, J.; Holcova, L. Use of waste ceramics in adsorption technologies. *Appl. Clay Sci.* **2016**, *134*, 145–152. [CrossRef]
16. Doušová, B.; Grygar, T.; Martaus, A.; Fuitová, L.; Koloušek, D.; Machovič, V. Sorption of As^{V} on aluminosilicates treated with Fe^{II} nanoparticles. *J. Colloid Interface Sci.* **2006**, *302*, 424–431. [CrossRef]
17. Jeong, Y.; Fan, M.; Singh, S.; Chuang, C.L.; Saha, B.; van Leeuwen, J.H. Evaluation of iron oxide and aluminium oxide as potential arsenic(V) adsorbents. *Chem. Eng. Process.* **2007**, *46*, 1030–1039. [CrossRef]
18. Misak, N.Z. Langmuir isotherm and its application in ion-exchange reactions. *React. Polym.* **1993**, *21*, 53–64. [CrossRef]
19. CEN 2002. CEN 2002. Characterization of Waste-Leaching-Compliance Test for Leaching of Granular Waste Material and Sludge Part 2. Comite Europeen de Normalisation 2002, EN 12457-2. Available online: https://www.sis.se/en/produkter/environment-health-protection-safety/wastes/solid-wastes/ssen124574/ (accessed on 1 October 2003).
20. Massiot, D.; Farnan, I.; Gautier, N.; Trumeau, D.; Trokiner, A.; Coutures, J.P. Ga-71 and Ga-69 Nuclear-Magnetic-Resonance Study of Beta-Ga_2O_3-Resolution of 4-Fold and 6-Fold Coordinated Ga Sites in Static Conditions. *Solid State Nucl. Magn. Reson.* **1995**, *4*, 241–248. [CrossRef]
21. Brus, J. Heating of samples induced by fast magic-angle spinning. *Solid State Nucl. Magn. Reson.* **2000**, *16*, 151–160. [CrossRef]
22. Walther, J.V. *Essentials of Geochemistry*, 2nd ed.; Jones and Bartlett Publishers Inc.: Sudbury, MA, USA, 2009; 486p.
23. Bohn, H.L.; Mc Neal, B.L.; O'Conner, G.A. *Soil Chemistry*, 2nd ed.; John Wiley and Sons: New York, NY, USA, 1985; 341p.
24. Um, W.; Papelis, C. Sorption mechanisms of Sr(II) and Pb(II) on zeolitized tuffs from the Nevada Test Site as a function of pH and ionic strength. *Am. Mineral.* **2003**, *88*, 2028–2039. [CrossRef]
25. Marcus, Y. Thermodynamics of Solvation of Ions. *J. Chem. Soc. Faraday Trans.* **1991**, *87*, 2995–2999. [CrossRef]
26. Ding, M.; de Jong, B.H.W.S.; Roosendaal, S.J.; Vredenberg, A. XPS studies on the electronic structure of bonding between solid and solutes: Adsorption of arsenate, chromate, phosphate, Pb^{2+}, and Zn^{2+} ions on amorphous black ferric oxyhydroxide. *Geochim. Cosmochim. Acta* **2000**, *64*, 1209–1219. [CrossRef]
27. Wen, X.; Du, Q.; Tang, H. Surface complexation model for the heavy metal adsorption on natural sediment. *Environ. Sci. Technol.* **1998**, *32*, 870–875. [CrossRef]
28. Fayon, F.; Farnan, I.; Bessada, C.; Coutures, J.; Massiot, D.; Coutures, J.P. Empirical Correlations between ^{207}Pb NMR Chemical Shifts and Structure in Solids. *J. Am. Chem. Soc.* **1997**, *119*, 6837–6843. [CrossRef]
29. Rajec, P.; Macasek, F.; Feder, M.; Misaelides, P.; Samajova, E. Sorption of caesium and strontium on clinoptilolite- and mordenite-containing sedimentary rocks. *J. Radioanal. Nucl. Chem.* **1998**, *229*, 49–56. [CrossRef]
30. Faghihian, H.; Marageh, M.G.; Kazemian, H. The use of clinoptilolite and its sodium form for removal of radioactive cesium and strontium from nuclear wastewater and Pb^{2+}, Ni^{2+}, Cd^{2+}, Ba^{2+} from municipal wastewater. *Appl. Radiat. Isotopes* **1999**, *50*, 655–660. [CrossRef]
31. Bergaoui, L.; Lambert, J.F.; Prost, R. Caesium adsorption on soil clay: Macroscopic and spectroscopic measurements. *Appl. Clay Sci.* **2005**, *29*, 23–29. [CrossRef]

8

Non-Destructive Testing of Technical Conditions of RC Industrial Tall Chimneys Subjected to High Temperature

Marek Maj [1,*], Andrzej Ubysz [1], Hala Hammadeh [2] and Farzat Askifi [3]

[1] Faculty of Civil Engineering, Wroclaw University of Science and Technology, Wroclaw 50-370, Poland; andrzej.ubysz@pwr.edu.pl
[2] Faculty of Engineering, Middle East University, Amman 11831, Jordan; hhamm2131@gmail.com
[3] Department of Structure Engineering, Faculty of Civil Engineering, Damascus University, Damascus PO Box 30621, Syria; frzataskifi@gmail.com
* Correspondence: marek.maj@pwr.edu.pl

Abstract: Non-destructive tests of reinforced concrete chimneys, especially high ones, are an important element in assessing their condition, making it possible to forecast their safe life. Industrial chimneys are often exposed to the strong action of acidic substances, They are negatively exposed to the condensation of the flue gases. Condensate affects the inside of the thermal insulation and penetrates the chimney wall from the outside. This is one reason for the corrosion of concrete and reinforcing steel. Wet thermal insulation settles, and drastically reduces its insulating properties. This leads to an increase in temperature in the reinforced concrete chimney wall and creates additional large variations in temperature fields. This consequently causes a large increase in internal forces, which mainly increase tensile and shear stresses. This results in the appearance of additional cracks in the wall. The acid condensate penetrates these cracks, destroying the concrete cover and reinforcement. Thermographic studies are very helpful in monitoring the changes in temperature and consequently, the risk of concrete and reinforcement corrosion. This simple implication between changes in temperature of the chimney wall and increasing inner forces as shown in this article is particularly important when the chimney cannot be switched off due to the nature of the production process. Methods for interpreting the results of thermovision tests are presented to determine the safety and durability of industrial chimneys.

Keywords: nondestructive testing; thermography; monitoring of structures; reinforced concrete chimney; corrosion processes; service life of a structure

1. Introduction

Industrial reinforced concrete chimneys are often exposed to a chemically aggressive environment. The combustion gases conveyed via the chimney undergo condensation inside it or are dissolved in precipitation, becoming strongly acidic liquids. The concrete/acid contact results in the corrosion of the concrete and after the concrete cover is penetrated, in the corrosion of the steel also. The two corrosion processes result in the rapid degradation of the chimney structure.

As regards chimneys, one should bear in mind not only the high cost and the technological challenges involved in their construction but also that they perform an essential role in the production processes and so cannot be put out of service for repairs. For example, in steelworks and coking plants, the damping of the furnace from which the combustion gases are conveyed to a chimney results in the destruction of the whole power unit. Therefore, nondestructive tests are vital for both assessing the current technical condition of the chimney and monitoring the degradation processes over its whole service life [1].

2. Causes and Effects of Reduced Effectiveness of Chimney Thermal Insulation

One of the major causes of the degradation of the chimney's reinforced concrete (RC) shell is the too rapid fall in the temperature of the combustion gas as it flows through the chimney flue. The chimney wall consists of the following three layers:

- A reinforced concrete shaft;
- An internal wall (e.g., made of fire brick) constituting the chimney's inner lining which is very resistant to high temperatures;
- A mineral wool layer, placed between the two walls, serving as thermal insulation.

The durability of a chimney to a considerable degree depends on the quality and longevity of the thermal insulation. Since the chimney cannot be taken out of the production process, it is highly important to constantly monitor the condition of the insulation. Particularly suitable for this purpose are non-destructive testing methods. If the RC shell were removed in places to expose the insulation, the places could become corrosion centres and a thermal bridge could form.

When designing the geometry and thermal insulation of a chimney, one should consider the combustion gas inlet and outlet temperature, the amount and rate of flow of the flue gas and its condensation temperature. Significant internal forces can also arise from a temperature difference [2].

The above parameters are determined by

- The amount of the exhausted gas;
- Changes in flue gas temperature due to changes in production technology;
- The daily changes of the physicochemical parameters of the chimney environment, such as the wind velocity, the atmospheric pressure, the air humidity, etc.

An important factor is a reduction in the insulating power of the chimney walls caused by changes in the physical properties of the insulation due to, e.g., the mineral wool getting damp [3]. As a result, the temperature of the flue gas in the chimney's upper part decreases, and after the critical temperature is reached, this causes excessive flue gas condensation on the chimney lining.

As the condensate penetrates through cracks in the lining, it makes the mineral wool damp, whereby the latter loses its insulating properties. Moreover, the damp mineral wool sinks, and as a result, areas devoid of thermal insulation are created. As the coefficient of thermal conductivity decreases, the temperature of the chimney's inner wall falls further, and so does the temperature of the flue gas. As a result, more and more flue gas condenses on the chimney's inner wall, whereby the degradation processes in the chimney's RC shell intensify (Figures 1–3).

Figure 1. Deposition of condensation on outer surface chimney wall along construction joints.

Figure 2. Condensation degrading of chimney wall.

Figure 3. Condensation dripping down from a chimney upper ring.

As condensation drips down the inner wall, it penetrates via cracks to the mineral wool and outside to the reinforcement of the concrete wall, causing intense corrosion of the concrete and the reinforcing steel (Figure 4).

Figure 4. Condensation penetrating chimney crown wall.

3. The Idea of a Thermographic Survey of Chimney Thermal Insulation

Thermographic surveys [4–9] have been used to detect thermal bridges in residential buildings for many years. Thermographic surveys are conducted using a camera which can take thermograms. In the case of chimneys, only thermograms of the external surface of chimney's wall are taken. The air temperature considerably affects the accuracy of thermographic surveys. The latter is more precise. The larger the thermal contrast between the surface of the chimney shell and the ambient air temperature the better the thermogram. Therefore, the best period for evaluating the condition of the thermal insulation in chimneys is in winter.

Passive thermography is the optimal method for the thermographic surveying of chimneys. Based on the thermographic images and reference readings, one can determine the temperatures on the surface of the chimney shell [3,10]. In this method, the image is obtained for a set scale range. Knowing the temperature of the flow gas flowing through the chimney and the temperature on the outer surface of the chimney shell, one can determine the actual thermal transmittance coefficient and compare it with its calculated values specified in the design documents. Based on the relative temperature differences, one can determine the degree of damage to the thermal insulation. The measurements were made by a company specializing in thermovision. The cameras were calibrated before measurements.

However, there may be a few percentages of error due to the curvature of the chimney surface, non-homogeneity of the surface, etc.

4. Thermographic Surveys of Chimney Thermal Insulation

The monitoring of the condition of the chimney's insulation is an important element in the assessment of the durability of the chimney. Based on such monitoring, one can forecast the actual service lifespan of the structure and systematically eliminate the causes of chimney shell degradation. The condition of the thermal insulation of industrial chimneys is examined using the classic thermographic method [11,12]. It is mainly the structure's envelope which is examined in this way to detect places where there are gaps in the insulation, or there is no insulation. In the considered case, the thermograms showed that temperature differences on the chimney's outer surface occurred along practically its whole height and on its circumference (latitudinally). According to the results [13] of the thermographic surveys of the 120 m high chimney with the lower and upper diameter of respectively $D_d = 7.16$ m and $D_g = 4.48$ m, the temperature difference in the particular points of the surface amounted to

- As much as 21.8 °C in the chimney's upper part;
- 15.9 °C in the chimney's middle part;

- 10.5–11.9 °C (being more uniform) in the chimney's lower part (Figure 5).

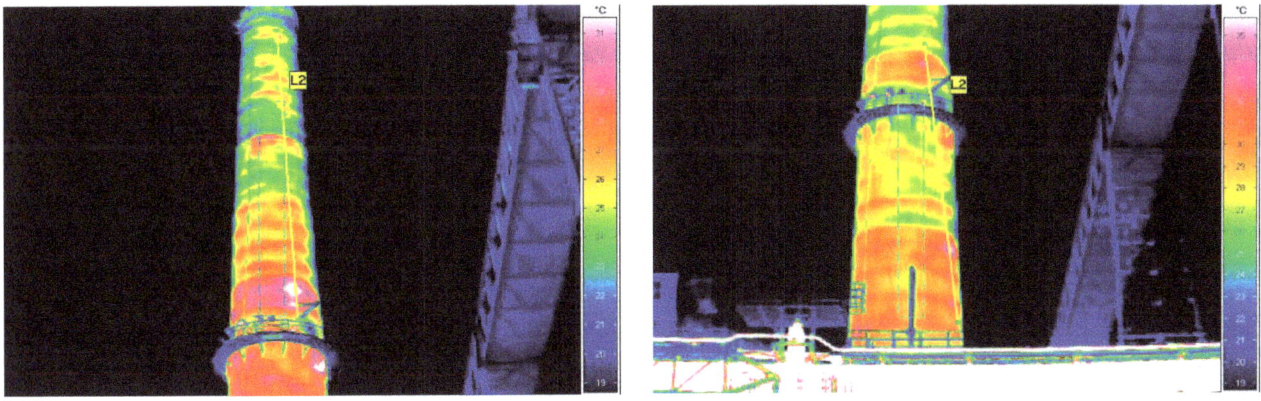

(a) (b)

Figure 5. Chimney thermograms revealing insulation imperfections: (**a**) No thermal insulation more than two metres above and under platform; (**b**) Varied temperature distribution in chimney shaft.

The lower part of the chimney does not show such a high degree of degradation as its middle and upper parts. The thickness of the outer reinforced concrete shell, which is considerably greater in the lower part of the chimney, is one of the determining factors.

To nondestructively investigate the condition of the thermal insulation, an analysis of temperature values v_e on the outer surface of the chimney's shell was carried out. The parameters for the analysis had been determined based on the known thermophysical characteristics of the materials, the chimney's geometry (specified in the design documents), and temperature measurements. The following parameters of the layered chimney wall were assumed for the analysis:

- Thermal conductivity:
    - fire brick = 1.30 [W/(m K)],
    - mineral wool = 0.05 [W/(m K)],
    - reinforced concrete = 1.74 [W/(m K)],
- Heat-insulating layer thickness:
    - fire brick layer—11.4 cm,
    - mineral wool layer—13 cm,
    - reinforced-concrete chimney shell wall—24 cm,
- Temperature:
    - inside chimney 200 °C,
    - external 12.3 °C.
- Temperature values v_e on the outer surface of the chimney shell are:
    - for the wall with mineral wool: v_e = 14.4 °C,
    - for the wall without a mineral wall: v_e = 32.4 °C.

Calculations and in situ measurements indicate that in the places where the temperature of the reinforced-concrete shell is the highest there is no mineral wool or the wool there has very poor insulating properties. The measured minimum temperatures of the shell in well-insulated places on average amounted to 17 °C while the maximum temperatures in the same thermograms on average amounted to 33 °C. The variation coefficient for the temperature on the chimney surface ranges from

0.13 to 0.25. The average decrease in thermal performance in the particular chimney wall surface areas ranges from 20% to 90%.

5. Thermal and Static Load Analysis

As a result of the nonuniform temperature distribution, caused by damage to the thermal insulation, additional internal forces, such as bending moments, shear stresses, and annular tensile and compressive forces, arise. The values of the forces can be traced by studying real cases of damaged chimneys.

5.1. Adopted Assumptions

The following were assumed: the velocity of flow of the flue gas in the chimney: v = 10 m/s, the operating temperature of the flue gas at the chimney outlet: t_{w2} = 180 °C, and the operating temperature of the flue gas at the chimney inlet: t_{w1} = 220 °C. The inlet of the combustion gas takes place through a connecting flue pipe located below the ground level.

The external design temperature was assumed based on the national annex to standard [14]: in winter T_{min} = −36 °C and in summer T_{max}= + 40 °C.

Since no detailed information was available, the emergency flue gas temperature was assumed as 20% higher than the typical one. The emergency temperature amounts to $t_{wa,1}$ = 220 °C and $t_{wa,2}$ = 270 °C, respectively, at the chimney outlet and inlet. The emergency temperature was used in the calculations.

5.2. Types of Thermal Effects in the Chimney

The linearly elastic models of chimney concrete walls are the simplest models that indicate the influence of thermal factors. The wall thickness of the reinforced concrete chimney is not large as in the case of silos [15] for hot materials. The model of linear temperature gradient used for these silos gives excessive bending moments. The wall of a reinforced concrete chimney subjected to repeated wind and thermal loads behaves linearly. Another problem is with cracks in the wall which reduce the stiffness of the chimney. Cracks in a chimney reduce the impact of bending moments but also expose the chimney to corrosion. The first load results from the difference between the temperatures on the surface of the chimney shaft. The load was calculated for an ideal situation, i.e., immediately after chimney erection—no degraded insulation, and for a situation when some of the insulation has degraded. The temperature load was introduced as the gradient of the temperatures in the chimney shaft.

The second load stems from the difference between the chimney operating temperature during chimney service life and the initial temperature, i.e., the temperature at which the chimney was erected.

5.3. Distribution of Temperature in Chimney Wall for Undamaged Insulation

The chimney wall consists of the following layers: the shaft, the thermal insulation, and the lining. The thicknesses and diameters were assumed according to Table 1. As the reference situation, the condition of the chimney immediately after its erection, i.e., with the continuous mineral wool and fibre brick insulation, was adopted. The following thermal conductivity coefficients were assumed: λ_b = 1.74 W/mK for the reinforced concrete wall, λ_{iz} = 0.05 W/mK for the insulation (mineral wool), and λ_{sz} = 1.30 W/mK for the fire brick.

Thermal transmittance coefficient k:

$$\frac{1}{k} = \frac{1}{\alpha_n} + \sum_i \left(\frac{t_i}{\lambda_i} \kappa_i \frac{r_z}{r_i} \right) + \frac{1}{\alpha_o}, \tag{1}$$

where $\alpha_n = 8 + v = 8 + 10 = 18\ \mathrm{W/m^2K}$—the inflow coefficient (the zone of the inner surface of the lining), where v—the mean velocity of the flue gas;

$\alpha_o = 24\ \mathrm{W/m^2K}$—the outflow coefficient (the outer surface of the shaft);

t_i—the thickness of layer i;
λ_i—the thermal conductivity coefficient of layer i;
r_i—the outside radius of layer i;
r_z—the outside radius of the shaft;
κ_i—correction coefficients taking into account wall curvature:
$\kappa_i = \left(\frac{r_z}{r_i}\right)^{0.57}$ where $i = b$ for the reinforced concrete shaft, $i = iz$ for the mineral wool, $i = sz$ for the fire brick.

Table 1. Geometric dimensions of the thermographically surveyed existing chimney.

Segment No.	H	Shaft Outside Radius	Chimney Wall Thickness	Shaft Inside Radius	Insulation Thickness	Lining Thickness
-	[m]	[m]	[m]	[m]	[m]	[m]
1	2.5	4.00	0.42	3.58	0.15	0.23
2	15	3.80	0.39	3.41	0.15	0.23
3	30	3.61	0.36	3.25	0.13	0.23
4	45	3.41	0.33	3.08	0.13	0.114
5	60	3.21	0.3	2.91	0.13	0.114
6	75	3.01	0.27	2.74	0.13	0.114
7	90	2.82	0.24	2.58	0.13	0.114
8	105	2.62	0.21	2.41	0.13	0.114
9	120	2.42	0.18	2.24	0.13	0.114

The temperature drop in a given layer is expressed by the formula

$$\Delta T_i = k\frac{t_i}{\lambda_i}\kappa_i\frac{r_z}{r_i}\Delta t, \tag{2}$$

where $\Delta t = t_w - t_z$.

For the inflow and outflow, the temperature drops are

$$\Delta T_n = k\cdot\frac{1}{\alpha_n}\Delta t \qquad \Delta T_o = k\cdot\frac{1}{\alpha_o}\Delta t. \tag{3}$$

The temperature at the boundary of each of the layers is

$$T_j = t_w - \frac{k}{\alpha_n}\cdot\Delta T - \sum_i \Delta T. \tag{4}$$

Figures 6 and 7 show temperature drops in the particular layers in the winter season and in the summer season. The largest temperature drop occurs in the full insulation layer, and it is about 10 times larger than in the reinforced concrete shaft. Because of the small thickness of the fire brick layer, the temperature drop in this layer is the smallest.

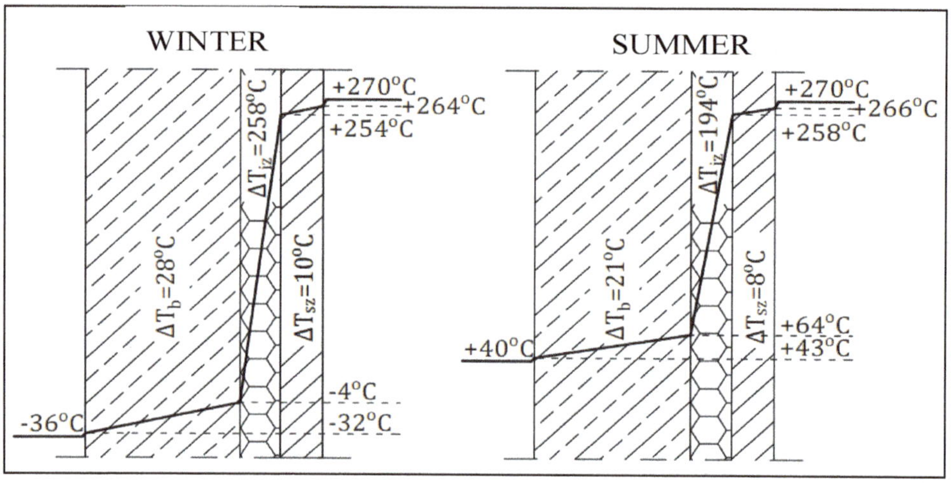

Figure 6. Temperature drops in particular layers for the winter and summer season.

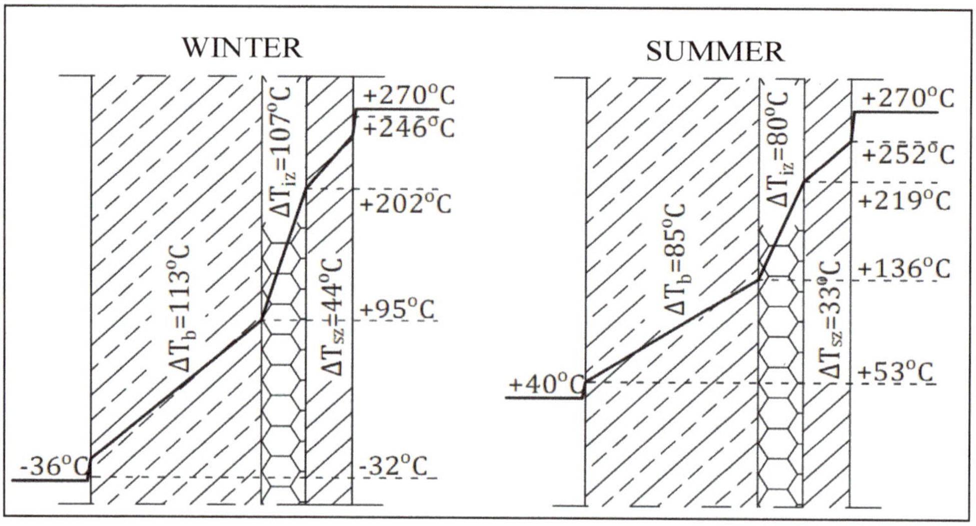

Figure 7. Temperature drops in particular layers of design chimney insulation in the first segment with damaged insulation.

5.4. Distribution of Temperature in Chimney Wall in Case of Insulation Discontinuity

Insulation discontinuities were assumed to occur in 2.5 high segments uniformly distributed along the whole height of the chimney. The mineral wool in the segments was assumed to lack the design insulating power.

The temperature drops in the particular layers for the decreased mineral wool insulating power are shown in Table 2. Since it is not possible to directly assess the degree to which the mineral wool's insulating power decreased, an approximate method was used. Calculations in which λ_{iz} was the unknown were carried out on the basis of the thermographic surveys of the existing chimney and the temperatures on its surface. The value of λ_{iz} was adjusted consistently with the actual temperatures on the surface of the investigated chimney. The results are presented in Table 3. The results apply to the 10-fold decrease in the insulating power of the mineral wool in the selected places ($\lambda_{iz} = 0.5$ W/mK).

The maximum temperature gradient in the reinforced concrete shell in the winter season increased by about 420% in comparison with the gradient calculated for the worst insulation case. For the summer season temperature, the gradient increased by about 405%. The temperature gradient which the insulation transfers is approximately equal to the gradient transferred by the chimney shaft under both the winter and summer temperature load. Such a large increase in temperature load can lead to the cracking of the reinforced concrete shell and to its damage. The temperature gradient values in the summer season and the temperature on the chimney's surface in summer and in winter and a figure (Figures 8–10) showing the temperature gradient values in each of the layers are presented below.

Table 2. Temperatures on the surface of the calculated chimney at an ambient temperature of 15 °C.

Seg. No.	H	Damaged Insulation					Full Insulation				
		$T_{sz,w}$	$T_{sz,z}$	$T_{iz,z}$	$T_{b,z}$	$T_{ext.}$	$T_{sz,w}$	$T_{sz,z}$	$T_{iz,z}$	$T_{b,z}$	$T_{ext.}$
-	[m]	[°C]	[°C]	[°C]	[°C]	[°C]	[°C]	[°C]	[°C]	[°C]	[°C]
1	10	250	213	124	30	15	265	256	41	19	15
2	20	246	208	118	30	15	261	252	39	19	15
3	30	241	203	112	30	15	257	248	38	19	15
4	40	234	192	118	32	15	251	240	42	20	15
5	50	230	187	111	33	15	247	236	40	20	15
6	60	223	175	113	35	15	241	227	44	21	15
7	70	218	169	106	35	15	237	223	41	21	15
8	80	212	160	103	37	15	232	216	42	22	15
9	90	207	153	95	37	15	228	212	39	22	15
10	100	204	150	94	37	15	224	208	38	21	15
11	110	200	148	92	36	15	220	204	38	21	15
12	120	193	142	89	35	15	212	197	37	21	15

Where $T_{sz,w}$—the temperature of the outside surface of the ceramic wall; $T_{sz,z}$—the temperature of the inside surface of the ceramic wall (contact with isolation); $T_{iz,z}$—the temperature of the inside surface of the concrete wall; $T_{b,z}$—the temperature of the outside surface of the concrete wall; T_{ext}—outside temperature.

Table 3. Differences between chimney operating temperature and chimney erection temperature.

Seg. No.	H	Winter		Summer	
-	[m]	Insulation		Insulation	
		Full	Damaged	Full	Damaged
		[°C]	[°C]	[°C]	[°C]
1	10	−28	28	44	86
2	20	−29	25	43	83
3	30	−30	22	42	81
4	40	−27	27	44	84
5	50	−28	23	43	81
6	60	−25	26	45	83
7	70	−27	22	44	79
8	80	−25	21	45	79
9	90	−27	17	44	75
10	100	−28	16	43	74
11	110	−28	15	43	73
12	120	−29	13	42	71

To relate the assumptions to the actual temperature loading of the chimney, the calculated gradients were compared with the temperatures appearing in the thermographic pictures of the existing chimney. The investigated real chimney is a 120 m high reinforced concrete structure serving a coke oven battery. The temperature of the combustion gas at the investigated chimney's inlet reaches 270 °C (maximally 340 °C) while the flue gas temperature at the outlet amounts to 220 °C (maximally 300 °C). The chimney is made of concrete C25/30 and reinforced with steel A-II. The chimney was divided into 9 segments, each about 15 m high. The chimney's insulation is made of semi-hard mineral wool boards and batts and its lining is made of fire brick. The thicknesses of the particular layers and their diameters are presented in Table 1.

The thermographic surveys revealed temperature differences on the chimney's surface, which indicates that the insulation was not uniform, there were gaps in it, and in places the insulation had slid down and was damp. The temperature differences can also be due to the degradation of the reinforced concrete chimney shaft. Since the measurements were carried out outside the chimney, the places with deteriorated thermal insulation are visible as dark red (the hottest places). The suspected places where the insulation is missing are visible in the thermograms as bands running at regular intervals around the circumference of the chimney. One can discern a certain pattern in insulation discontinuity: the bands are spaced at every 1 to 3 m. The insulation loss is most visible in the lower part of the chimney, and it is uniform there. In the thermographic picture, the surface temperature in this part reaches about 34 °C. In the segments situated higher one can see more distinct temperature differences on the outside surface, ranging from 22 to 34 °C.

Since the temperature of the flue gas in the existing chimney is comparable with the assumed emergency temperature in the chimney and also the diameters and thicknesses of the layers are similar, the temperature on the surface of the RC shaft, indicated by the thermographic surveys should be similar to the one yielded by the calculations. As indicated by the high background temperature, the surveys were carried out in the summer season. The ambient temperature was assumed to be 15 °C. The temperatures on the shaft surface calculated for the case with insulation and the case with insulation loss are comparable with the ones revealed by the thermographic surveys of the real chimney, as shown in the tables. For the assumed value of $\lambda_{iz} = 0.5$ W/mK, the temperature on the RC shell is comparable with the temperature indicated by the thermographic surveys.

Temperature loading in 2.5 m high bands for alternately full insulation and damaged insulation was assumed. The values were assumed according to Tables 2 and 3.

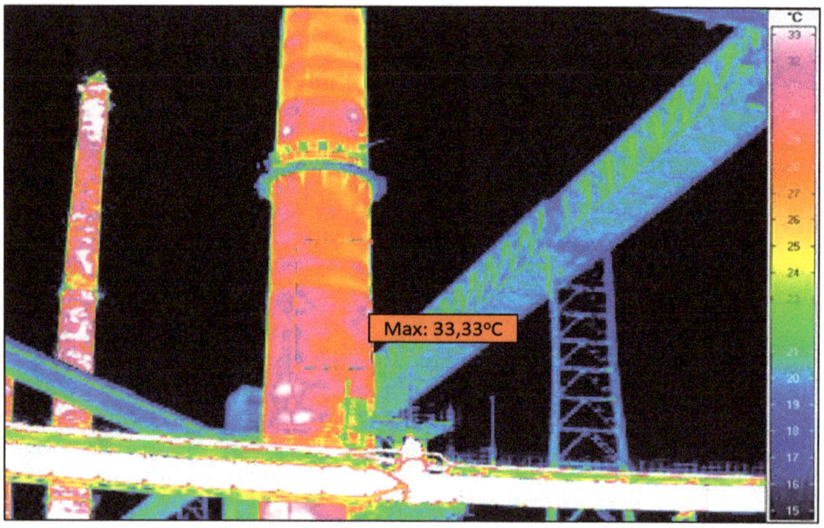

Figure 8. Thermographic picture of the lower part of chimney shaft.

Figure 9. Thermographic picture of the middle part of chimney shaft.

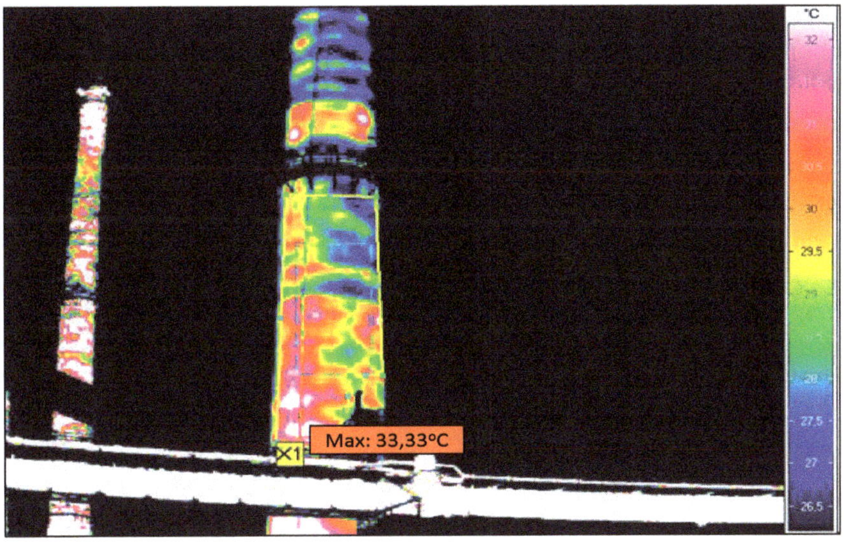

Figure 10. Thermographic picture of the upper part of chimney shaft.

5.5. Surface Temperature Load

Since the chimney works at a temperature different than the one prevailing during its erection, a temperature load generating internal forces due to the chimney's limited freedom of deformation was assumed. Two load cases, the summer season load and the winter season load, were considered. In each of the cases, a load at full insulation and a load at damaged insulation were analysed.

The mean temperature in the wall can be calculated from the formula

$$\overline{T} = T_{b,wew} - \frac{\Delta T_b}{2}, \tag{5}$$

where $T_{b,wew}$—the temperature on the inside surface of the reinforced concrete shaft; ΔT_b—the temperature drop in the reinforced concrete shaft.

The temperature at which the chimney had been erected was assumed as $T_o = 10\ ^oC$.

The difference between the chimney operating temperature and the chimney erection temperature was calculated from the formula

$$\Delta T = \overline{T} - T_o, \tag{6}$$

5.6. Increase in Bending Moments Due to the Increase in the Temperature Gradient

The difference in the temperature distribution across the thickness of the reinforced concrete wall between the case with damage insulation and the case with undamaged insulation amounts to

$T = 113 - 28 = 85°$

Such a big gradient causes additional bending moment and additional shear force.

The model of these calculations is presented in Figure A1.

Calculations of this influence are included in Appendix A

Additional thermally induced internal forces.

Let us consider the chimney as a cylindrical shell in which the unexpandable (well thermally insulated) band can be regarded as an element resembling a clamping tendon. Circumference *L* of the surrounding area with damaged thermal insulation is increased by *L* due to higher temperature *T*. The tendon with the introduced compressive force reduces the circumference to its size before the thermal expansion.

Using such a static load model one gets additional bending moment *M* and additional shear force Q (model is presented on Figure A2. The calculation of these additional factors is presented in Appendix B.

A few remarks

The presented computing models do not average their results. They assume quite a sharp transition from one temperature field to another without a transition area. In fact, the intermediate area can smooth calculations outcomes.

6. Conclusions Emerging from Investigations of Thermal Insulation of Chimneys

The method of thermographic surveys, used to monitor thermal insulation in housing, also finds an important application in the investigation of the durability and failure hazard of industrial structures in which the condensation of chemically aggressive gases is likely to occur. This particularly applies to chimneys where invasive tests are not preferred because of the consequences of taking samples from the structure.

Owing to thermographic surveys, one can monitor the hazards leading to the degradation of the chimney structure.

It is possible to indicate the area where flue gas condensate penetrates through cracks and leaks into the inner fire-brick lining into the insulation layer made of mineral wool. The condensate penetrating through cracks in the RC shaft to the structural reinforcement intensifies the corrosion of the steel and the concrete. As a result of the decrease in insulating power the temperature in the chimney RC core rises, whereby the bending moments and the tensile forces increase and the reinforced concrete shell cracks under the additional stresses which had not been taken into account in the design of the walls [2].

Thermograms can show increased shear and tensile stresses that can cause these large cracks in concrete.

Soaked by condensate or damaged mineral wool areas are visible on thermograms.

The calculations presented in the paper indicate the possibility of assessing the values of shear stresses from thermal interactions.

The authors took part in the design of chimney repairs with large diagonal cracks caused by high shear stresses.

In the case of chimneys operating in the uninterruptable process mode, special measures are required to improve the condition of the insulation. The repair of the reinforced concrete chimney consisting solely in injection filling the external cracks [10] stops the corrosion of the reinforcing steel, but it does not improve the insulation of the chimney.

Author Contributions: Conceptualisation, M.M. and A.U.; Methodology, M.M.; Software, H.H. and F.A.; Validation, M.M., A.U., H.H. and F.A.; Formal Analysis, M.M.; Investigation, M.M. and A.U.; Resources, M.M., A.U., H.H. and F.A.; Data Curation, M.M. and A.U.; Writing—Original Draft Preparation, M.M. and A.U.; Writing—Review and Editing, M.M., A.U., H.H. and F.A.; Visualisation, M.M., H.H. and F.A.; Supervision, A.U.

Acknowledgments: Developed as part of the research project "Industrialized construction process (Construction 4.0). Technological and methodological conditions of application of selected composite elements in civil engineering". This project is carried out by the Wroclaw University of Technology together with the Peoples' Friendship University of Russia in Moscow, Research Project PWr-RUDN 2017 no. 45WB/0001/17 Industrialized Construction Process.

Appendix A

$M_t = * T * E_c J_c / t = 0.00001*85*33{,}000{,}000*1*0.242/12 = 134.5$ kNm

The cracking moment for the concrete

$M_{cr} = f_{ctm} W_1$ (1)

where $E_c = 33$ GPa and $f_{ctm} = 2.9$ Mpa—the mean tensile strength of the concrete (C30/37),

A_c—the cross-sectional area $A = 1*0.24 = 0.24$ m^2, $W_1 = bt^2/6$, $b = 1$ m, $t = 0.24$ m, the bending section modulus for phase I.

$M_{cr} = 2.9*1*0.242/6 = 117$ kNm

Shear stress at damaged insulation/undamaged insulation boundary

The difference between the elongations of the adjacent rings depends on (for diameter $D = 6.68$ m):
$(\pi D) = \pi * D* * T = 3.14*6.68*0.00001*85 = 20.98*0.00001*85 = 0.018$ m
The annular strain is
$= *(\pi D)/(\pi * D) = 0.018/20.98 = 0.000858$.
The annular force generated due to strain locking is
$N = A_c * \; *E_c = 0.24*0.000858*33{,}000{,}000 = 6795.4$ kN/m
The force is blocked by horizontal surfaces with width t, whereby shear stress $_c$ arises
$_c = N/(2*t) = 14.2$ MPa
at the mean shear strength of the concrete
$f_{c\,m} = 0.5 * f_{ctm} = 0.5*2.9 = 1.45$ MPa.

The above bending moments due to gradient T, and shear forces are additional quantities, whereby the forces assumed in the design are exceeded.

Calculations were carried out for simplified static load diagrams. The values of the bending moments and shear stresses indicate high additional stresses in the concrete.

As the horizontal band in the concrete wall with damaged insulation heats up, the band is "pushed" outside more strongly, whereby a bending moment arises.

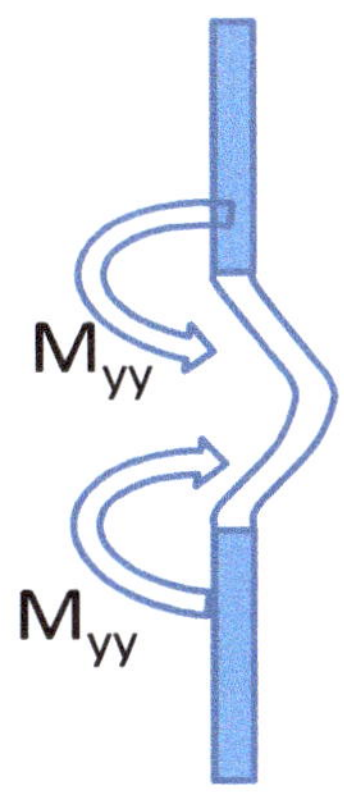

Figure A1. Change in length of cylindrical shell results in the meridional bending moment.

As this displacement is blocked, annular force $N = 6795.4$ kN/m is generated.

Appendix B

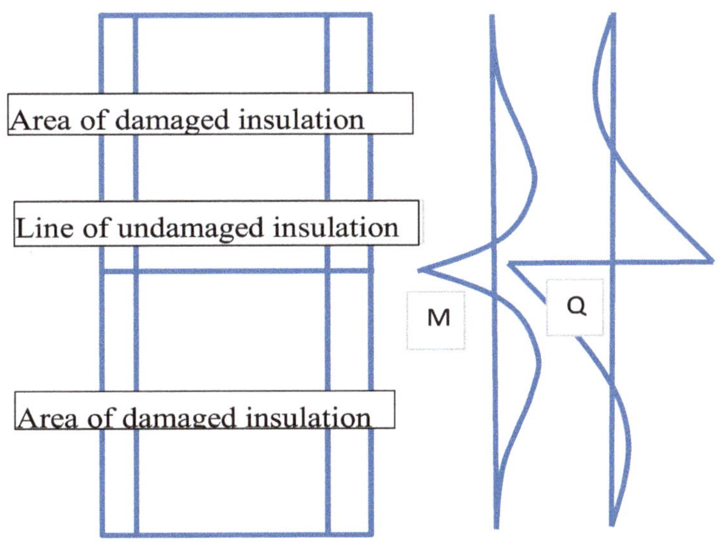

Figure A2. Model of influence damaged on the undamaged area of insulation.

In this case, calculations of the internal forces for design value T obtained from the thermograms are as follows ([16]):

$E_c = 33$ GPa
t – thickness = 0.24 m
= 0.2
$T_{design} = 15°$.
$D_{diameter} = 6.68$ m
r =D/2 = 3.5 m
L= 20.986 m
L = 0.003 m
N (force in shell, generated by T) = 1188 kN/m
$_p = (12*(1 - {}^2)/(D^2h^2))^{(1/4)}*1.42$ ([9,16])
/ (4 $_p$) = 0.55 m
$F_{des} = N/r = 339.43$ kN/m
$M = F_{des}/(4*{}_p) = 59.70$ kNm/m
$Q = F_{des}/2 = 169.71$ kN/m

The above values added to the stress resulting from the permanent loads and the variable loads may contribute to horizontal cracking.

References

1. Dörr, R.; Noakowski, P.; Breddermann, M.; Leszinski, H.; Potratz, S. Verstärkung eines Stahlbetonschornsteins. *Beton- und Stahlbetonbau* **2004**, *99*, 670–674.
2. Schnell, J.; Kautsch, R.; Noakowski, P.; Breddermann, M. Verhalten von hochbaudecken bei zugkräften aus zwang: Einfluß von kriechen, betonfestigkeit, temperaturdifferenz, plattendicke und spannweite—Auswirkung auf schnittgrössen, stahlspannung, rissbreite, druckzonenhöhe und durchbiegung. *Beton- und Stahlbetonbau* **2005**, *100*, 406–415. [CrossRef]
3. Jäger-Cañás, A.; Pasternak, H. Influence of closely spaced ring-stiffeners on the axial buckling behavior of cylindrical shells. *Ce/Papers* **2017**, *1*, 928–937. [CrossRef]
4. Bobkiewicz, J. Badania termograficzne budynków jako podstawa skutecznej termomodernizacji. *Pomiary Autom. Robotyka* **2003**, *7*, 23–25. (In Polish)
5. Jaworski, J. *Termografia Budynków. Wykorzystanie Obrazów Termalnych w Diagnostyce Budynków*; Dolnośląskie Wydawnictwo Edukacyjne: Wrocław, Poland, 2000.
6. Nowak, H. *Zastosowanie Badań Termowizyjnych w Budownictwie*; Oficyna Wydawnicza Politechniki Wrocławskiej: Wrocław, Poland, 2012.
7. Pichniarczyk, P.; Zduniewicz, T. Wykorzystanie w budownictwie metody termowizji w podczerwieni. *Izolacje* **2010**, *15*, 21–23. (In Polish)
8. Żurawski, J. Termowizja jako weryfikacja jakości prac izolacyjnych. *Izolacje* **2008**, *11–12*, 19–23.
9. Sendkowski, J.; Tkaczyk, A.; Tkaczyk, Ł. Termowizja i termografia w diagnostyce kominów przemysłowych. Przykłady, możliwości. *Prz. Bud.* **2013**, *84*, 21–25.
10. Kaminski, M.; Maj, M.; Ubysz, A. Chimney cracked reinforced concrete walls as a problem of durability exploitation. In Proceedings of the Fifth International Conference on Structural Engineering (SEMC 2013), Cape Town, South Africa, 2–4 September 2013.
11. Horváth, L.; Iványi, M.; Pasternak, H. Thermovision: An efficient tool for monitoring of steel members. *IABSE Symposium Rio De Janerio* **1999**, 142–143.
12. Świderski, W. Metody i techniki termografii w podczerwieni w badaniach nieniszczących materiałów kompozytowych (Methods and techniques of infrared thermography in non-destructive testing of composite materials). *Probl. Tech. Uzbroj. (Probl. Armament Technol.)* **2009**, *38*, 75–92.
13. Maj, M.; Ubysz, A. Some methods of repairing of chimney cracked reinforced concrete wall. *TISNOB Pozn.* **2015**, 213–224.

14. Oddziaływania na konstrukcje. Część 1-5: Oddziaływania ogólne. Oddziaływania termiczne. *PN-EN 1991-1-5 Eurokod 1.*
15. Brown, C.J.; Nielsen, J. *Silos Fundamentals of Theory Behavior and Design*; E & FN Spon: London, UK, 1998.
16. Timoshenko, S.; Woinowsky Krieger, S. *Theory of Plates and Shells*, 2nd ed.; McGraw-Hill Book Company: New York, NY, USA, 1959; pp. 466–532.

9

Application of Nanoindentation and 2D and 3D Imaging to Characterise Selected Features of the Internal Microstructure of Spun Concrete

Jarosław Michałek *, Michał Pachnicz and Maciej Sobótka

Faculty of Civil Engineering, Wrocław University of Science and Technology, 50-370 Wrocław, Poland; michal.pachnicz@pwr.edu.pl (M.P.); maciej.sobotka@pwr.edu.pl (M.S.)

* Correspondence: jaroslaw.michalek@pwr.edu.pl

Abstract: The spinning of concrete is a process in which concrete mixture is moulded and compacted under the action of the centrifugal force arising during the fast rotational motion of the mould around its longitudinal axis. As a result of the spinning of the liquid concrete mixture, an element annular in cross section, characterised by an inhomogeneous layered wall structure, is produced. The heavier constituents tend towards the cross-section wall's outer side, while the lighter components tend towards its inner side. The way in which the particular constituents are distributed in the element's cross section is of key importance for the macro properties of the manufactured product. This paper presents procedures for investigating spun concrete and interpreting the results of such investigations, which make it possible to characterise the microstructure of the concrete. Three investigative methods were used to assess the distribution of the constituents in the cross section of the element: micro-computed tomography (μCT), 2D imaging (using an optical scanner) and nanoindentation. A procedure for interpreting and analysing the results is proposed. The procedure enables one to quantitatively characterise the following features of the microstructure of spun concrete: the mechanical parameters of the mortar, the aggregate content, the pore content, the cement paste content, the aggregate grading and the size (dimensions) of the pores. Special attention is devoted to the determination of the variation of the analysed quantities in the cross section of the element. The result of the application of the investigative procedures is presented for an exemplary spun concrete element. The proposed procedures constitute a valuable tool for evaluating the process of manufacturing spun concrete elements.

Keywords: spun concrete; micro-computed tomography; nanoindentation; deconvolution; mathematical morphology

---

1. Introduction

The main aim of this study was to assess the capabilities of three different investigative methods, namely nanoindentation, micro-computed tomography (μCT) and 2D optical scanning, as applied to evaluate selected properties of the microstructure of spun concrete. In particular, it was attempted to use these methods to describe the internal microstructure of concrete, e.g. local strength parameters, pore space morphology, spatial distribution of aggregate and cement paste. The performance of utilised methods was demonstrated on the example of spun concrete samples. This type of concrete has an internal structure different from the commonly used cast in place concrete. Due to the production process, the spun concrete is characterised by a layered structure across the wall of the annular cross section. Therefore, the parameters describing the internal microstructure vary within the cross-section. Proposed methodology for determining aforementioned parameters can be a practical tool helpful, for example, in the development of the optimal technology for the production of spun concrete elements.

The spinning of concrete is a process in which a concrete mixture is shaped and compacted under the action of a normal (radial) force arising during the fast (500–700 rpm) rotational motion of the mould around its longitudinal axis. Because of its peculiarities, this method can be used to produce exclusively hollow elements. As a result of spinning, the concrete mixture introduced into the mould consolidates and retains the acquired shape as the process continues. The moulding pressure produced by spinning is not uniform across the layer being consolidated. The moulding pressure on the inner layer of the product wall is close to zero, while, on its outer side, it reaches a maximum value. A graph of the pressure across the product wall can be presented in the form of a triangle.

The generated normal pressure is initially taken by the water present between the particles of the concrete mixture. As a result, the hydrostatic pressure increases and under this pressure the water begins to displace (through the forming filtration channels directed radially towards the mould axis) from one layer to another. At the same time, the concrete mixture consolidates. The displacing water takes the finest concrete mixture constituents (mainly cement) with it. In the course of this process, the cohesion of the mixture increases and the mechanical bond between the mixture constituents acquires some strength and in certain conditions a state of equilibrium is reached, whereby the filtration stops. Since at this moment the moulding pressure is fully transferred to the solid particles of the already considerably compacted concrete mixture, the ultimate effect of the compaction will depend on the composition of the concrete mixture (to a large extent on the aggregate grading).

Under such a distribution of pressure during spinning, water is not carried away from the cross-section wall thickness equally (the element's wall layers being characterised by different w/c values and thus by different porosity). Consequently, the mixture layers closer to the mould axis begin to consolidate only after the limit density of the outer layer is reached while the moulding pressure increases. This means that, unlike in the case of vibrated concrete, the structure of spun concrete across the element wall is inhomogeneous and layered. The structure is characterised by the fact that the heavier constituents (large particles) tend towards the cross section's outer side, while the lighter components (cement paste) tend towards its inner side. As a result, the outer layer can become highly compact and after the concrete sets it can acquire high strength and resistance to chemical and mechanical impacts. The inner layer will consist of highly consolidated cohesive cement paste and after setting it can become highly impervious and resistant to the impact of flowing water. According to Ref. [1], the 2–4 mm thick inner layer can contain >25% more cement than the intermediate layers.

The deformation and strength characteristics of the layered structure of spun concrete have been the subject of only a few studies. Mainly macroscopic investigations of spun concrete are carried out. Marquardt was one of the first researchers [1] who described the structure of spun concrete used for the production of pipes. He experimentally found that the larger is the difference between the specific weight of the concrete mixture constituents, the faster and more completely does the mixture fractionate. For the concrete mixture, he recommended well sorted mixed aggregate particles with similar petrographic properties and a maximum diameter of 15 mm. He also described the variation in cement content across the wall of a pipe made of spun concrete. He found that cement across 83% of the wall thickness is quite uniformly distributed and an increase in cement content takes place in only the 2–5 mm thick inner layer of the cross section. To reduce cross-sectional lamination, he proposed to use layered spinning at lower mould rotational speeds. After spinning each of the layers and before feeding the next concrete mixture batch, one should pour out the spun out water from the mould and smooth the inner layer.

Achverdov, investigating the cross section of a spun concrete pipe, found [2] that the 2–3 mm thick inner layer can contain >25% more cement than the intermediate layers. The cement content in the inner layer of the cross section depends on the kind and amount of cement and on the initial water content in the concrete mixture. He also stated that the principles used for selecting a filler particle size distribution for vibrocompacted regular concretes cannot be applied to spun concretes. The total fine aggregate content should be considerable and in the case of pipes, the amount of 0/2 mm sand should constitute 40–50% by weight of the total filler content. Achverdov (similarly to Marquardt [1]) noted that, to obtain higher integrity and a higher qualitatively structure of spun concrete, it is necessary to use a multilayer moulding system. In the case of multilayer compaction, two rotation phases (the initial one and the proper one) are used for each of the layers and, when the moulding of a particular layer is finished, the centrifuge is stopped and the water drained from the mixture is removed.

Dilger, Ghali and Krishna Mohan Rao, when investigating spun concrete poles [3,4], found that, because the cross section's inner layer has a high water and cement content, the shrinkage inside the pole's cross section is greater than on its outside, which results in shrinkage cracks developing vertically and extending deep into the pole wall. They also found that segregation can be reduced or even eliminated through a proper concrete mixture design precluding an excess of fine-grain fractions in the concrete mixture. This means that only mortar necessary to fill the gaps between coarse aggregate particles should be supplied. Because of the high compaction energy generated during spinning, the amount of mortar is considerably smaller than in the case of normal concrete mixtures. Based on this research, it was determined that the ratio of the mass of the sand in the mixture to the mass of the coarse aggregate should not exceed 25–30%. However, because of such a low fine-grained parts content in the mixture, the latter is difficult to mix and compact in the centrifuge.

Adesiyun, Kamiński, Kubiak and odo [5,6] studied the influence of such parameters as sand equivalent (25–50%), cement content (410–530 kg/m^3), plasticiser amount (0–2%), rotational speed (400–700 rpm) and spinning time (5–11 min) on the structure of spun concrete. They conducted tests on 230 mm high annular spun concrete samples with the inside diameter of 45–60 mm. The test results confirm the conclusions drawn in Ref. [3,4], concerning the difference in shrinkage between the inner and outer surface of the wall of the tested cross section. They used computer image analysis, which makes it possible to extract the information contained in the image of the cross section of the element wall, to describe the structure of spun concrete. To precisely describe the structure, the sample wall thickness was divided into twenty 2.25–3.0 mm wide strips (depending on the thickness of the cross section wall). They carried out tests for different combinations of the parameters mentioned above. Using computer image analysis, they graphically presented the distribution of aggregate, cement paste and air voids across the wall for the particular samples. They found that the aggregate content decreases from the spun concrete sample's outer zone to its outer zone, whereas the cement paste content changes in the opposite direction (in the inner zone, the cement paste content amounts to almost 100%). The air content in the wall cross section is higher for the inner layers than for the outer ones. Composition segregation occurs in all samples. The sample with the lowest sand equivalent value (25%), the lowest speed (400 rpm) and the shortest spinning time (5 min) has the least fractionated composition. The test result confirms the ones reported in [2–4].

The above investigations of spun concrete mainly focus on the macroscopic description and strength and deformation tests of the sample (usually the whole sample). It is only in [5,6] that modern methods based on computer image analysis are reported, whereby the peculiar structure of spun concrete across the cross section wall can be more precisely described. However, examples reported in the literature indicate that such methods can be used to determine spatial distributions of air void [7–9] aggregate [7,10] or fibres [11–13]. Furthermore, size and shape of the mentioned constituents of cementitious composites can be characterised [14]. Thus, desiring to probe even deeper into the structure of spun concrete, the present authors used innovative (as applied to this field) investigative methods, such as nanoindentation, micro-computed tomography (μCT) and 2D optical scanning. Thanks to the use of these methods, the following features of the inner concrete microstructure were successfully quantitatively determined: the distribution of concrete components, the variation of the mortar mechanical parameters across the cross section wall and the spatial distribution of pores.

2. Preparation of Samples for Tests

A sample of the concrete mixture used for the manufacture of spun concrete power poles was prepared for these investigations. The sample was made in a little mould with an inside diameter of 150 mm and a height of 300 mm (Figure 1), attached to the steel mould used for manufacturing spun concrete power poles in one of the precast concrete plant in Poland [15]. The steel mould together with the attached little mould was placed in a centrifuge (Figure 2) and subjected to spinning for 8 min at the maximum speed of 600 rpm. After concrete mixture spinning, the excess of evaporable water was removed from the little mould and the latter, together with the moulded sample, was transferred to a steam box. After about 4 h, concrete spinning steam was fed gradually (the temperature rising at a rate not higher than 10 °C/h) into the steam box. The temperature in the steam box did not exceed 60–70 °C and the concrete was cured for 8 h. Then, the supply of steam to the steam box was turned off to allow the moulds to naturally cool down to a temperature below 40 °C. Subsequently, the sample was extracted from the mould and was left in laboratory conditions for two weeks. An about 10 mm thick slice (Figure 3), in which the characteristic structure of spun concrete is visible, was cut out from the sample to be used in the tests.

Figure 1. Little mould attached to steel mould used for manufacturing spun concrete power poles.

Figure 2. Steel mould with attached little mould is being placed in centrifuge.

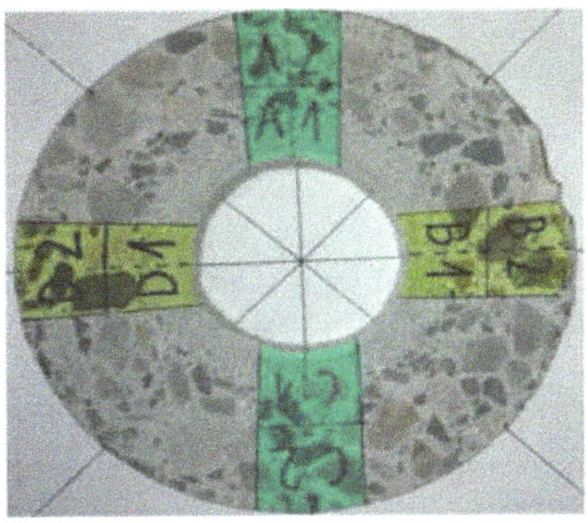

Figure 3. About 10 mm slice cut out of principal sample. Division of slice into smaller samples for tests proper is shown.

Prior to the tests proper, smaller samples to be tested in the particular devices were marked off on the slice (Figure 3). Samples A1–D2 (Figure 3) were tested in the nanoindenter, while the remaining part of the cross section was examined using a computer microtomograph and an optical scanner. The smaller samples were cut out using a high-speed diamond saw made by Struers Labotom-5 (Struers, Shanghai, China) (Figure 4). A series of preliminary tests was carried out on the prepared samples to determine: the indentation parameters (the force, the spacing of test points, and the necessary number of tests), the parameters of the scanning in the microtomograph (the radiation intensity, the exposure time and the filters used) and the optical scanning parameters (the way of preparing the surface and the method of imaging the tested surface). Based on the preliminary test results, a test plan was adopted. The aim of the tests was to determine the following three parameters of the tested cross section:

- the distribution of aggregate across the wall of the cross section;
- the variation of the mortar's mechanical parameters across the wall of the cross section; and
- the distribution of pores across the wall of the cross section.

Figure 4. The slice mounted in the chamber of precision saw.

The identified quantities were related to conventional coordinate R in the radial direction (perpendicular to the axis of rotation of the spun concrete element), where coordinate $R = 0$ applies to the peripheral edge of the element.

3. Analysis of Aggregate Distribution Across Wall

3.1. Preparation of Samples and Scanning

The investigations presented in this section consisted in analysing the images obtained from the optical scanning of the cross section of the spun concrete element. The investigative procedure included the following steps:

1. Cutting out samples
2. Preparing their surface for scanning
3. Scanning
4. Segmenting aggregate from obtained images
5. A morphometric analysis of the aggregate based on its binary image

First, samples were cut out of the tested element and their surfaces were levelled so that they could be scanned. The most effective method of preparing the surface of the samples was sought to obtain the best result of aggregate segmentation in the next step. Different ways of grinding, etching and dyeing (using various inks and dyes (including fluorescent ones)) the matrix were tried. Ultimately, the most effective of the ways was found to be grinding, etching with 10% hydrochloric acid solution and repeated dyeing with acrylic ink and grinding again until a flat surface was obtained owing to the filling of the etched cement matrix volume with the ink. Grinding took place in a Struers Labo-Pol-5 grinding (Struers, Shanghai, China) and polishing machine using an MD Piano disc (Struers, Shanghai, China).

Scanning with a resolution of 600 dpi (which in pixel size terms amounts to 42.33 mm/pix) was carried out in a Brother MFC L5750DW device (Brother, Bridgewater, NJ, USA). The scanning result for an exemplary sample is shown in Figure 5.

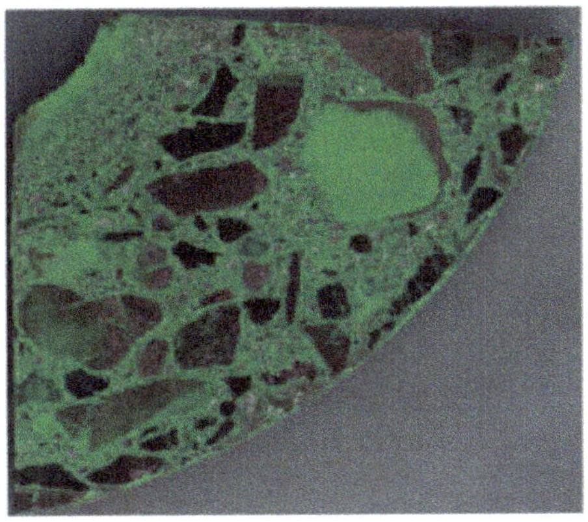

Figure 5. Scan of spun concrete element cross section after matrix dyeing (scale 1:1).

3.2. Segmentation of Aggregate

The next step consisted in aggregate segmentation. Without going into details (presented below), the aim of segmentation was to obtain a binary image in which white pixels indicate the area occupied by aggregate, while pixels in the remaining area (constituting the background) are black. The aggregate segmentation procedure proposed below is based on image processing methods. GIMP 2.10.4, ImageJ 1.52e (Fiji distribution) and CTAn 1.17.1.7 + were used for this purpose. Similar to in the case of the preparation of samples for scanning, the segmentation procedure was developed through many trials. The procedure deemed the best consisted of the following steps:

1. Gaussian blurring (smoothing) with a radius of 1 pixel (3 times)
2. The manual "retouching" of large aggregate fragments that were found to be susceptible to etching (Figure 6)
3. The segmentation of the "dark" aggregate
4. The segmentation of the "light" aggregate
5. The product of the images from Step 3 and Step 4
6. The removal of "pores"

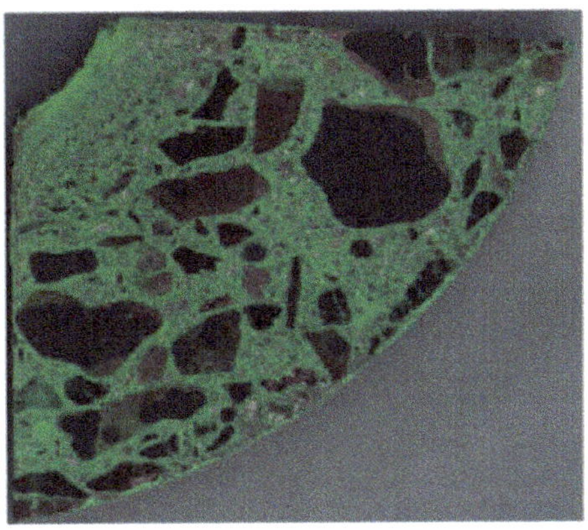

Figure 6. Image of sample after segmentation procedure Step 2.

To segment the "dark" (i.e., darker than the dye used in the selected RGB channel) aggregate (Step 3), first a mask was created from the green channel (G) of the image and by superimposing it on the white background a new image was created, this time in shades of grey (Figure 7). Then, thresholding was applied, assigning the value of 1 to the pixels with a brightness of 0–87/255 and the value of 0 to the other pixels (Figure 8).

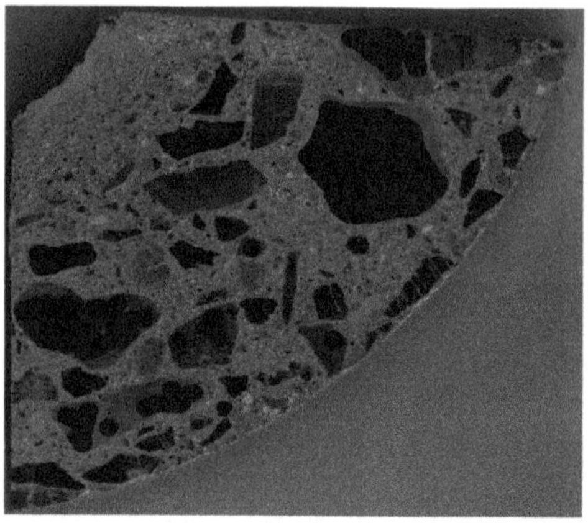

Figure 7. Mask created from green channel (G).

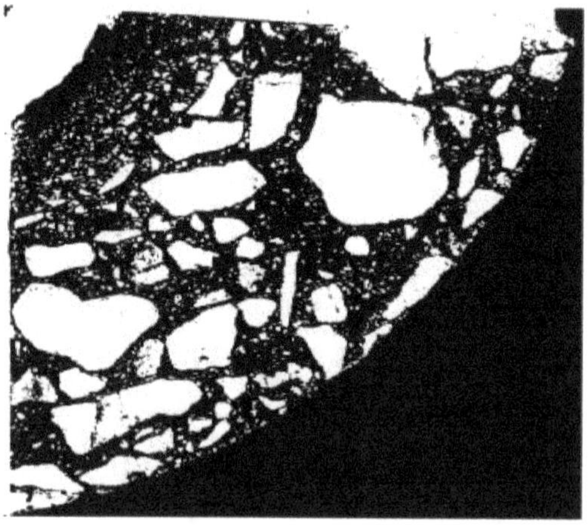

Figure 8. Image of sample after segmentation procedure Step 3.

The "light" aggregate (Step 4) was segmented similarly as in Step 2, this time superimposing red channel (R) and blue channel (B) masks (Figure 9). Then, thresholding according to the 20–255/255 range was applied. The result of this operation is shown in Figure 10.

Figure 9. Superimposition of masks created from red (R) and blue (B) channels.

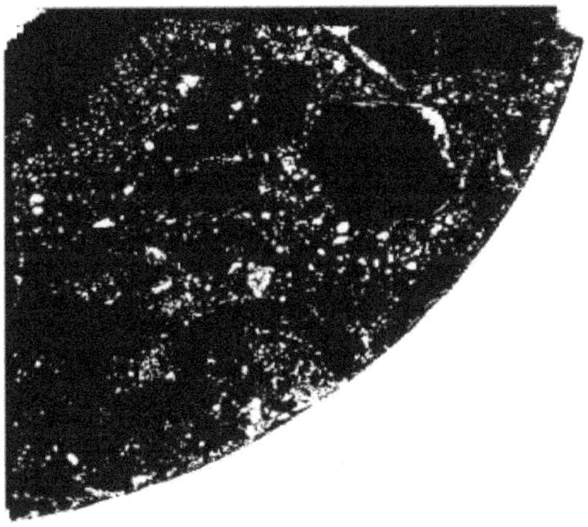

Figure 10. Image of sample after segmentation procedure Step 4.

The product of the images from Step 3 and Step 4 (Step 5) is shown in Figure 11. The removal of pores (i.e., black areas completely surrounded by the white area) (Step 6) and the use of a mask confining the area to the surface of the scanned sample are shown in Figure 12.

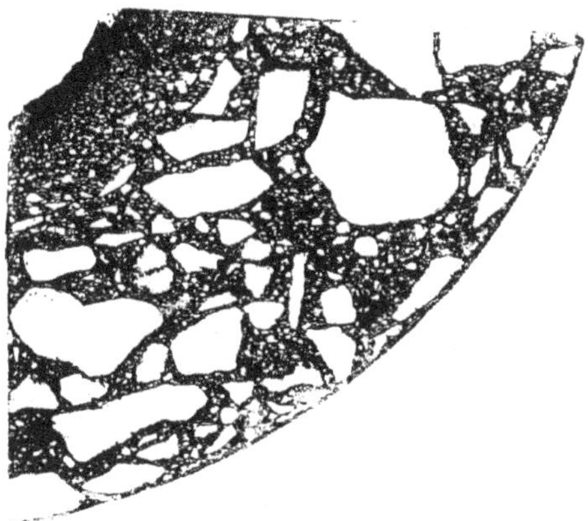

Figure 11. Step 5—product of images from Steps 3 and 4.

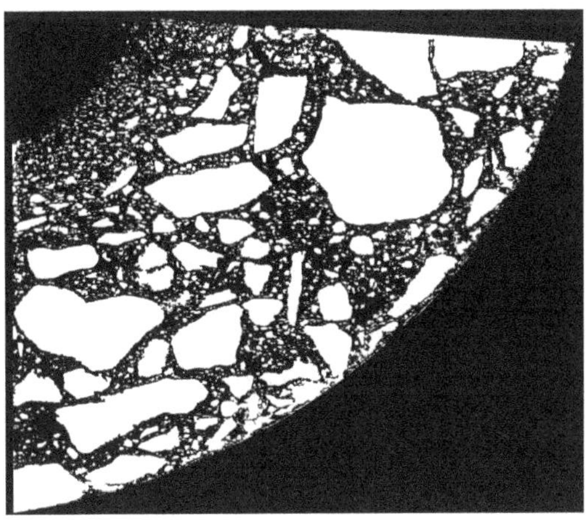

Figure 12. Step 6—the removal of "pores".

3.3. Morphometric Analysis

3.3.1. Basic Terms and Measures

Generally, the so-called morphologic microstructure measures are used to characterise the spatial distribution of the constituents of composite materials [16]. As part of this research, the aggregate content was determined and a procedure for calculating this quantity as a function of coordinate R was proposed. Moreover, the so-called local structure thickness was used to describe the fractionation of aggregate across the wall. The methodology presented below is based on the fact that a binary image can be understood as a discrete approximation of the indicator function (as defined in [17–19]) on a uniform grid of rows and columns of pixels. Then, the fractional content (of aggregate in this case) in a certain considered image area Ω can be defined as the sum of pixels in the constituent, divided by the sum of pixels in area Ω.

$$\phi = \frac{\sum\limits_{\Omega(i,j)} I^{(k)}(i,j)}{\sum\limits_{\Omega(i,j)} 1} \tag{1}$$

where $I^{(k)}$ is the binary indicator function of constituent k, i is the row number, j is the column number, and

$$\Omega(i,j) = \{(i,j) : \mathbf{x}(i,j) \in \Omega\} \tag{2}$$

stands for the number of pixels in area Ω.

To describe the variation in aggregate content as a function of element wall thickness R, a series of subareas, constituting a narrow circumferential band characterised by a fixed value of coordinate R in a range from $R - 1/2{\cdot}\Delta R$ to $R + 1/2{\cdot}\Delta R$, was selected (Figure 13).

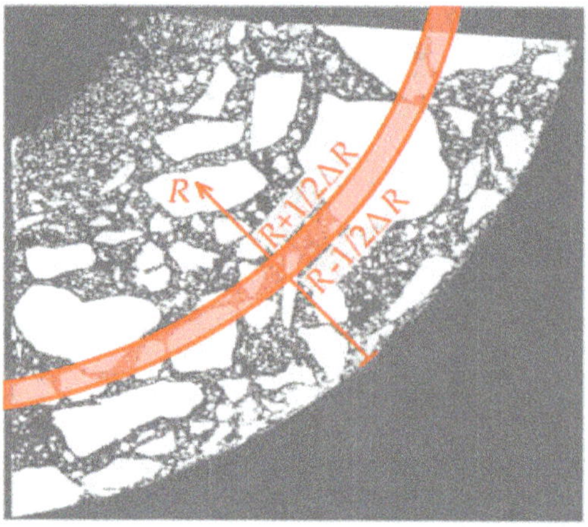

Figure 13. Way of selecting circumferential band.

Then, the value of fractional content $\phi(R)$ in the band whose centre line is described by coordinate R can be defined as:

$$\phi(R) = \frac{\sum\limits_{\Omega_R(i,j)} I^{(k)}(i,j)}{\sum\limits_{\Omega_R(i,j)} 1} \tag{3}$$

where

$$\Omega_R(i,j) = \left\{(i,j) : R(i,j) \in \left[R - \frac{1}{2}\Delta R, R + \frac{1}{2}\Delta R\right)\right\} \tag{4}$$

runs through the set of pixels contained in the considered band. The linear dimension of a single pixel, i.e., $\Delta R = 42.3$ μm, was assumed as the basic value of ΔR in the analysis.

Local structure thickness is a scalar field defined in the area occupied by the considered constituent. In general, the procedure for determining local structure thickness consists in filling the area occupied by the considered constituent with circles with the possibly largest diameter. Then, the local structure thickness in given point x is defined as the largest diameter of the circle that is fully contained in the considered constituent and at the same time contains point x [20,21]. This is shown schematically in Figure 14.

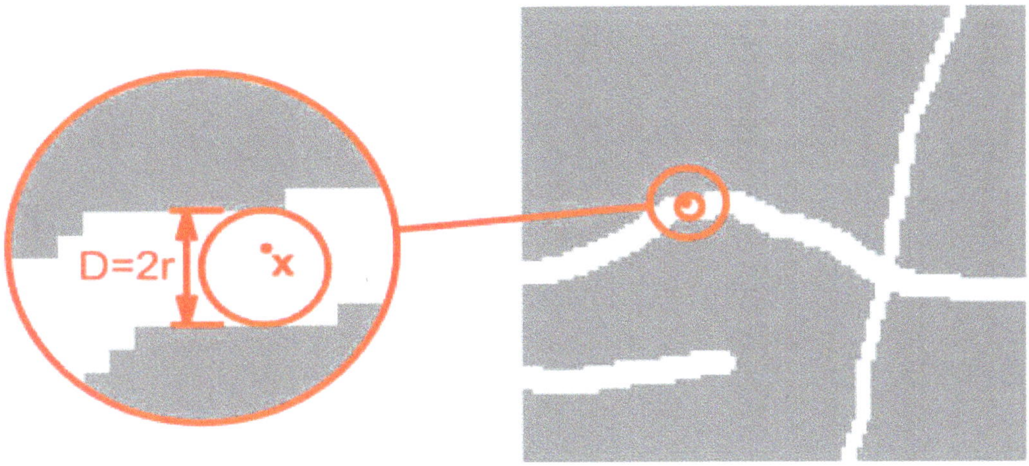

Figure 14. Element width in point x, as maximum diameter of inscribed circle [21].

Calculations are first performed for the whole image, whereby a map of the local structure thickness of considered constituent, $D^{(k)}(i,j)$, is obtained. A similar procedure as for the fractional content was used to characterise the variation in aggregate size across the concrete element wall (i.e., relative to coordinate R). This time the average size of the aggregate in circumferential band is defined as:

$$D(R) = \frac{\sum\limits_{\Omega_R(i,j)} D^{(k)}(i,j)}{\sum\limits_{\Omega_R(i,j)} I^{(k)}(i,j)} \quad (5)$$

i.e., as the weighted average of all the $D^{(k)}(i,j)$ values of the pixels found simultaneously in band and in the aggregate. It should be noted here that the decided advantage of this way of determining aggregate size (as local structure thickness) over other methods consisting, e.g., in counting pixels, is the fact that the result in this case does not depend on an accidental "merger" of a few aggregate particles into one geometric object in the analysed image.

3.3.2. Results

The authors' own procedure written in the Wolfram Language in Mathematica was used for the calculations. Selected morphological transformations were performed in the programs GIMP, ImageJ (Fiji distribution) and Bruker CTAn. A graph of the variation in aggregate content in the considered sample is shown in Figure 15.

A graph of mean aggregate size is shown in Figure 16, while a map of local structure thickness is shown in Figure 17b. Moreover, Figure 17c shows the sum of the images presenting the map of aggregate thickness, the applied ROI mask and the outermost circumferential bands taken into account in the analysis.

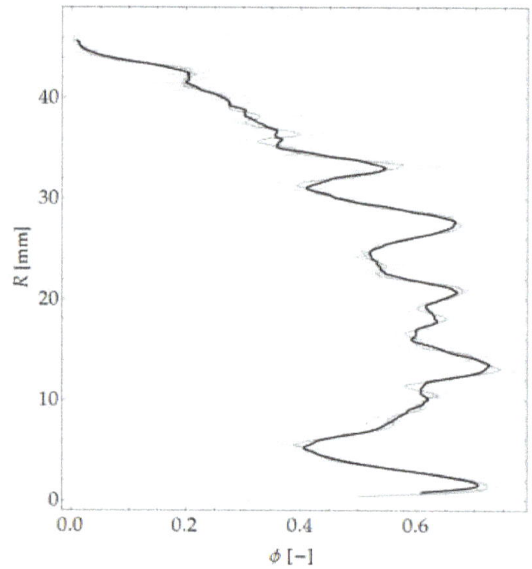

Figure 15. Aggregate content across element thickness.

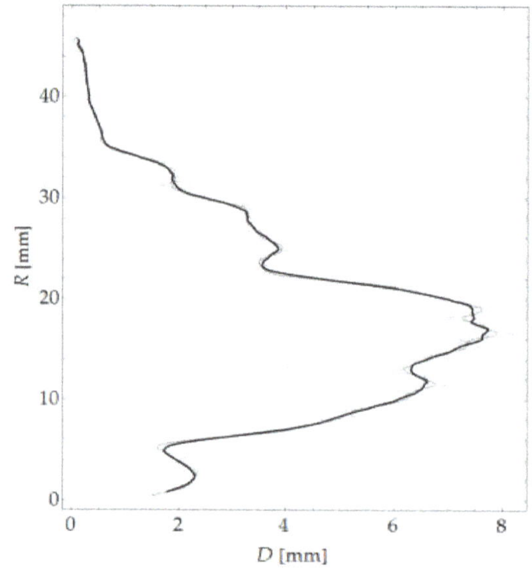

Figure 16. Mean aggregate size versus coordinate *R*.

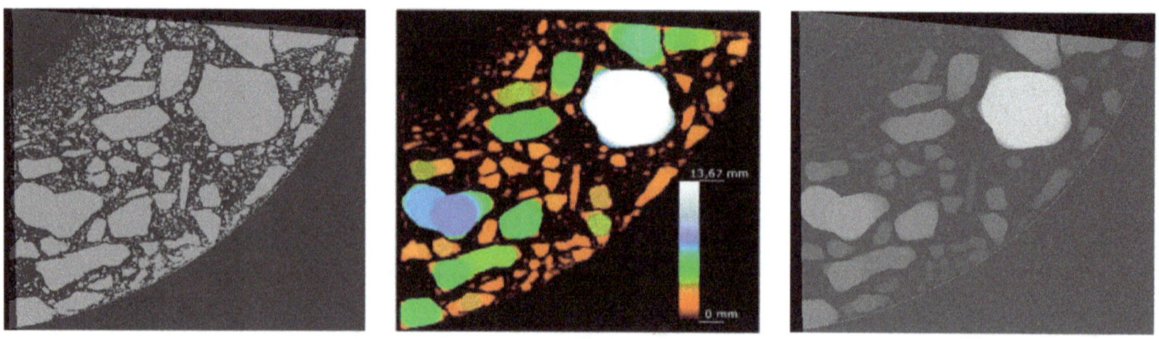

Figure 17. Results of applied morphological transformations: (**a**) image of sum of analysed binary image, applied mask defining ROI and outermost circumferential bands taken into account in analysis; (**b**) map of local aggregate size; and (**c**) image of sum of local aggregate size, applied map defining ROI and outermost circumferential bands taken into account in analysis.

4. Analysis of Variation of Mortar Mechanical Parameters Across Wall

The nanoindentation method [22] was used to determine the mechanical parameters of the mortar, understood within the particular analysis as cement paste together with fine aggregate, i.e. finer than sand particles. Hardness (HIT), indentation modulus (MIT) and surface aggregate content (ϕ) in the mortar were determined on the basis of the test results.

The concrete samples were specially prepared for tests in the nanoindenter. The preparation included: embedding concrete pieces in epoxy resin (Figure 18a) and levelling and polishing the surface tested (Figure 18b). The grinding and polishing procedure was individually fitted to each sample on the basis of the authors' experience in the testing of concrete samples.

(a) (b)

Figure 18. Samples for testing in nanoindenter: (**a**) samples embedded in epoxy resin; and (**b**) samples with surface polished for testing.

The proper preparation of the samples was verified by evaluating the images obtained from the optical microscope (Figure 19) and carrying out a series of trial indentations. The absence of artefacts (cracks or streaks) in the images indicated that the surface had been properly prepared for testing [23,24].

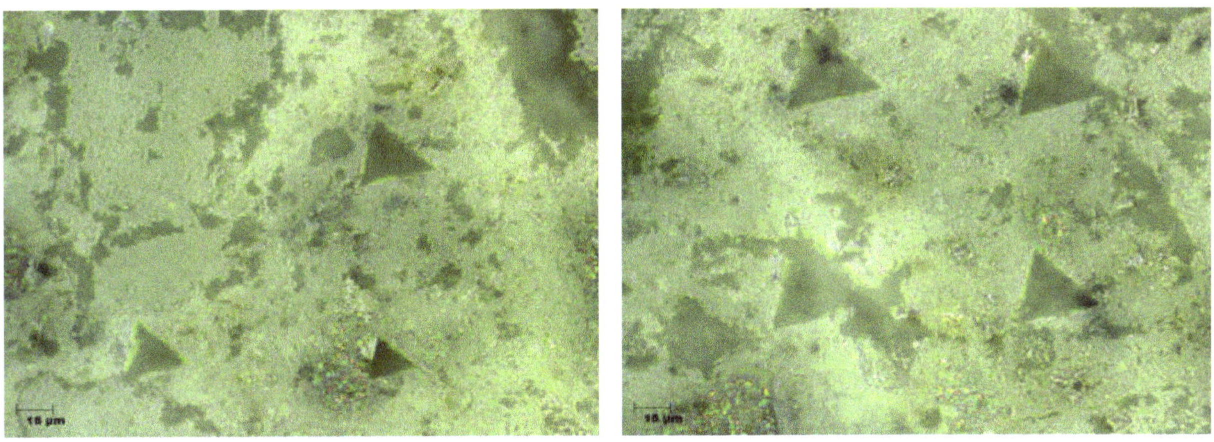

Figure 19. Exemplary photos of surface with trial indentations.

The indentation test was carried out in accordance with the standard procedure [25]. An indenter with known geometry and known mechanical parameters was pressed into the tested material and simultaneously the characteristic values of the applied load and the depth of penetration of the indenter tip were measured. In the course of the test, force F continuously grew, whereby penetration depth h continuously increased. Hence, the F-h dependence (Figure 20a) could be determined, based on which the maximum penetration depth (h_{max}) of the indenter tip and the range of elastic W_e and plastic W_p deformations of the material could also be determined (Figure 20b) [22,26,27].

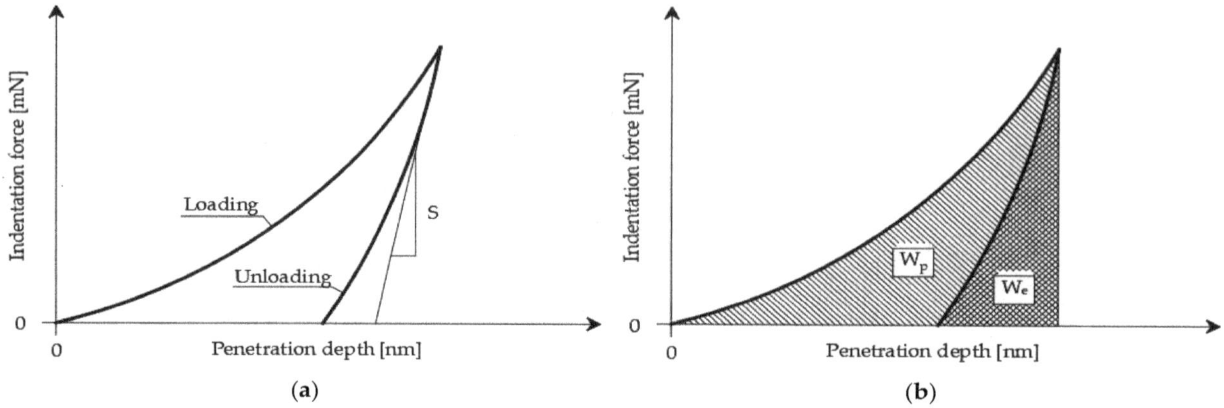

Figure 20. Graphical representation of standard indentation test parameters: (**a**) exemplary F-h dependence obtained during nanoindentation test; and (**b**) way of determining plastic behaviour (W_p) and elastic behaviour (W_e).

The basic parameters determined by the test are hardness (H_{IT}) and indentation modulus (M_{IT}). Hardness is defined as follows:

$$H_{IT} = \frac{F_{max}}{A} \tag{6}$$

where A is a projection of the indenter contact surface onto the surface of the sample. This quantity is usually determined as a function of maximum penetration depth h_{max} [28,29].

Indentation modulus M_{IT} is calculated using the Sneddon solution [30] describing the pressing of an axially symmetric rigid cone into an elastic half-space. Then, M_{IT} is defined as follows:

$$M_{IT} = \frac{1}{2}\frac{dF}{dh}\frac{\sqrt{\pi}}{\sqrt{A}} \tag{7}$$

Based on trials, a loading procedure comprising two forces (10 mN and 250 mN) was adopted. The pattern of indenter loading during each test is shown in Figure 21.

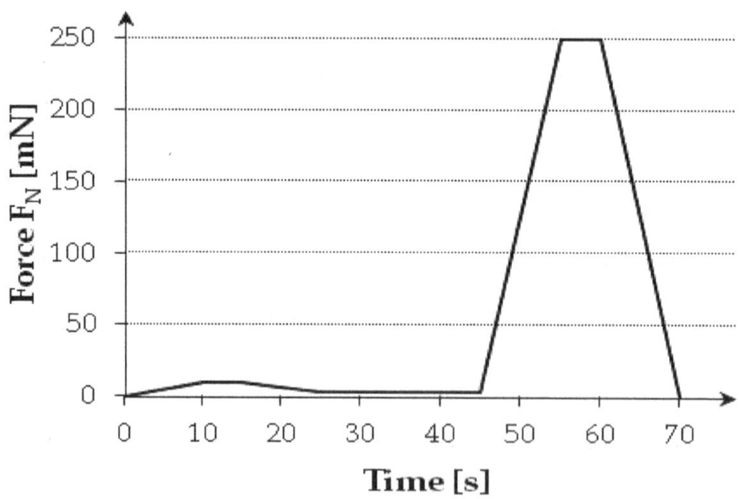

Figure 21. Loading function in each of the tests.

The grid indentation technique (GIT) was used to determine the parameters of the sample surface [31]. According to the assumptions of GIT, penetration depth h, interindentation spacing l

(mesh dimension) and number of carried out tests N should satisfy the following conditions (see, e.g., [32]):

$$3h < D < l(N)^{1/2} \tag{8}$$

$$d < \Omega < h \tag{9}$$

$$R_q < 3h \tag{10}$$

where D represents the characteristic dimension of nonuniformity on a given scale, d is the maximum dimension of inclusions in the tested constituent, Ω is the characteristic dimension of the so-called representative elementary volume [33,34] and R_q is the mean square deviation of surface roughness.

Using the CSM TTX-NHT nanoindenter with the Berkovich tip [35], a series of mortar hardness measurements on seven measurement grids (Figure 22) was carried out. One hundred indentations were made at every 50 μm on each of the grids.

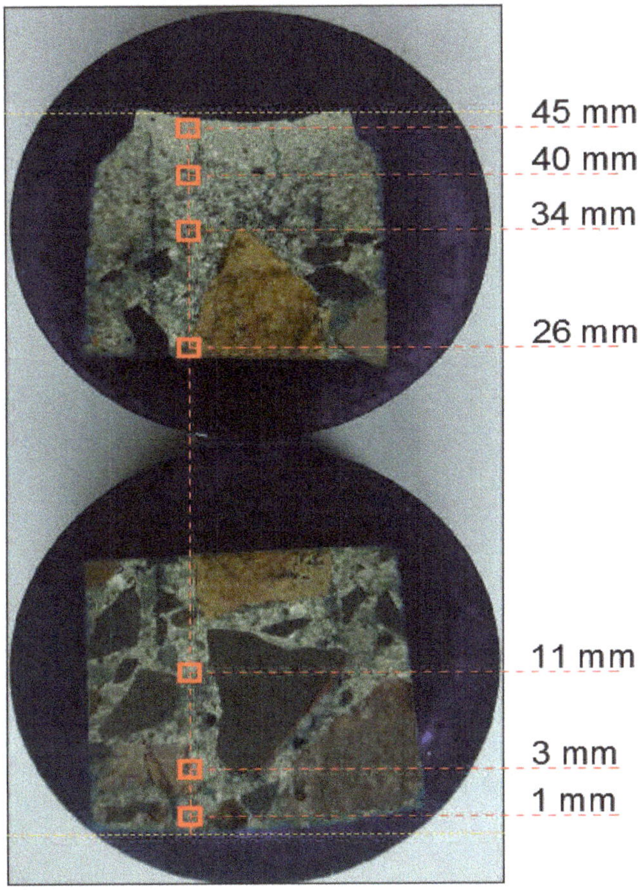

Figure 22. Arrangement of particular test grids—schematic representation.

The results of the nanoindentation tests were processed using a Mathematica script written by the authors. The mean value and the standard deviation of hardness HIT and indentation modulus MIT for the indentation force of 250 mN were determined for each of the grids. Because of the high inhomogeneity of the tested material, it was necessary to considerably increase the number of measurements to determine the parameters for the load level of 10 mN. The determined values are presented in Table 1 and in the diagrams of the variation of the mechanical parameters along radius R (Figure 23). The solid line marks the calculated mean value, while the broken lines correspond to the (upper and lower) boundaries of the values.

Table 1. Mean values and standard deviations of measured hardness H_{IT} and indentation modulus M_{IT}.

R	μM_{IT}	σM_{IT}	μH_{IT}	σH_{IT}
(mm)	(GPa)	(GPa)	(GPa)	(GPa)
1	46.26	23.86	4.49	5.00
3	45.52	25.83	4.29	5.07
11	66.72	30.00	7.87	5.67
26	50.76	27.12	5.75	5.28
34	52.09	23.75	5.01	4.84
40	40.04	21.53	3.26	4.26
45	23.66	12.31	1.00	1.53

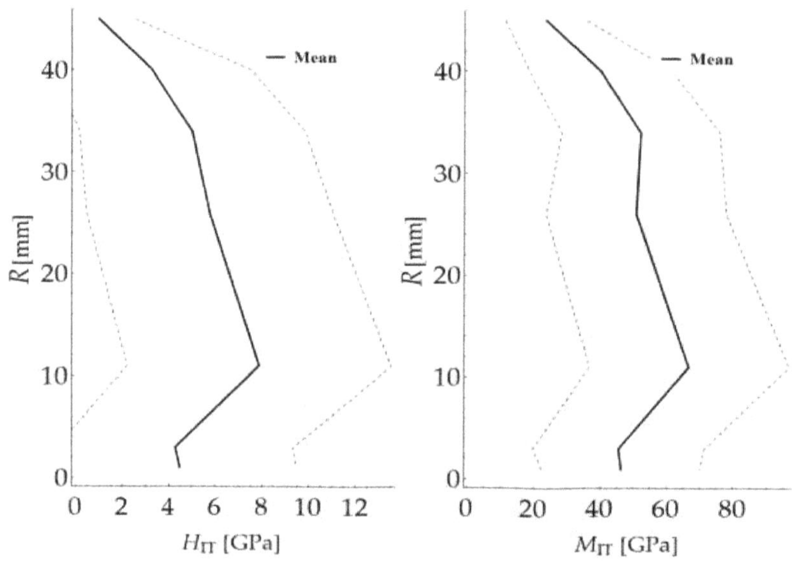

Figure 23. Variation in hardness HIT and indentation modulus MIT across wall of cross section.

In addition, for the above measurements, the segmentation of the mortar constituents was carried out assuming two material phases (cement paste and aggregate). The deconvolution technique was used for this purpose [31,36]. According to this method, each individual indentation is considered to be an independent random event, and its results (M_{IT} and H_{IT}) are considered to be random variables. The values of the cumulative distribution function for the measured M_{IT} and H_{IT} values (respectively, F_M and F_H) can be calculated as follows:

$$\left.\begin{aligned} F_M\left(M_{IT(i)}\right) &= \tfrac{i}{N} - \tfrac{1}{2N} \\ F_H\left(H_{IT(i)}\right) &= \tfrac{i}{N} - \tfrac{1}{2N} \end{aligned}\right\} \text{for } i \in [1, N] \tag{11}$$

Assuming that the distribution of the mechanical parameters of the particular constituents can be described using Gaussian distributions, expressed by mean values $\mu_j^{M_{IT}}$ and $\mu_j^{H_{IT}}$ and standard deviations $\sigma_j^{M_{IT}}$ and $\sigma_j^{H_{IT}}$, for, respectively, indentation modulus M_{IT} and hardness H_{IT}, the cumulative distribution function for each segmented constituent has the form:

$$F\left(X_{IT(i)}; \mu_j^{X_{IT}}, \sigma_j^{X_{IT}}\right) = \frac{1}{\sigma_j^{X_{IT}}} \frac{1}{\sqrt{2\pi}} \int_{-\infty}^{X_{IT(i)}} \exp\left(\frac{-\left(u - \mu_j^{X_{IT}}\right)^2}{2\left(\sigma_j^{X_{IT}}\right)^2}\right) du; \text{ z } X_{IT} = (M_{IT}, H_{IT}) \tag{12}$$

The sought values $\left\{f_j, \mu_j^{M_{IT}}, \sigma_j^{M_{IT}}, \mu_j^{H_{IT}}, \sigma_j^{H_{IT}}\right\}, j = 1, n$ are determined by minimising the difference between the cumulative distribution function for the experimental results and the one assumed in the form of a Gaussian distribution, i.e.,

$$\min\left[\sum_{i=1}^{N}\left(\sum_{j=1}^{n} f_j F\left(M_{IT(i)}; \mu_j^{M_{IT}}, \sigma_j^{M_{IT}}\right) - F_M\left(M_{IT(i)}\right)\right)^2 + \sum_{i=1}^{N}\left(\sum_{j=1}^{n} f_j F\left(H_{IT(i)}; \mu_j^{H_{IT}}, \sigma_j^{H_{IT}}\right) - F_H\left(H_{IT(i)}\right)\right)^2\right] \tag{13}$$

Thanks to this approach, one can estimate the aggregate content in the mortar. An exemplary result of segmentation for one measurement grid is shown in Figures 24 and 25.

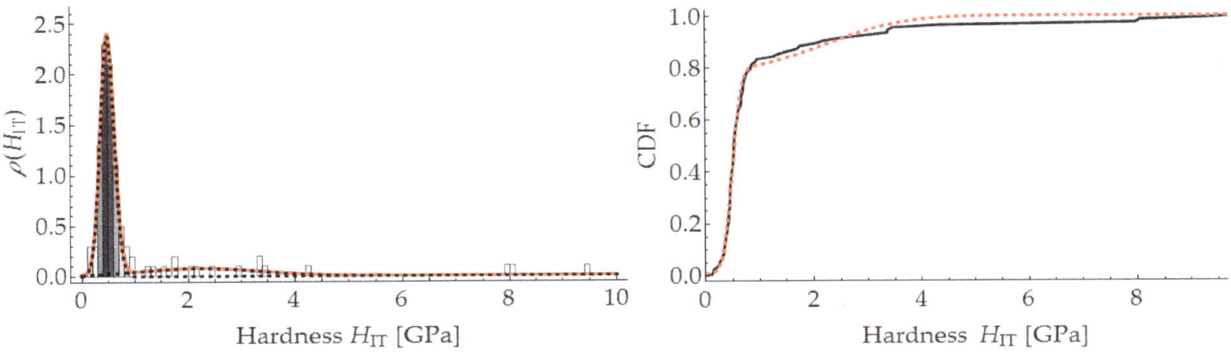

Figure 24. Segmentation of mortar components through deconvolution based on hardness histogram.

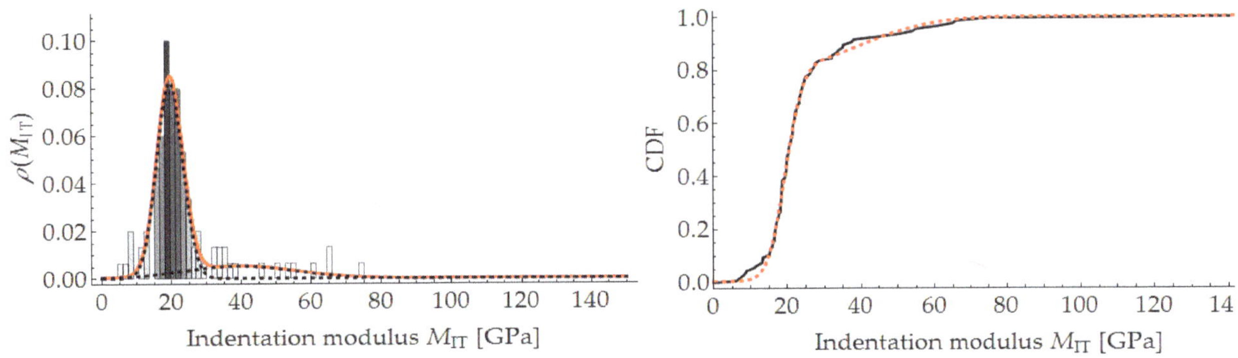

Figure 25. Segmentation of mortar constituents through deconvolution based on indentation modulus histogram.

The segmentation results for the whole sample are presented in Table 2 and in the diagrams of the variation of the mechanical parameters along radius R (Figure 26).

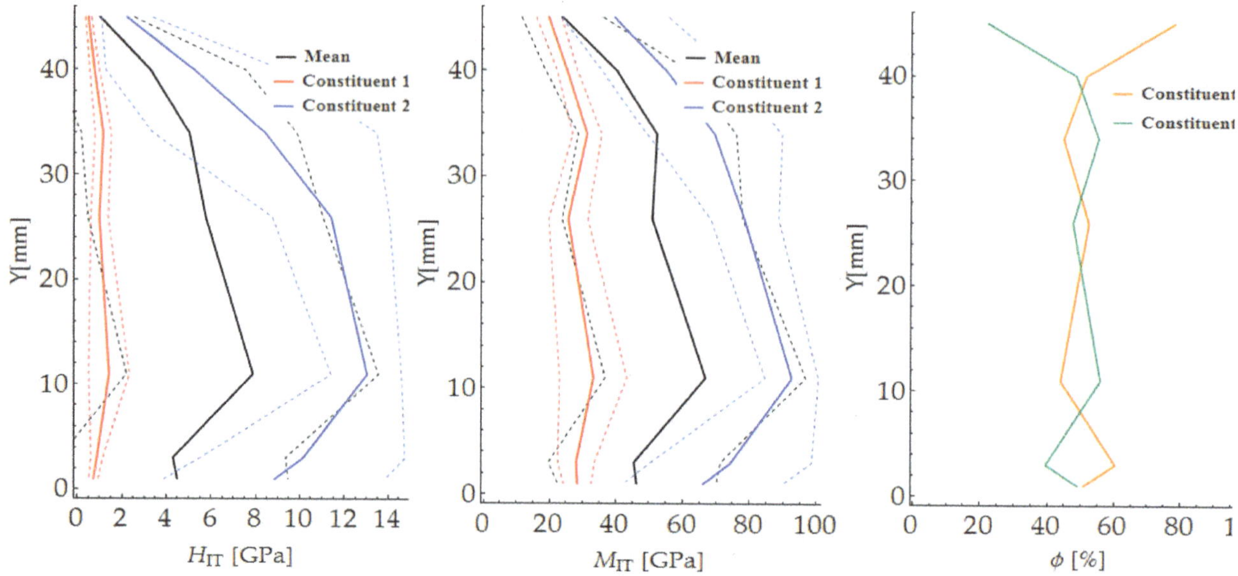

Figure 26. Variation in hardness H_{IT}, indentation modulus E_{IT} and constituent content across wall.

Table 2. Hardness H_{IT}, indentation modulus E_{IT} and fractional content ϕ of tested material constituents.

R (mm)	Constituent	μM_{IT} (GPa)	σM_{IT} (GPa)	μH_{IT} (GPa)	σH_{IT} (GPa)	ϕ (%)
1	1	28.42	4.16	0.74	0.21	50.90
	2	66.13	23.69	8.87	4.98	49.10
3	1	28.05	5.53	0.89	0.26	60.40
	2	74.51	24.12	10.16	4.64	39.60
11	1	33.23	10.17	1.43	0.91	44.10
	2	92.63	7.99	13.02	1.62	55.90
26	1	25.54	5.88	0.99	0.39	52.40
	2	78.44	10.14	11.41	2.66	47.60
34	1	31.12	4.38	1.16	0.36	44.80
	2	69.36	20.21	8.42	5.04	55.20
40	1	25.06	3.04	0.77	0.22	51.50
	2	54.80	19.31	5.23	3.97	48.50
45	1	19.52	3.76	0.49	0.13	78.00
	2	39.42	16.14	2.21	1.15	22.00

5. Analysis of Distribution of Pores Across Wall

The non-destructive technique of micro-computed tomography (μCT) was used to characterise pore space variation. The adopted approach was similar to those presented in [37–40]. The sample was trimmed to a rectangular prism whose dimensions ensured the analytical resolution of 10 μm/pix. Then, the rectangular prism was mounted on a base and placed in the microtomograph chamber (Figure 27).

Figure 27. Sample mounted on base and placed in microtomograph chamber.

Scanning was carried out using the Bruker Skyscan 1172 scanner. The scanning consists in subjecting the prepared sample to a series of exposures and reconstructing the material structure from the obtained projections (Figure 28).

Figure 28. Exemplary projection of material structure.

The scanning parameter values shown in Table 3 were selected on the basis of trial scans and the earlier tests of the samples.

Table 3. Selected major scanning parameters.

Parameter	Value
Source Voltage (kV)	100
Source Current (μA)	100
Image Pixel Size (μm)	9.88
Filter	Al + Cu
Exposure (ms)	1100
Rotation Step (deg)	0.24
Frame Averaging	ON (8)
Random Movement	ON (10)
Use 360 Rotation	YES
Geometrical Correction	ON

The NRecon program based on the Feldkamp algorithm [41,42] was used for image reconstruction. The set of reconstruction parameters is shown in Table 4.

Table 4. Selected major reconstruction parameters.

Parameter	Value
Post-alignment	−1.00
Pixel Size (μm)	9.87947
Object Larger than FOV	OFF
Ring Artefact Correction	18
Beam Hardening Correction (%)	10
Threshold for defect pixel mask (%)	50
CS Static Rotation (deg)	5.00
Minimum for CS to Image Conversion	0.000
Maximum for CS to Image Conversion	0.040

The reconstructed structure of the tested sample is shown in Figure 29. To quantitatively and qualitatively evaluate the material structure, one should additionally determine the "volume" of interest (VOI) in the reconstructed model. A rectangular region of interest, marked red in the exemplary cross sections of the sample (Figure 30), was adopted.

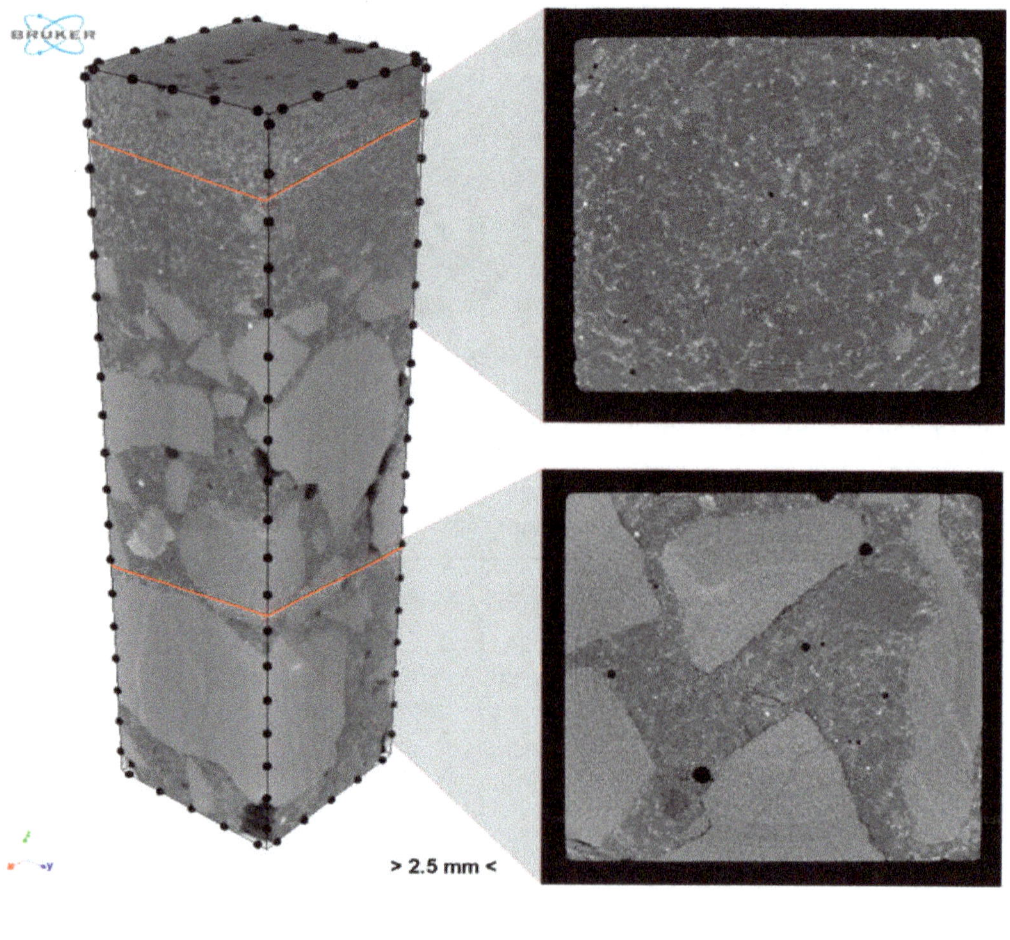

Figure 29. Reconstructed structure of sample with exemplary cross sections.

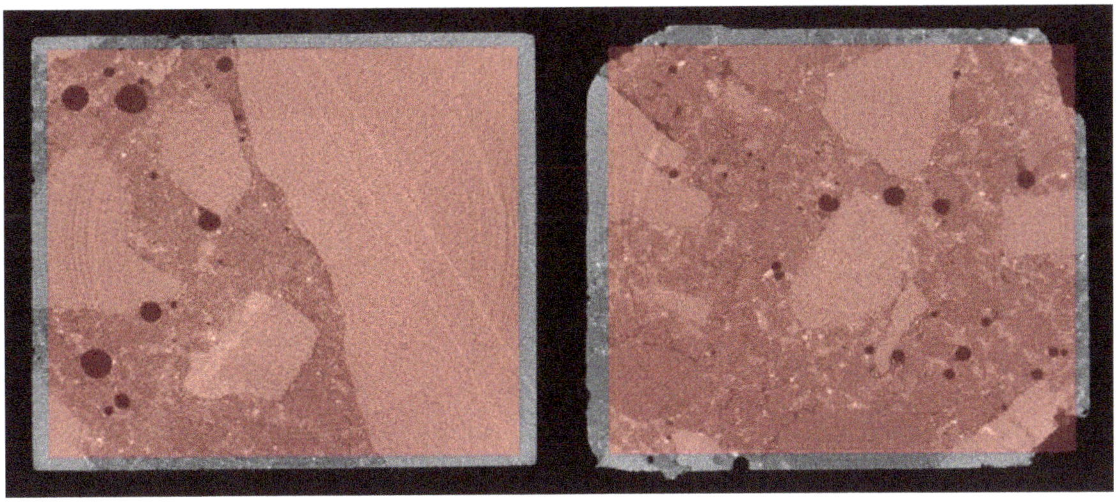

Figure 30. Region of interest (ROI) for selected cross sections.

The obtained results were analysed using the Bruker software (CTAn and CTVox) and a Mathematica script written by the authors.

To estimate the porosity of the tested material, it was necessary to segregate the pore space from the images. For this purpose, the images were binarised (using thresholding) for the particular cross sections. An exemplary result of the segmentation for selected cross sections is shown in Figure 31.

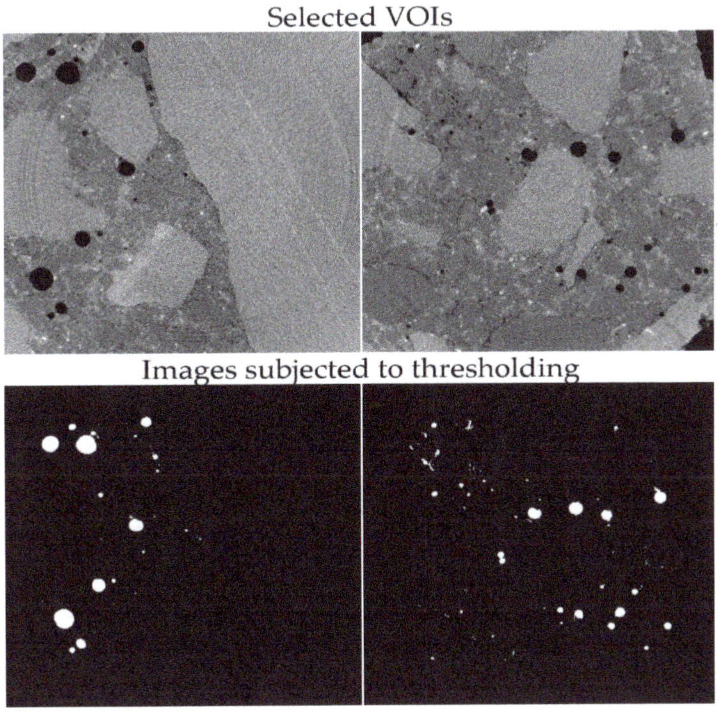

Figure 31. Segmentation of pores for selected cross sections.

The isolated pore space could be presented in a 3D model and quantitatively analysed (Figure 32). The graph in Figure 32 represents porosity versus sample height. Figure 33 shows the spatial distribution and a graph of pore local thickness versus sample height. It should be noted that pore thickness is understood here similarly as described in Section 3.3. However, this time, the calculations were performed as three-dimensional and local thickness was defined as the maximum diameter of a sphere contained in the pore volume.

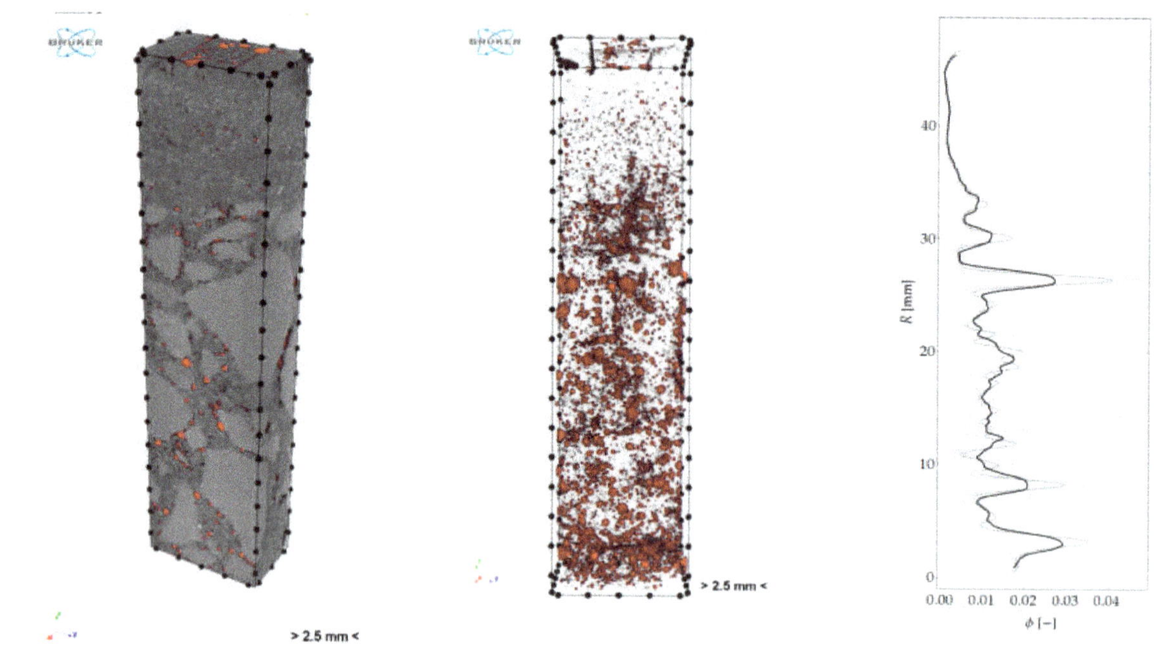

Figure 32. Cross section through 3D model of concrete sample with detailed pores, and pore space within VOI.

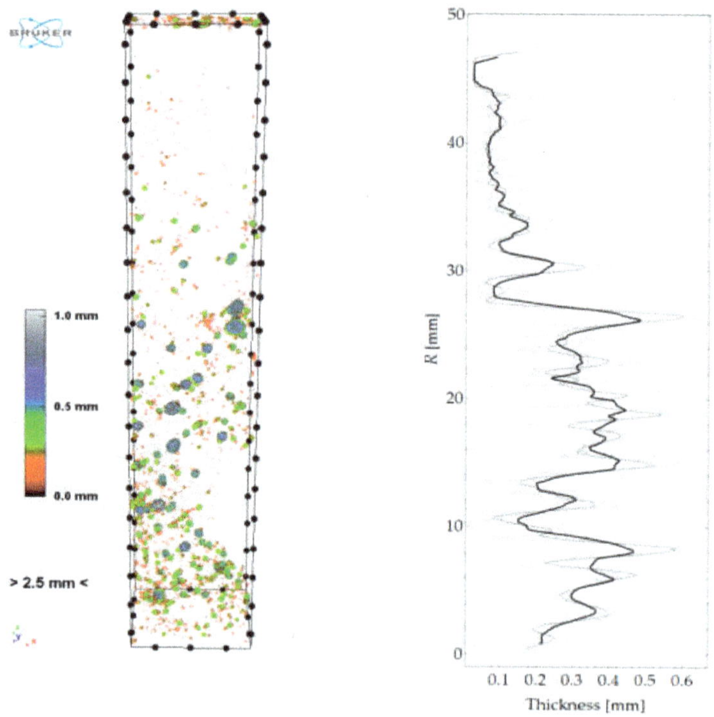

Figure 33. Spatial distribution of pore "thickness" and graph of mean thickness along sample height.

Considering the results obtained from 2D and 3D scanning and knowing the aggregate and pore content, the cement paste content was calculated. A graph of this quantity is shown in Figure 34. In this case, it was assumed that the analysed material consisted of three constituents, i.e., aggregate, mortar and pores.

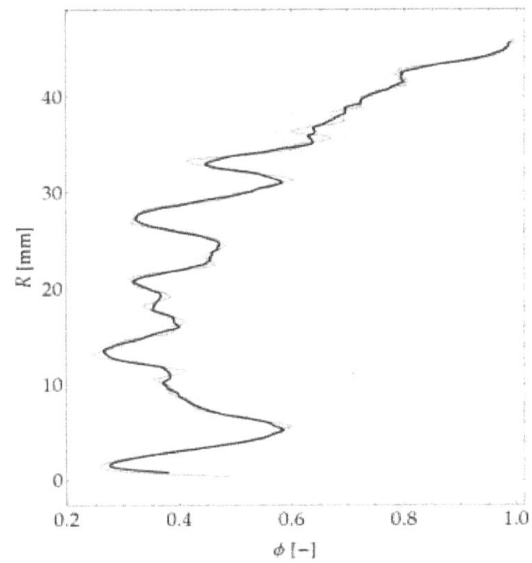

Figure 34. Graph of mortar content along sample height.

6. Discussion

A test methodology comprising three different testing methods, namely nanoindentation, micro-computed tomography (μCT) and two-dimensional optical scanning, is presented. The methodology was used to describe selected features of the inner microstructure of spun concrete, such as the distribution of its constituents, the variation of the mechanical parameters of the mortar across the cross-section wall, and the spatial distribution of pores. Tests were carried out on a sample moulded from the mixture used for the manufacture of spun concrete power poles in one of the precast concrete plants in Poland.

The variation of the aggregate content and the mean aggregate size across the wall of the cross section of a spun concrete sample was characterised using an aggregate segmentation procedure based on image processing methods, and a morphometric analysis employing mathematical morphology. The aggregate segmentation method is similar to the computer image analysis method used in [5,6]. The obtained results are satisfactory, providing not only a qualitative, but also quantitative description of the distribution of aggregate across the wall of a spun concrete cross section.

Thanks to the use of the nanoindentation method, it was possible to determine the mechanical parameters of the mortar (cement paste and fine aggregate) across the wall of the cross section of a spun concrete sample. The hardness (HIT), the indentation modulus (MIT) and the surface concrete content (ϕ) in the tested mortar were determined. Because of the size of the nanoindenter tip and the indentation forces used, testing was limited to places where there was no coarse aggregate in the sample concrete. This means that acquired information about the mechanical characteristics of spun concrete across the wall of the cross section is not full. However, in addition to the measurements of the mechanical parameters, the authors succeeded in segmenting the mortar constituents (cement paste and aggregate) using the deconvolution technique. Based on the obtained results, it was possible to trace the distribution of cement paste (and thus of cement) across the wall of the cross section. The findings reported by Marquardt [1] that cement is quite uniformly distributed across the wall and that only in the inner layer of the cross section its content increases were confirmed.

The variation in pore space was characterised using micro-computed tomography (μCT). Thanks to this method, it was possible to spatially (in 3D) describe the distribution of pores in the sample as well as their distribution across the wall of the cross section. Mainly closed air voids were found to occur. The results of the analyses are surprising and they do not corroborate the literature results (e.g., [5,6]), where it is suggested that the air content in the cross section of the wall is rather higher in the inner layers than in the outer layers.

The investigative methodology comprising nanoindentation, micro-computer tomography (μCT) and two-dimensional optical scanning seems to be useful for evaluating selected features of the internal microstructure of spun concrete. The methodology allows one to gain an insight into the interior of concrete and it yields qualitatively and quantitatively valuable results. The developed procedures for investigating the internal structure of spun concrete can be successfully used, e.g., to evaluate the process of manufacturing spun concrete elements, by verifying the porosity of the produced material, the fractionation of the aggregate, and the cement content, taking into account the variation of the parameters across the wall.

Taken together, the main contribution of the work to the field is the development of a procedure for the quantitative description of selected microstructure parameters. It should be emphasised that special attention was paid to the evaluation of the variability of these parameters in the cross section of the element. In the future perspective, the developed procedure will serve as a base for upcoming studies planned by the authors. For instance, the procedure allows for a comparative assessment of elements manufactured under different technological regimes. Thus, the methodology proposed in the article may prove to be a useful tool in the optimisation of manufacturing process concerning, i.e. selection of time and speed of mould spinning. Moreover, the procedure can be utilised in the analysis of the spun concrete microstructure changes along the traction pole, for which the diameter decreases towards its top.

Author Contributions: Conceptualisation, J.M.; methodology, J.M., M.P., and M.S.; software, M.P. and M.S.; validation, J.M., M.P., and M.S.; formal analysis, J.M.; investigation, J.M.; resources, J.M., M.P., and M.S.; data curation, M.P. and M.S.; writing—original draft preparation, J.M., M.P., and M.S.; writing—review and editing, J.M., M.P., and M.S.; visualisation, M.P. and M.S.; supervision, J.M.; project administration, J.M.; and funding acquisition, J.M. and M.S.

References

1. Marquardt, E. Geschleuderte Beton-und Eisenbetonrohre. *Die Bautechnik* **1930**, *40*, 587–602.
2. Achverdov, I.N. Novoe v technologii železobetonnych centrifugirovannych rastrubnych trub. *Beton i železobeton* **1961**, *5*, 195–200.
3. Dilger, W.H.; Ghali, A.; Krishna Mohan Rao, S.V. Improving the Durability and Performance of Spun-Cast Concrete Poles. *PCI J.* **1996**, *41*, 68–90. [CrossRef]
4. Dilger, W.H.; Krishna Mohan Rao, S.V. High Performance Concrete Mixtures for Spun-Cast Concrete Poles. *PCI J.* **1997**, *42*, 82–96. [CrossRef]
5. Adesiyun, A.; Kamiński, M.; Kubiak, J.; odo, A. Laboratory test on the properties of spun concrete. In Proceedings of the Third Interuniversity Research Conference, Szklarska Poręba, Poland, 22–26 November 1994; pp. 3–8.
6. Adesiyun, A.; Kamiński, M.; Kubiak, J.; odo, A. *Investigation of Spun-Cast Concrete Structure, School of Young Research Methodology of Concrete Structures*; Dolnośląskie Wydawnictwo Edukacyjne: Wrocław, Poland, 1996; pp. 13–19.
7. Wong, R.C.K.; Chau, K.T. Estimation of air void and aggregate spatial distributions in concrete under uniaxial compression using computer tomography scanning. *Cem. Concr. Res.* **2005**, *35*, 1566–1576. [CrossRef]
8. Skarżyński, .; Tejchman, J. Experimental investigations of fracture process in concrete by means of X-ray micro-computed tomography. *Strain* **2016**, *52*, 26–45. [CrossRef]
9. Du Plessis, A.; Olawuyi, B.J.; Boshoff, W.P.; Le Roux, S.G. Simple and fast porosity analysis of concrete using X-ray computed tomography. *Mater. Struct.* **2016**, *49*, 553–562. [CrossRef]
10. Ostrowski, K.; Sadowski, .; Stefaniuk, D.; Wałach, D.; Gawenda, T.; Oleksik, K.; Usydus, I. The Effect of the Morphology of Coarse Aggregate on the Properties of Self-Compacting High-Performance Fibre-Reinforced Concrete. *Materials* **2018**, *11*, 1372. [CrossRef]
11. Schabowicz, K.; Jóźwiak-Niedźwiedzka, D.; Ranachowski, Z.; Kudela, S.; Dvorak, T. Microstructural characterization of cellulose fibres in reinforced cement boards. *Arch. Civ. Mech. Eng.* **2018**, *18*, 1068–1078. [CrossRef]

12. Ponikiewski, T.; Gołaszewski, J.; Rudzki, M.; Bugdol, M. Determination of steel fibres distribution in self-compacting concrete beams using X-ray computed tomography. *Arch. Civ. Mech. Eng.* **2015**, *15*, 558–568. [CrossRef]
13. Ponikiewski, T.; Katzer, J.; Bugdol, M.; Rudzki, M. Steel fibre spacing in self-compacting concrete precast walls by X-ray computed tomography. *Mater. Struct.* **2015**, *48*, 3863–3874. [CrossRef]
14. Garboczi, E.J. Three-dimensional mathematical analysis of particle shape using X-ray tomography and spherical harmonics: Application to aggregates used in concrete. *Cem. Concr. Res.* **2002**, *32*, 1621–1638. [CrossRef]
15. Kubiak, J.; odo, A.; Michałek, J. Produkcja wirowanych żerdzi elektroenergetycznych w formach nieotwieranych podłużnie. *Mater. Bud.* **2015**, *1*, 40–41. [CrossRef]
16. ydżba, D.; Rajczakowska, M.; Stefaniuk, D.; Kmita, A. Identification of the carbonation zone in concrete using X-ray microtomography. *Stud. Geotech. Mech.* **2014**, *36*, 47–54. [CrossRef]
17. Torquato, S. *Random Heterogeneous Materials: Microstructure and Macroscopic Properties*; Springer Science & Business Media: Berlin, Germany, 2013.
18. ydżba, D. Zastosowania metody asymptotycznej homogenizacji w mechanice gruntów i skał. *Prace Naukowe Instytutu Geotechniki i Hydrotechniki Politechniki Wrocławskiej. Monografie* **2002**, *74*, 274.
19. Różański, A. Sur la représentativité, la taille minimale du VER et les propriétés effectives de transport des matériaux composites aléatoires. Ph.D. Thesis, Lille 1 University, Villeneuve-d' Ascq, France.
20. Hildebrand, T.; Rüegsegger, P. A new method for the model-independent assessment of thickness in three-dimensional images. *J. Microsc.* **1997**, *185*, 67–75. [CrossRef]
21. Cała, M.; Cyran, K.; Kawa, M.; Kolano, M.; ydżba, D.; Pachnicz, M.; Rajczakowska, M.; Różański, A.; Sobótka, M.; Stefaniuk, D.; et al. Identification of Microstructural Properties of Shale by combined Use of X-ray Micro-CT and Nanoindentation Tests. *Proced. Eng.* **2017**, *191*, 735–743. [CrossRef]
22. Oliver, W.C.; Pharr, G.M. An Improved Technique for Determining Hardness and Elastic Modulus Using Load and Displacement Sensing Indentation Experiments. *J. Mater. Res.* **1992**, *7*, 1564–1583. [CrossRef]
23. Miller, M.; Bobko, C.; Vandamme, M.; Ulm, F.-J. Surface Roughness Criteria for Cement Paste Nanoindentation. *Cem. Concr. Res.* **2008**, *38*, 467–476. [CrossRef]
24. Struers-Ensuring Certainity. Available online: https://www.struers.com/en/Knowledge/Grinding-and-polishing# (accessed on 2 November 2019).
25. *ISO 14577-1:2015, Metallic Materials-Instrumented Indentation Test for Hardness and Materials Parameters-Part 1: Test Method ISO Central Secretariat*; rue de Varembé, 1211: Geneva, Switzerland, 2015.
26. Doerner, M.F.; Nix, W.D. A Method for Interpreting the Data from Depth-Sensing Indentation Instruments. *J. Mater. Res.* **1986**, *1*, 601–609. [CrossRef]
27. Fischer-Cripps, A.C. Illustrative Analysis of Load-Displacement Curves in Nanoindentation. *J. Mater. Res.* **2007**, *22*, 3075–3086. [CrossRef]
28. Fischer-Cripps, A.; Nicholson, D. Nanoindentation. Mechanical Engineering Series. *Appl. Mech. Rev.* **2004**, *57*, B12. [CrossRef]
29. Oliver, W.C.; Pharr, G.M. Measurement of Hardness and Elastic Modulus by Instrumented Indentation: Advances in Understanding and Refinements to Methodology. *J. Mater. Res.* **2004**, *19*, 3–20. [CrossRef]
30. Sneddon, I.N. Boussinesq's problem for a rigid cone. *Math. Proc. Camb. Philos. Soc.* **1948**, *44*, 492–507. [CrossRef]
31. Constantinides, G.; Ravi Chandran, K.S.; Ulm, F.-J.; Van Vliet, K.J. Grid Indentation Analysis of Composite Microstructure and Mechanics: Principles and Validation. *Mater. Sci. Eng. A* **2006**, *430*, 189–202. [CrossRef]
32. Bobko, C.; Ulm, F.-J. The Nano-Mechanical Morphology of Shale. *Mech. Mater.* **2008**, *40*, 318–337. [CrossRef]
33. Kanit, T.; Forest, S.; Galliet, I.; Mounoury, V.; Jeulin, D. Determination of the Size of the Representative Volume Element for Random Composites: Statistical and Numerical Approach. *Int. J. Solids Struct.* **2003**, *40*, 3647–3679. [CrossRef]
34. ydżba, D.; Różański, A. Microstructure Measures and the Minimum Size of a Representative Volume Element: 2D Numerical Study. *Acta Geophys.* **2014**, *62*, 1060–1086. [CrossRef]
35. Berkovich, E.S. Three Faceted Diamond Pyramid for Micro-Hardness Testing. *Ind. Diam. Rev.* **1951**, *11*, 129.
36. Sorelli, L.; Constantinides, G.; Ulm, F.-J.; Toutlemonde, F. The Nano-Mechanical Signature of Ultra High Performance Concrete by Statistical Nanoindentation Techniques. *Cem. Concr. Res.* **2008**, *38*, 1447–1456. [CrossRef]

37. Rajczakowska, M.; Stefaniuk, D.; ydżba, D. Microstructure Characterization by Means of X-Ray Micro-CT and Nanoindentation Measurements. *Stud. Geotech. Mech.* **2015**, *37*, 75–84. [CrossRef]
38. Al-Raoush, R.; Papadopoulos, A. Representative Elementary Volume Analysis of Porous Media Using X-Ray Computed Tomography. *Powder Technol.* **2010**, *200*, 69–77. [CrossRef]
39. Cnudde, V.; Cwirzen, A.; Masschaele, B.; Jacobs, P.J.S. Porosity and Microstructure Characterization of Building Stones and Concretes. *Eng. Geol.* **2009**, *103*, 76–83. [CrossRef]
40. Peyton, R.L.; Haeffner, B.A.; Anderson, S.H.; Gantzer, C.J. Applying X-ray CT to Measure Macropore Diameters in Undisturbed Soil Cores. *Geoderma* **1992**, *53*, 329–340. [CrossRef]
41. Feldkamp, L.A.; Davis, L.C.; Kress, J.W. Practical Cone-Beam Algorithm. *J. Opt. Soc. Am. A JOSAA* **1984**, *1*, 612–619. [CrossRef]
42. Rodet, T.; Noo, F.; Defrise, M. The Cone-Beam Algorithm of Feldkamp, Davis, and Kress Preserves Oblique Line Integrals. *Med. Phys.* **2004**, *31*, 1972–1975. [CrossRef]

10

Variation in Compressive Strength of Concrete across Thickness of Placed Layer

Jarosław Michałek

Faculty of Civil Engineering, Wrocław University of Science and Technology, 50-370 Wrocław, Poland; jaroslaw.michalek@pwr.edu.pl

Abstract: Is the variation in the compressive strength of concrete across the thickness of horizontally cast elements negligibly small or rather needs to be taken into account at the design stage? There are conflicting answers to this question. In order to determine if the compressive strength of concrete varies across the thickness of horizontally cast elements, ultrasonic tests and destructive tests were carried out on core samples taken from a 350 mm thick slab made of class C25/30 concrete. Special point-contact probes were used to measure the time taken for the longitudinal ultrasonic wave to pass through the tested sample. The correlation between the velocity of the longitudinal ultrasonic wave and the compressive strength of the concrete in the slab was determined. The structure of the concrete across the thickness of the slab was evaluated using GIMP 2.10.4. It was found that the destructively determined compressive strength varied only slightly (by 3%) across the thickness of the placed layer of concrete. Whereas the averaged ultrasonically determined strength of the concrete in the same samples does not vary across the thickness of the analyzed slab. Therefore, it was concluded that the slight increase in concrete compressive strength with depth below the top surface is a natural thing and need not be taken into account in the assessment of the strength of concrete in the structure.

Keywords: concrete slabs and floorings; horizontal casting; compressive strength; ultrasonic tests

---

1. Introduction

The view that the compressive strength of concrete varies across the thickness of horizontally cast elements (concrete slabs, floorings, etc.) is seldom expressed in the literature on the subject. Opinions as to the significance of this variation are widely divided, as the following survey of literature indicates.

The research published by Stawiski [1–3] provided the direct incentive for this study of the distribution of concrete compressive strength along the height of horizontally cast elements. On the basis of ultrasonic tests of core samples taken from concrete, Stawiski found the compressive strength of the concrete to be lower in the top zone than in the bottom zone by as much as 40–50% [1–3]. The variation in compressive strength along the height of the cross section was approximately linear. The fall in ultrasonic wave velocity at the sample's top surface is ascribed by Stawiski [1] to the surface weakening effect connected with concrete consolidation resulting in the segregation of concrete components. The main factor responsible for the decrease in concrete compressive strength is considered to be porosity, which very strongly affects ultrasonic wave velocity. Also the inadequate curing of fresh concrete, damage to the structure of concrete caused by corrosion, and mechanical damage to the top surface of the concrete which can arise in the course of the service life of the element are also possible factors.

Stawiski [3] proposed to introduce (besides the grade of concrete) strength gradient ∇fc into the evaluation of concrete in horizontally cast elements (e.g., floor toppings). On the basis of his research [3] Stawiski pointed out that in, e.g., an approximately 15 cm thick element the strength gradient of the concrete at the depth of 10 cm from the bottom amounted to 0.7 MPa/cm, whereas in the layers situated closer to the top surface it varied markedly (−3.0, −4.5, −8.0 MPa/cm). Therefore, Stawiski calls for [3] defining allowable variations in concrete compressive strength, e.g., $\nabla f_c \leq 1.0$ MPa/cm. The increase of 1.0 MPa/cm in the strength of concrete in the lower situated layers relative to the top layer suggested by Stawiski [3] seems to be very large.

On the basis of ultrasonic tests of the compressive strength across the thickness of samples taken from cut out pieces of 40, 45 and 60 mm thick floorings made of cement mortars, Hoła, Sadowski and Hoła A. [4] found the strength was not identical and varied across the thickness. The lowest strength was in the top zone, the highest in the bottom zone, while in the middle zone, it was close to the destructively determined compressive strength. In the considered case, the strength gradient of the mortar across the thickness of the flooring amounted to 6–7 MPa/cm.

Petersons in [5] found the compressive strength of the lower situated layers to be higher than that of the top layer, but only by 10–20%. No further increase in concrete strength was observed in the layers situated below 300 mm. The difference in compressive strength between the top surface and the bottom surface in slabs was ascribed to the inadequate curing of the concrete [5].

In monograph [6], Dąbrowski, Stachurski and Zieliński found that the deeper situated layers of concrete had higher strength than the surface layer. Below 80 cm, this increase in strength stabilized at the level of approximately 10%. In the authors' opinion [6], this is due to the well-known property of concrete—it reaches higher strength when hardening under a moderate pressure—and that is why this phenomenon does not occur in samples of low height.

Yuan, Ragab, Hill and Cook [7] found that the compressive strength of concrete along the height of the placed layer did not vary significantly. Suprenant [8] found that the compressive strength of concrete in slabs varied minimally, and only in a small upper part of the element. The most marked variation in concrete strength along element height has been observed in walls and beams. This is mainly due to the greater static pressure exerted by the concrete situated above.

Neville [9] found that the slight increase in concrete compressive strength below the top surface was a natural thing, but need not be taken into account. When testing a reinforced concrete wall and beam by means of the ultrasonic method, Watanabe, Hishikawa, Kamae and Namiki [10] found the compressive strength of concrete in samples taken from the lower part of the element was slightly higher than in samples taken from its upper part. They treated this as a natural thing which did not need to be taken into account.

Neville [11] ascribed the variation in the compressive strength of concrete along the height of the sample to the presence of retained water, occurring during concrete bleeding.

According to standard [12], the compressive strength of concrete in a structure can be lower in the top layer than in the bottom layer by as much as 25%. Concrete characterized by lower compressive strength usually occurs to a depth of 300 mm or to 20% of the height of the cross section, depending on which of the values is lower. According to standard [13], the range of variation in the compressive strength of concrete in a structure can differ between the particular portions of the structure. The variation is random and often forced (by, e.g., the relative density, the degree of compaction, the curing conditions, etc.).

Therefore, the questions arises: Is the variation in the compressive strength of concrete across the thickness of horizontally cast elements negligibly small or rather needs to be taken into account at the design stage?

2. Description of Author's Investigations

2.1. *Ultrasonic Tests of Concrete*

In order to verify the phenomenon of concrete compressive strength variation across the thickness of horizontally cast elements, samples with diameter d = 100 mm height h = 350 mm (Table 1), taken from a specially cast slab made of concrete C25/30 (the grade of the concrete was determined using concrete cubes cast when casting the slab) were subjected to ultrasonic tests. CEMII/BS-32.5 cement

(270 kg/m^3), fly ash additive (60 kg/m^3), plasticizer (2.43 kg/m^3), water (170 kg/m^3) and 1879 kg/m^3 of natural aggregate (sand 0/2 mm −40%, gravel 2/8 mm −26%, gravel 8/16 mm −34%) were used in the concrete mix for the slab construction. The latter had been compacted by means of an immersion vibrator and cured for 28 in the laboratory conditions defined in standard [14]. The slab had been exposed to variable weather conditions for two years. Samples (01–06 in Figure 1) were drilled out of the slab perpendicularly to its top surface. For reference purposes specimens (07–12) were drilled out of the slab parallel with its top surface. Prior to the tests, the actual dimensions of the samples and their weight were determined (Table 1).

Table 1. The mean dimensions of core samples and their weight.

Sample Number	$d_{śr}$	$h_{śr}$	m	V	ρ
	mm		g	cm^3	g/cm^3
01	98.7	351.7	6158.5	2691	2.29
02	98.5	351.4	6167.5	2680	2.30
03	98.6	351.5	6138.0	2682	2.29
04	98.6	351.4	6176.5	2682	2.30
05	98.5	350.0	6121.5	2669	2.29
06	98.6	350.2	6121.0	2675	2.29
07	98.6	350.0	6122.0	2672	2.29
08	98.7	351.3	6172.5	2688	2.30
09	98.5	351.4	6168.2	2678	2.30
10	98.5	350.0	6122.4	2667	2.30
11	98.6	351.5	6140.0	2684	2.29
12	98.5	351.4	6137.6	2678	2.29

Figure 1. The samples to be tested using ultrasonic device.

The measuring points spaced at every 1 cm (Figures 1 and 2) were marked on the sides of the core samples. Velocity C_L of ultrasonic wave passage through concrete was measured in two perpendicular directions. No concrete/probe coupling material was used. The probes were set perpendicularly to the tested side surface of the sample (Figure 2). The distributions of velocity C_L of longitudinal ultrasonic wave passage through the concrete were obtained from the tests (Tables 2 and 3, and Figures 3 and 4).

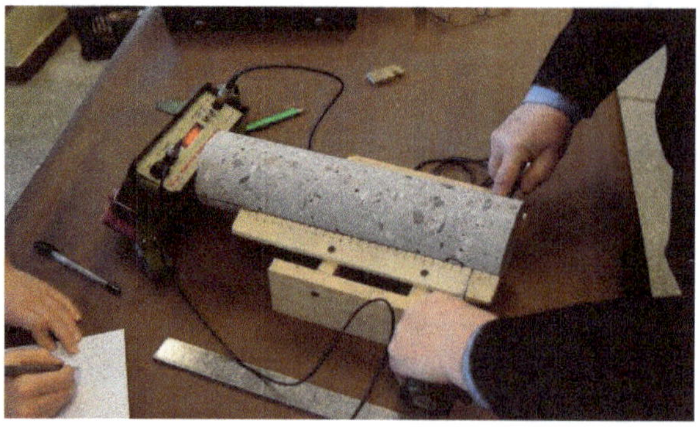

Figure 2. The measurement of velocity C_L of longitudinal ultrasonic wave.

Table 2. Velocities C_L ultrasonic wave propagation, measured in two perpendicular directions in core samples taken perpendicularly to top surface of slab.

Measuring Place Distance from Sample Top	Sample 01		Sample 02		Sample 03		Sample 04		Sample 05		Sample 06	
	C_{L1}	C_{L2}	C_{L1}	C_{L2}	C_{L1}	C_{L2}	C_{L1}	C_{L2}	C_{L1}	C_{L2}	C_{L1}	C_{L2}
cm	km/s											
0.5	3.19	3.39	3.49	3.18	3.32	3.44	3.40	3.31	3.36	3.36	3.23	3.29
1.5	3.18	3.35	3.48	3.02	3.28	3.38	3.29	3.26	3.35	3.30	3.22	3.33
2.5	3.30	3.41	3.49	3.48	3.31	3.37	3.40	3.36	3.28	3.33	3.27	3.36
3.5	3.58	3.63	3.62	3.57	3.43	3.54	3.57	3.62	3.33	3.54	3.55	3.65
4.5	3.60	3.68	3.60	3.61	3.49	3.52	3.53	3.62	3.53	3.52	3.61	3.63
5.5	3.51	3.66	3.64	3.54	3.66	3.62	3.66	3.62	3.62	3.67	3.57	3.61
6.5	3.60	3.68	3.70	3.52	3.57	3.67	3.68	3.44	3.47	3.66	3.60	3.63
7.5	3.53	3.59	3.62	3.48	3.64	3.58	3.58	3.60	3.53	3.62	3.62	3.58
8.5	3.46	3.60	3.72	3.61	3.57	3.50	3.70	3.60	3.54	3.61	3.56	3.58
9.5	3.53	3.57	3.74	3.57	3.26	3.62	3.71	3.58	3.44	3.59	3.60	3.65
10.5	3.61	3.59	3.70	3.57	3.57	3.63	3.54	3.53	3.47	3.58	3.67	3.79
11.5	3.58	3.59	3.62	3.66	3.60	3.62	3.62	3.56	3.50	3.58	3.58	3.62
12.5	3.56	3.65	3.62	3.65	3.55	3.65	3.60	3.58	3.52	3.62	3.41	3.62
13.5	3.32	3.63	3.61	3.60	3.57	3.49	3.65	3.61	3.57	3.57	3.57	3.70
14.5	3.38	3.65	3.72	3.66	3.55	3.63	3.70	3.70	3.63	3.62	3.66	3.76
15.5	3.45	3.67	3.73	3.57	3.65	3.63	3.44	3.53	3.61	3.65	3.73	3.68
16.5	3.60	3.66	3.73	3.58	3.62	3.51	3.72	3.59	3.50	3.61	3.62	3.65
17.5	3.53	3.68	3.72	3.53	3.53	3.62	3.66	3.53	3.54	3.61	3.65	3.70
18.5	3.68	3.67	3.64	3.49	3.62	3.53	3.70	3.59	3.54	3.66	3.61	3.70
19.5	3.58	3.72	3.66	3.59	3.47	3.48	3.65	3.57	3.62	3.66	3.63	3.62
20.5	3.59	3.68	3.63	3.47	3.57	3.39	3.64	3.70	3.43	3.70	3.62	3.66
21.5	3.62	3.57	3.57	3.58	3.53	3.56	3.62	3.54	3.54	3.62	3.48	3.65
22.5	3.40	3.61	3.62	3.61	3.55	3.45	3.70	3.50	3.54	3.61	3.58	3.70
23.5	3.45	3.64	3.62	3.65	3.68	3.59	3.59	3.61	3.40	3.37	3.58	3.65
24.5	3.35	3.62	3.51	3.62	3.52	3.58	3.64	3.70	3.48	3.53	3.62	3.65
25.5	3.55	3.66	3.65	3.48	3.53	3.50	3.53	3.65	3.46	3.57	3.64	3.57
26.5	3.38	3.71	3.69	3.61	3.57	3.32	3.48	3.65	3.44	3.61	3.58	3.58
27.5	3.54	3.55	3.62	3.55	3.54	3.43	3.58	3.63	3.37	3.58	3.54	3.49
28.5	3.54	3.61	3.52	3.57	3.56	3.59	3.62	3.66	3.57	3.66	3.62	3.65
29.5	3.61	3.65	3.63	3.58	3.60	3.59	3.68	3.69	3.48	3.54	3.57	3.57
30.5	3.57	3.62	3.64	3.57	3.57	3.57	3.55	3.62	3.57	3.53	3.61	3.54
31.5	3.45	3.61	3.58	3.52	3.45	3.48	3.57	3.29	3.57	3.50	3.27	3.51
32.5	3.24	3.20	3.52	3.25	3.13	3.34	3.30	3.42	3.48	3.26	3.43	3.34
33.5	3.33	3.25	3.57	3.22	3.28	3.20	3.51	3.42	3.54	3.41	3.42	3.42
34.5	3.39	3.29	3.59	3.23	3.47	3.04	3.62	3.49	3.49	3.49	3.40	3.49

Table 3. Velocities C_L of longitudinal ultrasonic wave propagation, measured in two perpendicular directions in core samples taken parallel with the top surface of the slab.

Measuring Place Distance from Sample Top	Sample 07		Sample 08		Sample 09		Sample 10		Sample 11		Sample 12	
	C_{L1}	C_{L2}	C_{L1}	C_{L2}	C_{L1}	C_{L2}	C_{L1}	C_{L2}	C_{L1}	C_{L2}	C_{L1}	C_{L2}
cm	km/s											
0.5	3.31	3.42	3.60	3.40	3.34	3.60	3.60	3.35	3.30	3.28	3.50	3.38
1.5	3.27	3.33	3.50	3.28	3.25	3.30	3.32	3.34	3.29	3.26	3.40	3.31
2.5	3.55	3.47	3.31	3.60	3.36	3.56	3.41	3.51	3.32	3.41	3.45	3.33
3.5	3.56	3.45	3.25	3.46	3.59	3.42	3.62	3.53	3.52	3.54	3.43	3.59
4.5	3.54	3.49	3.54	3.42	3.45	3.55	3.50	3.51	3.40	3.41	3.46	3.56
5.5	3.55	3.46	3.46	3.60	3.40	3.50	3.48	3.44	3.41	3.55	3.42	3.36
6.5	3.57	3.50	3.41	3.45	3.51	3.44	3.41	3.44	3.28	3.38	3.46	3.37
7.5	3.49	3.50	3.57	3.42	3.51	3.53	3.44	3.45	3.39	3.50	3.56	3.43
8.5	3.45	3.43	3.56	3.60	3.47	3.54	3.55	3.41	3.49	3.51	3.51	3.49
9.5	3.55	3.43	3.60	3.52	3.44	3.57	3.48	3.51	3.50	3.44	3.45	3.50
10.5	3.40	3.47	3.49	3.50	3.45	3.50	3.40	3.45	3.41	3.49	3.39	3.39
11.5	3.56	3.43	3.46	3.51	3.54	3.46	3.41	3.55	3.36	3.46	3.44	3.50
12.5	3.55	3.38	3.42	3.50	3.54	3.46	3.42	3.47	3.52	3.55	3.43	3.43
13.5	3.57	3.43	3.38	3.46	3.51	3.56	3.37	3.40	3.39	3.50	3.55	3.51
14.5	3.48	3.47	3.41	3.55	3.57	3.51	3.38	3.40	3.51	3.55	3.46	3.52
15.5	3.59	3.47	3.55	3.51	3.43	3.59	3.42	3.41	3.48	3.58	3.44	3.45
16.5	3.54	3.45	3.56	3.46	3.42	3.46	3.38	3.38	3.41	3.45	3.56	3.40
17.5	3.50	3.40	3.53	3.35	3.57	3.56	3.49	3.44	3.40	3.38	3.41	3.39
18.5	3.47	3.44	3.46	3.47	3.49	3.45	3.40	3.36	3.42	3.39	3.38	3.40
19.5	3.42	3.43	3.41	3.55	3.45	3.43	3.40	3.46	3.42	3.50	3.43	3.54
20.5	3.40	3.48	3.47	3.46	3.44	3.50	3.44	3.43	3.38	3.40	3.38	3.55
21.5	3.45	3.50	3.50	3.39	3.42	3.42	3.44	3.56	3.48	3.44	3.44	3.49
22.5	3.40	3.41	3.41	3.42	3.46	3.42	3.40	3.44	3.50	3.43	3.42	3.43
23.5	3.43	3.43	3.41	3.45	3.54	3.41	3.41	3.49	3.33	3.50	3.37	3.41
24.5	3.50	3.38	3.53	3.50	3.38	3.42	3.49	3.42	3.52	3.38	3.51	3.50
25.5	3.42	3.50	3.51	3.38	3.40	3.57	3.46	3.40	3.46	3.41	3.45	3.53
26.5	3.56	3.38	3.45	3.36	3.44	3.37	3.45	3.25	3.55	3.45	3.49	3.44
27.5	3.54	3.45	3.51	3.47	3.39	3.31	3.30	3.36	3.36	3.43	3.41	3.46
28.5	3.42	3.37	3.55	3.46	3.40	3.38	3.48	3.36	3.37	3.37	3.50	3.55
29.5	3.50	3.43	3.46	3.34	3.43	3.36	3.44	3.35	3.43	3.35	3.53	3.50
30.5	3.56	3.50	3.42	3.43	3.45	3.45	3.44	3.53	3.50	3.37	3.37	3.32
31.5	3.57	3.40	3.55	3.33	3.44	3.43	3.46	3.50	3.40	3.37	3.37	3.45
32.5	3.40	3.34	3.40	3.53	3.40	3.41	3.50	3.56	3.40	3.45	3.45	3.44
33.5	3.50	3.40	3.59	3.51	3.47	3.45	3.42	3.49	3.51	3.40	3.44	3.47

A Unipan Materials Tester Type 543 with point-contact exponential probes [15] and a frequency of 40 kHz (Figure 2) was used to measure the time taken for the longitudinal ultrasonic wave to pass through the tested sample. Prior to the tests, the instrument had been calibrated to determine the time taken for the ultrasonic wave to pass through the probes alone (t_0 = 36.6 μs). The details of the operation of exponential heads with point-to-point contact with the examined surface are described in detail in the paper [15].

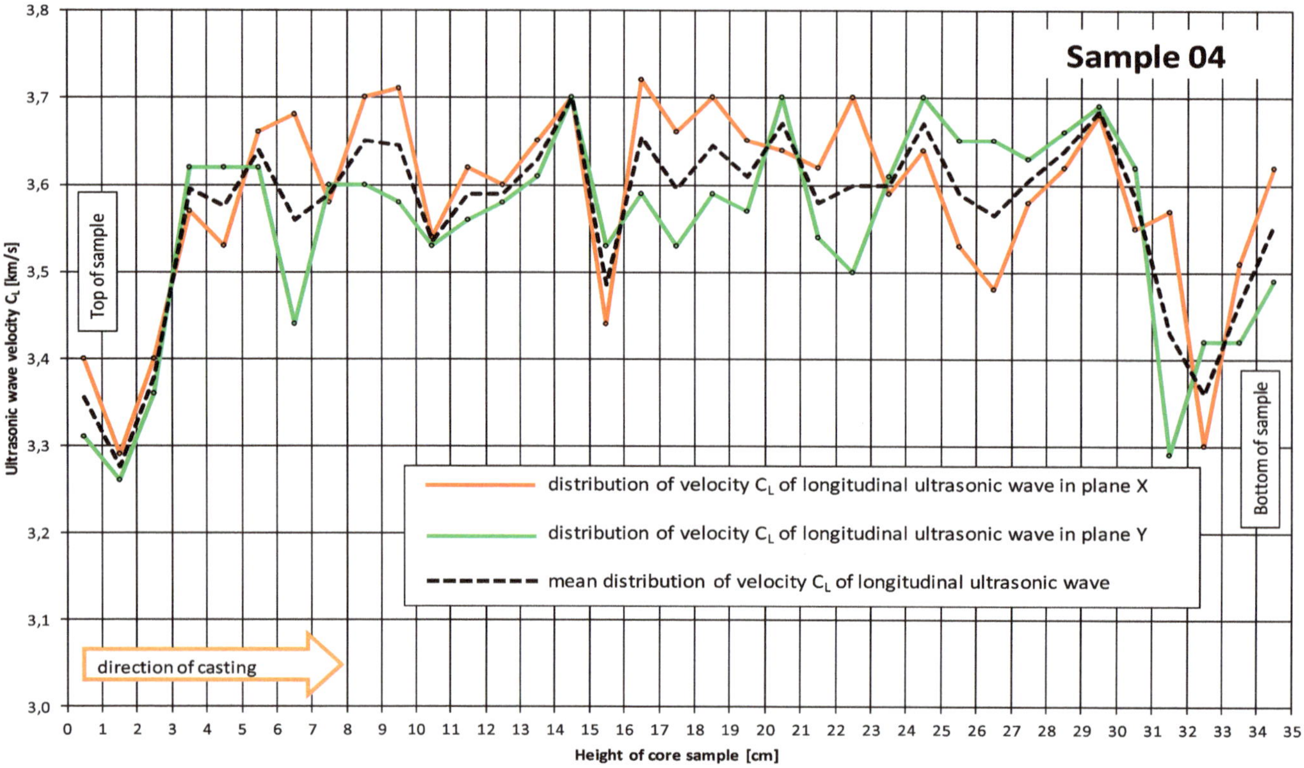

Figure 3. Exemplary distribution of velocity C_L of longitudinal ultrasonic wave along the height of sample 04 taken perpendicularly to the top surface of the slab.

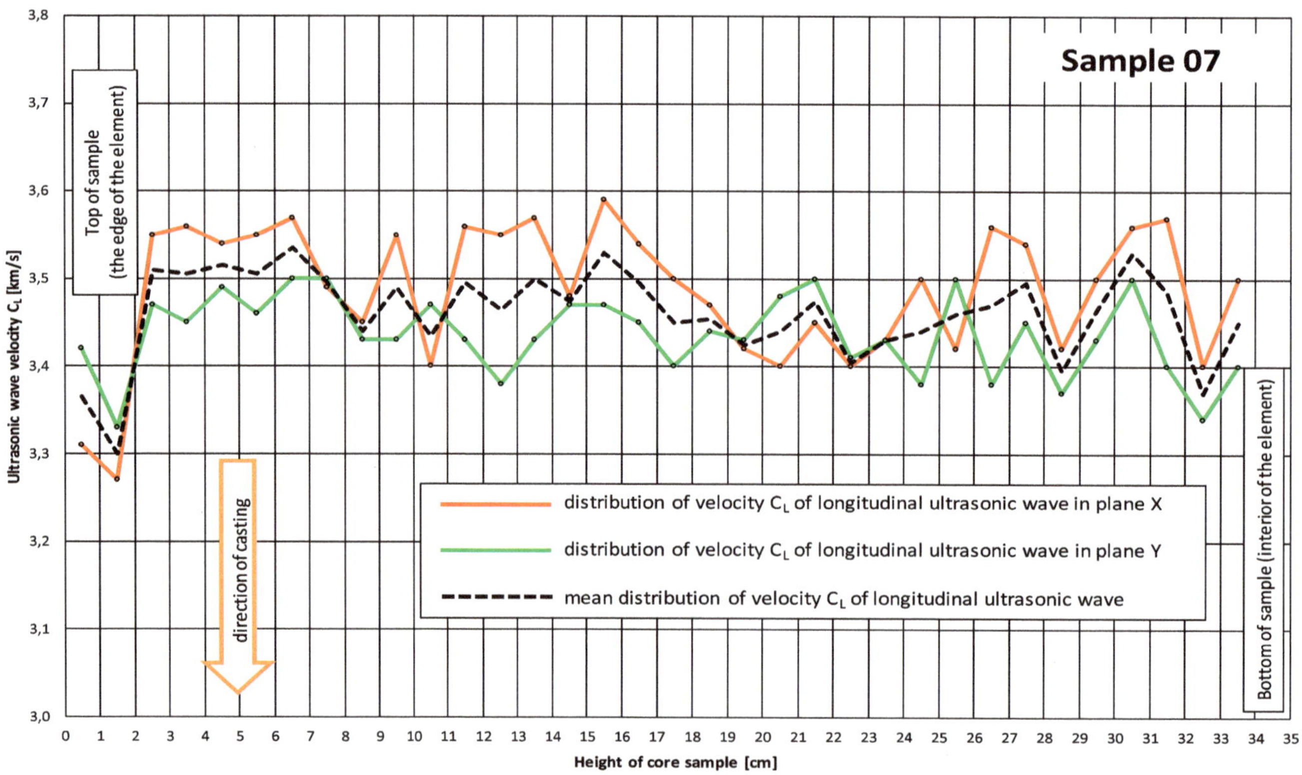

Figure 4. Exemplary distribution of velocity C_L of longitudinal ultrasonic wave along height of sample 07 taken parallel with the top surface of the slab.

Tables 2 and 3 and Figures 3 and 4 showed that the distributions of velocity C_L of the longitudinal ultrasonic wave along the height of the core samples with d ≅ 100 mm were quite constant (except for the top layers). The mean velocity C_L of ultrasonic wave propagation in the samples (01–06) taken perpendicularly to the top surface of the slab was C_{L1} = 3.54 km/s, a standard deviation s_{CL1} = 0.13 km/s and a variation coefficient ν_{CL1} = 3.67%. In the case of the samples (07–12) taken parallel with the top of the slab, the following were obtained: C_{L2} = 3.45 km/s, s_{CL2} = 0.072 km/s and ν_{CL2} = 2.09%. On the basis of the obtained velocities C_L of the longitudinal ultrasonic wave in the range: C_L = 3.5–4.5 km/s, it can be assessed [16] whether the quality of the concrete as good, whereas the velocities in the range of 3.0–3.5 km/s indicated dubious quality.

The observed slight fluctuations of velocity C_L of the longitudinal ultrasonic wave in the inner layers of the concrete are due to local concrete defects (e.g., air voids) or local strengthening with larger aggregate grains. Lower velocities C_L of the longitudinal ultrasonic wave were registered in the samples (01–06) taken perpendicularly to the top surface of the slab (Figures 3 and 5). In the samples (07–12) taken parallel with the top of the slab, a decline in velocity CL of the longitudinal ultrasonic wave was observed only at the edge constituting the side edge of the slab (Figures 4 and 6).

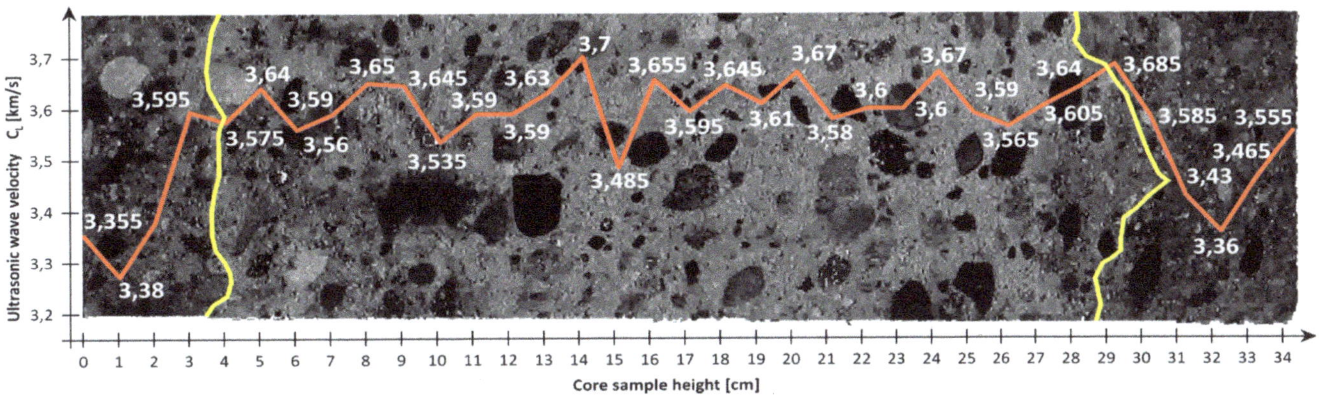

Figure 5. The structure of concrete of the sample taken perpendicularly to the top surface of the slab (slab top surface on left). The image also shows distribution of the mean velocity C_L of longitudinal ultrasonic wave (red line) and disturbance zone at the top and bottom surface of the slab (yellow line).

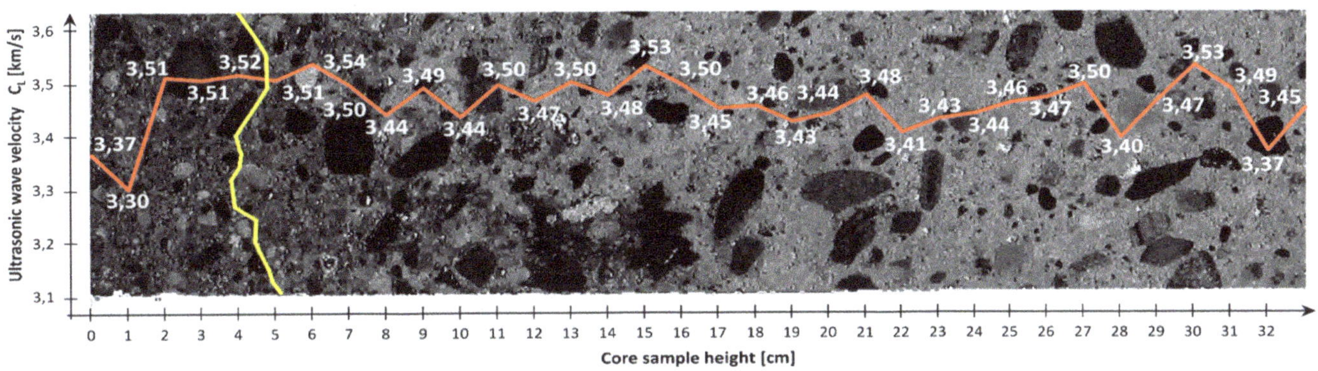

Figure 6. The structure of concrete of the sample taken parallel with the top surface of the slab (slab top surface on left). The image also shows distribution of the mean velocity C_L of longitudinal ultrasonic wave (red line) and zone of disturbance at the lateral surface of the slab (yellow line).

Initially, it was though that the falls in velocity C_L of the longitudinal ultrasonic wave were due to the sample end effect. Ultimately, it was decided that this phenomenon in samples 01–06 was caused from the top by bleeding [17] and concrete sedimentation, and from the bottom by the improper vibration of the concrete by means of the immersion vibrator (the vibrator was not fully immersed in the freshly placed concrete). In samples 07–12, the fall in velocity C_L of the longitudinal ultrasonic wave can be caused by the wall effect [9,10]. Figures 5 and 6 show the structure of the concrete along the height of the samples taken perpendicularly to and parallel with the top surface of the slab. The image of the structure of the samples in Figures 5 and 6 was prepared in GIMP 2.10.4 using the filter: LCHH(ab) component with a contrast of 50%. In Figure 5 the altered structure of the concrete is visible in the sample's upper part (an approximately 30–40 mm thick layer) and lower part (an approximately 30–50 mm thick layer). In these places, reduced velocities of the longitudinal wave velocity were observed. In Figure 6, the altered structure of the concrete can be seen in an approximately 20–80 mm thick layer located at the side wall of the slab. Also in this layer, falls in the velocity of the longitudinal ultrasonic wave were recorded.

2.2. *Ultrasonic Tests of Concrete*

For further investigations the core samples (Table 1) were cut into smaller samples (Figure 7) with height h = 100 mm (h/d = 1). In the terminal samples (e.g., 01/1—the top of the sample, 01/3—the bottom of the sample) the actual end faces were left unchanged (or were slightly trimmed to make them level). The middle sample (e.g., 01/2) was cut to the required size of 100 mm. Then, the end faces were prepared by grinding for compressive strength tests (Figure 8).

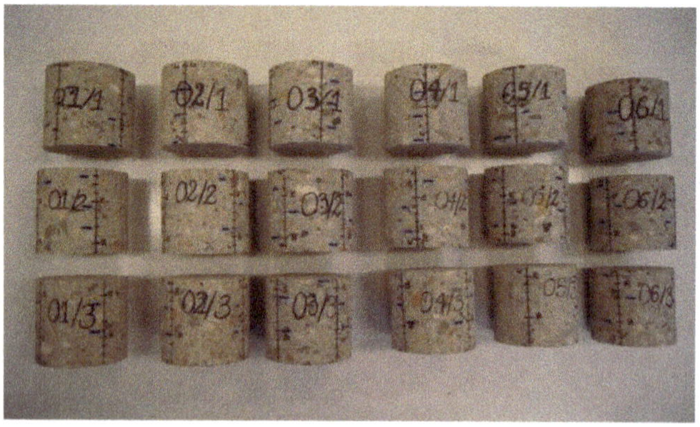

Figure 7. Core samples with height/diameter ratio h/d = 1, obtained from samples 01–06.

Figure 8. One of the core samples during grinding of its end surface.

The compressive strength tests were carried out in conformance with standard [13] in the ZD100 strength testing machine (Figure 9a) satisfying the requirements of standard [18]. All the samples showed the same type of failure (Figure 9b). The parameters of the samples and the test results are presented in Table 2.

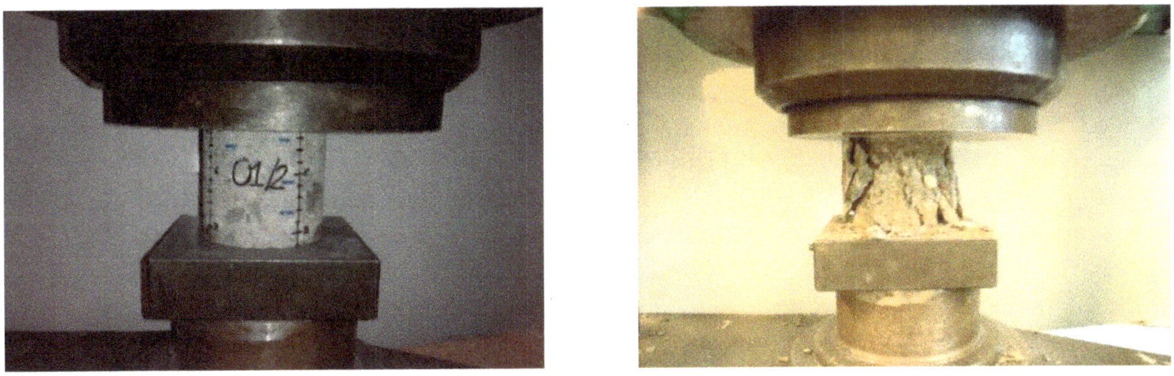

(a) (b)

Figure 9. The sample in strength testing machine: (**a**) during loading, (**b**) after failure.

The concrete strength values (Table 4) yielded by the tests carried out on the100 mm high samples were used to evaluate the class of the concrete and the variation of strength along the core sample height (h = 350 mm) and were correlated with the results obtained using the ultrasonic method. According to standard [19], the result of concrete compressive strength tests carried out on cylindrical specimens with diameter d = 100 mm and height h = 100 mm, cut out of a structure directly corresponded to the strength of concrete determined on 150 × 150 × 150 mm standard cubes ($f_{ck,is} = f_{ck,is,cube}$).

Table 4. The results of concrete compressive strength tests.

Samp. No.	d_m	h_m	m	A_c	V_c	ρ	F_{is}	f_{is}	Samp. No.	d_m	h_m	m	A_c	V_c	ρ	F_{is}	f_{is}
	mm		g	cm^2	cm^3	g/cm^3	kN	MPa		mm		g	cm^2	cm^3	g/cm^3	kN	MPa
01/1	98.7	99.9	1725	76.51	764	2.26	236	30.85	07/1	98.6	100.0	1760	76.36	764	2.30	226	29.60
01/2	98.7	99.8	1743	76.51	764	2.28	240	31.37	07/2	98.6	99.9	1751	76.36	763	2.30	231	30.25
01/3	98.7	99.8	1765	76.51	764	2.31	242	31.63	07/3	98.6	99.8	1746	76.36	762	2.29	232	30.38
02/1	98.5	100.0	1724	76.20	762	2.26	236	30.97	08/1	98.7	99.9	1755	76.51	764	2.30	230	30.06
02/2	98.5	99.8	1761	76.20	760	2.32	238	31.23	08/2	98.7	99.9	1761	76.51	764	2.30	235	30.71
02/3	98.5	99.9	1762	76.20	761	2.31	243	31.89	08/3	98.7	99.9	1757	76.51	764	2.30	237	30.98
03/1	98.5	99.8	1711	76.20	760	2.25	233	30.58	09/1	98.5	99.7	1744	76.20	760	2.30	234	30.71
03/2	98.5	99.9	1741	76.20	761	2.29	241	31.63	09/2	98.5	99.9	1740	76.20	761	2.29	238	31.23
03/3	98.6	99.6	1768	76.36	761	2.32	248	32.48	09/3	98.5	100.0	1760	76.20	762	2.31	243	31.89
04/1	98.5	99.9	1731	76.20	761	2.27	237	31.10	10/1	98.5	99.6	1754	76.20	759	2.31	239	31.36
04/2	98.5	99.9	1755	76.20	761	2.31	240	31.50	10/2	98.5	99.8	1742	76.20	760	2.29	245	32.15
04/3	98.5	99.8	1779	76.20	760	2.34	240	31.50	10/3	98.5	99.8	1748	76.20	760	2.30	248	32.55
05/1	98.5	99.9	1717	76.20	761	2.26	234	30.71	11/1	98.6	99.7	1741	76.36	761	2.29	237	31.04
05/2	98.5	99.9	1745	76.20	761	2.29	240	31.50	11/2	98.6	99.7	1747	76.36	761	2.29	232	30.38
05/3	98.5	99.9	1769	76.20	761	2.32	242	31.76	11/3	98.6	99.9	1750	76.36	763	2.29	238	31.17
06/1	98.5	99.9	1716	76.20	761	2.25	240	31.50	12/1	98.5	99.9	1766	76.20	761	2.32	240	31.50
06/2	98.5	99.9	1741	76.20	761	2.29	242	31.76	12/2	98.5	99.8	1754	76.20	760	2.31	243	31.89
06/3	98.5	99.9	1762	76.20	761	2.31	245	32.15	12/3	98.5	100.1	1750	76.20	763	2.29	245	32.15

Standard [19] states that due to drilling, which undoubtedly can slightly damage the core's material, the strengths of core samples determined in-situ are usually lower than the strengths of the standard samples. For this reason, it is allowed to use a correction factor of 0.85, understood as a ratio of the in-situ characteristic compressive strength to the characteristic compressive strength determined on the standard samples. As a result, the concrete compressive strength values coming directly from strength tests are increased.

The mean compressive strength of the concrete, determined on the 18 samples taken perpendicularly to the top surface of the slab, amounted to $f_{m(18),is}$ = 31.45 MPa (the minimum value $f_{is,lowest}$ = 30.58 MPa). The mean standard deviation amounted to s = 0.49 MPa. The coefficient k_1 = 1.48 was assumed. The characteristic compressive strength of the concrete in the structure ($f_{ck,is}$) was determined on the basis of standard [13], from the condition: $f_{ck,is} = \min(f_{m(18),is} - k_1 \times s; f_{is,lowest} + 4)$ = (30.72 MPa, 34.58 MPa). Thus the characteristic cube compressive strength of the concrete determined on samples taken perpendicularly to the top surface of the slab amounted to $f_{ck,is} = f_{ck,is,cube}$ = 30.72 MPa.

The mean compressive strength of the concrete determined on the 18 samples taken parallel with the top surface of the slab amounted to $f_{m(18),is}$ = 31.11 MPa (the minimum value $f_{is,lowest}$ = 29.60 MPa). The mean standard deviation amounted to s = 0.81 MPa. The characteristic cube compressive strength of the concrete, determined on the samples taken parallel with the top surface of the slab, amounted to $f_{ck,is}$ = (29.91 MPa, 33.60 MPa) = 29.91 MPa. The cube strength determined on the samples taken parallel to the top surface of the slab was 3% lower than the strength determined on the samples taken perpendicularly to the top surface of the slab. This confirms the observation that the strength of core samples drilled out horizontally is lower (on average by 8% [12,13]) than that of core samples drilled out vertically.

On the basis of the obtained $f_{ck,is,cube}$ values, the actual strength class of the concrete in the structure is estimated to be $f_{ck,is,cube}$ = 30.72 MPa and 29.91 MPa, respectively. When the correction factor of 0.85 is applied, this gives the concrete strength class respectively $f_{ck,is,cube}$ = 36.1 MPa and 35.2 MPa, which corresponds to at least concrete class C25/30.

2.3. *Scaling of Correlation Curve*

On the basis of the measurements, the mean longitudinal ultrasonic wave passage velocities C_L were correlated with the mean compressive concrete strengths f_{is}. When calculating the mean longitudinal ultrasonic wave passage velocity C_L, the values from the areas of disturbances near the ends of the samples were rejected. The correlation curve was scaled according to the procedure described in standard [13] (version 2). In accordance with [16], a hypothetical base regression curve for ordinary concrete, i.e., $f_{CL,b} = 2.39C_L^2 - 7.06C_L + 4.2$ for C_L = 2.4–5.0 km/s, was adopted. Then, the differences δf between experimental compressive strength f_{is} and the strength obtained from base curve f_{CL}, as well as the mean value $\delta f_{m(n)}$ of the differences and standard deviation s were determined. The shift of the base correlation curve was calculated from the relation $\Delta f = \delta f_{m(n)} - k_1 \cdot s$ for coefficient k_1 = 1.48 [13]. Ultimately, the corrected correlation curve $f_{CL} = f_{CL,b} + \Delta f$ has the form $f_{CL} = 2.39C_L^2 - 7.06C_L + 25.09$. The obtained curve only slightly differs from the curves determined separately for samples 01 ÷ 06 and 07–12.

The obtained correlation was evaluated using two accuracy characteristics, i.e., the correlation coefficient $\eta > 0.75$ and the mean square relative deviation $\nu_k \leq 12 \leq$ %. The correlation coefficient amounted to:

$$\begin{gathered}\eta = [0.25 \times \sum(f_{CL,I} - f_{CL(36),v})^2]^{1/2} \div [0.25 \times \sum(f_{is} - f_{m(36),is})^2]^{1/2} \\ \eta = [0.25 \times 15.34]^{1/2} \div [0.25 \times 16.33]^{1/2} = 0.97 > 0.75\end{gathered} \tag{1}$$

and the mean square relative deviation to:

$$\begin{gathered}\nu_k = 100 \times \{[1/(n-1)] \times \sum[(f_{CL,i} - f_{is})/f_{CL,i}]^2\}^{1/2} \\ \nu_k = 100 \times [(1/35)] \times 0.12555]^{1/2} = 5.99\% < 12\%.\end{gathered} \tag{2}$$

Thus, it can be said that a good correlation between the mean longitudinal ultrasonic wave passage velocities C_L and the mean concrete compressive strengths f_{is} was obtained.

The class of the concrete in the structure was determined on the basis of the concrete compressive strength values obtained from the correlation curve $f_{CL} = 2.39 \times C_L^2 - 7.06C_L + 25.09$ for the mean longitudinal ultrasonic wave passage velocities C_L. The mean compressive strength of the concrete determined using the ultrasonic method amounted to $f_{CL(36),is}$ = 29.80 MPa (minimal $f_{CL,is,lowest}$ = 28.80 MPa) and the mean standard deviation to s = 0.66 MPa. Hence, the characteristic compressive strength of the concrete amounted to $f_{ck,is} = f_{ck,is,cube} \leq (29{,}80 - 1.48 \times 0.66, 28.80 + 4) = (28.82\ \text{MPa}, 32.80\ \text{MPa}) = 28.82\ \text{MPa}$. When the correction factor of 0.85 was taken into account, concrete class C25/30 was obtained. The concrete class determined on the basis of the compressive strength values obtained from the correlation curve confirmed the destructively determined class of the concrete.

3. Analysis of Test Results

With the corrected correlation curve: $f_{CL} = 2.39 \cdot C_L^2 - 7.06 \cdot C_L + 25.09$, it was possible to trace the variation of the compressive strength of the concrete along the height of the analyzed samples. Figures 10 and 11 show the variations of concrete compressive strength along the height of the core samples respectively, and perpendicular to (Figure 10) and parallel with (Figure 11) the top surface of the slab. Further, the results of the destructive tests and the averaged results of the ultrasonic tests for the particular samples with a height h = 100 mm are included in the diagrams.

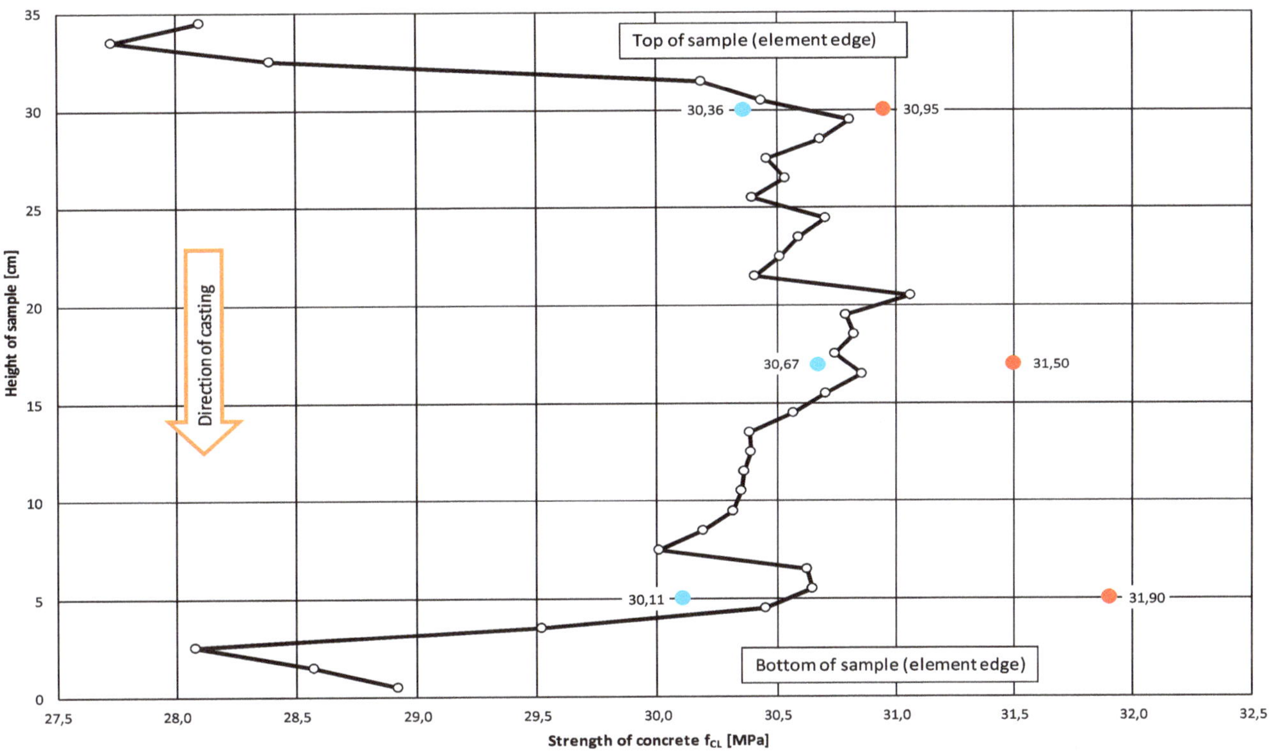

Figure 10. The mean distribution of compressive strength f_{CL} of concrete along the height of the sample taken perpendicularly to top surface of slab. The diagram includes the mean concrete strengths for the top, middle and bottom samples, obtained from ultrasonic tests (marked blue) and destructive tests (marked red).

An analysis of the concrete compressive strength values for the particular samples 01–06 (Figure 10) taken perpendicularly to the top plane of the slab indeed showed a slight increase (by 3%) in this strength in the sample's lower part relative to its upper part. A similar phenomenon (also an increase by approximately 2.6%) was observed for samples 07–12 (Figure 11) taken parallel with the top plane of the slab. However, it should be noted that the compressive strength values were strongly averaged for the samples and included the effect of various factors connected with the destructive test itself.

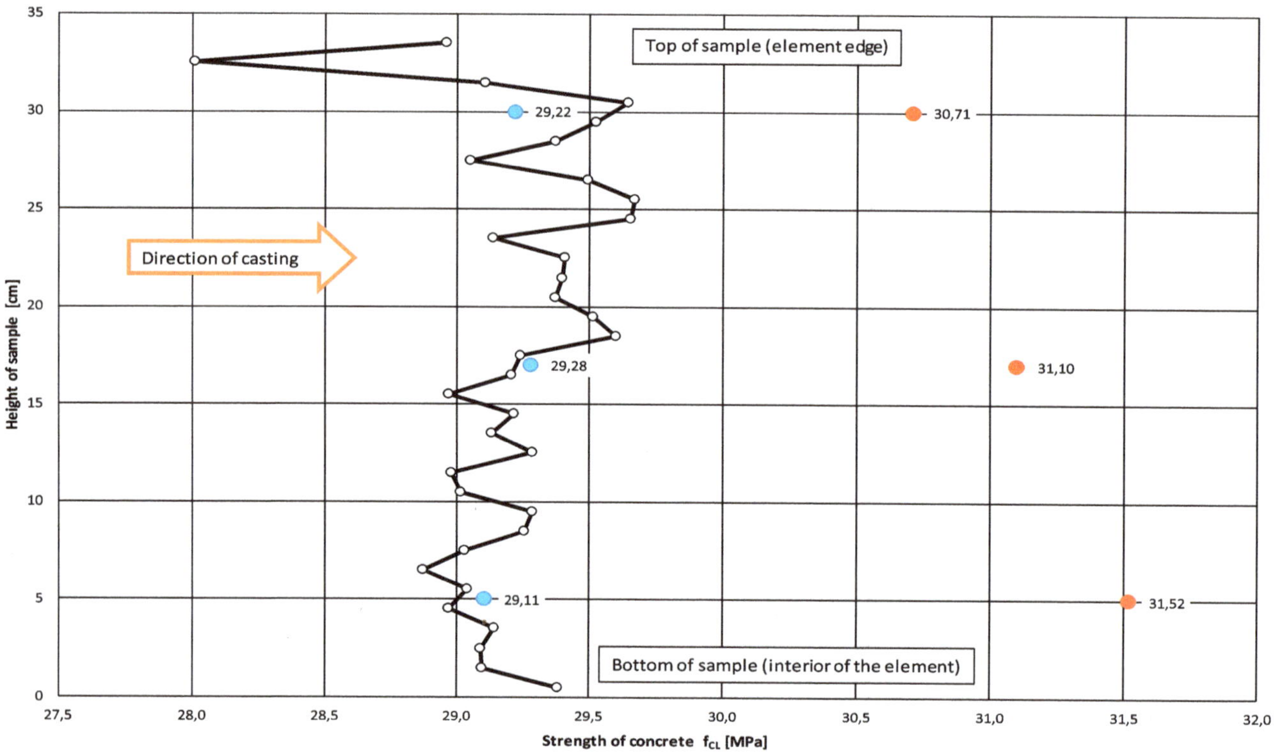

Figure 11. The mean distribution of compressive strength f_{CL} of concrete along the height of the sample taken parallel with the top surface of the slab. The diagram includes the mean concrete strengths for the top, middle and bottom samples obtained from ultrasonic tests (marked blue) and destructive tests (marked red).

The averaged compressive strength results obtained from the ultrasonic measurements showed (Figures 10 and 11), however, that there was no increase in the compressive strength of the concrete along the height of the sample. This applies to the samples taken both perpendicularly to and parallel with the top plane of the slab.

The ultrasonic tests indicate that the variation in the compressive strength of concrete along the height of the sample is minimal and random. It can even be considered as negligible. The obtained results do not corroborate Stawiski's theses [1–3], but confirm the results reported in [7–10].

The decreases in the compressive strength of the concrete occurring at the ends of the samples taken perpendicularly to the top plane of the slab (samples 01–06) and at the edge constituting the side edge of the slab for the samples taken parallel with the top plane of the slab (samples 07–12) were found to be interesting.

4. Conclusions

The following conclusions emerge from the investigations of the compressive strength of concrete carried out on core samples taken perpendicularly to and parallel with the top surface of the approximately 35 cm thick element, using different testing methods (the ultrasonic method and the destructive method):

The concrete compressive strength destructively determined along the height of the placed layer of concrete changed slightly (by 3%—samples 01–06) with a depth below the top surface of the element. The averaged concrete strength determined on the basis of the ultrasonic tests of the same samples did not vary across the thickness of the analyzed slab.

1. The obtained compressive strength increments across the thickness of the placed layer of concrete do not corroborate Stawiski's theses [1–3], but confirm the results reported in, [7–9]. Therefore, there can be agreement with Neville's statement [10] that the slight increase in concrete compressive strength with depth below the top surface is a natural thing and need not to be taken into account in the evaluation of the strength of concrete in the structure.
2. The concrete compressive strength determined on core samples only slightly depends on the depth of where the sample came from (provided the ingredients of the concrete do not segregate as it is being placed and compacted).
3. The use of the ultrasonic method for testing concrete with point-contact exponential probes showed the variation in concrete strength along the height of the core sample could be quite accurately evaluated and areas of lower quality concrete could be indicated. This was mainly from the thick layer (approximately 30–40 mm) extending from the top edge and the thick layer (approximately 30–50 mm) extending from the bottom edge of the samples 01–06 taken perpendicularly to the upper plane of the element. Further, from the thick layer (approximately 20–80 mm) extending from the edge constituting the side plane of the slab for the samples taken parallel with the top plane of the element (samples 07–12). The decline in the strength of the concrete in the upper part of samples 01–06 is caused by the bleeding of water from the concrete mixture (the bleeding phenomenon [17]) and the sedimentation of the latter. While in the lower part of the samples, it is due to the improper vibration of the placed layer of concrete mixture. The decrease in concrete strength at the side edge of the slab in the case of samples 07–12 can be due to the wall effect [9,10].
4. From the point of view of the assessment of the concrete structure, supplementary tests on the slab in the future should be carried out using ultrasonic tomography [20,21].
5. The ultrasonic method of testing concrete by means of point-contact exponential probes enables the accurate assessment of the quality of concrete (the segregation of concrete components, porosity, density, strength, etc.) along the height of a core sample drilled out perpendicularly to the placed layer.

References

1. Stawiski, B. *Ultrasonic Testing of Concrete and Mortar Using Point Probes*; Wroclaw University of Technology: Wroclaw, Poland, 2009; p. 154. (In Polish)
2. Stawiski, B. The heterogeneity of mechanical properties of concrete in formed constructions horizontally. *Arch. Civ. Mech. Eng.* **2012**, *12*, 90–94. [CrossRef]
3. Stawiski, B. Concrete strength gradients in industrial floors. *Mater. Bud.* **2017**, *543*, 22–24. (In Polish)
4. Hoła, J.; Sadowski, L.; Hoła, A.M. The effect of failure to comply with technological and technical requirements on the condition of newly built cement mortar floors. *Proc. Inst. Mech. Eng. Part L J. Mater. Des. Appl.* **2019**, *233*, 268–275. [CrossRef]
5. Petersons, N. Should standard cube test specimens be replaced by test specimens taken from structures? *Mater. Struct.* **1968**, *1*, 425–435. [CrossRef]
6. Dąbrowski, K.; Stachurski, W.; Zieliński, J.L. *Concrete Constructions*; Wydawnictwo Arkady: Warsaw, Poland, 1982. (In Polish)
7. Yuan, R.L.; Ragab, M.; Hill, R.E.; Cook, J.E. Evaluation of core strength in high-strength concrete. *Concr. Int.* **1991**, *13*, 30–34.
8. Suprenant, B.A. *Core Strength Variation of in-Place Concrete*; The Aberdeen Group: Boston, MA, USA, 1995.
9. Neville, A. Core tests: Easy to perform, not easy to interpret. *Concr. Int.* **2001**, *11*, 59–68.
10. Watanabe, S.; Hishikawa, K.; Kamae, K.; Namiki, S. Study on estimation of compressive strength of concrete in structure using ultrasonic method. *J. Struct. Constr. Eng.* **2016**, *81*, 191–198. [CrossRef]
11. Neville, A.M. *Properties of Concrete*; Polski Cement Sp. z o.o.: Cracow, Poland, 2000. (In Polish)
12. PN-EN 12390-2:2011. *Testing Hardened Concrete. Part 2. Making and Curing Specimens for Strength Tests*; Polish Committee for Standardization: Warsaw, Poland, 2011.

13. PN-EN 13791:2008. *Assessment of In-Situ Compressive Strength in Structures and Precast Concrete Components*; Polish Committee for Standardization: Warsaw, Poland, 2008.
14. PN-EN 12390-3:2011. *Testing Hardened Concrete. Part 3: Compressive Strength of Test Specimens*; Polish Committee for Standardization: Warsaw, Poland, 2008.
15. Gudra, T.; Stawiski, B. Non-destructive strength characterization of concrete using surface waves. *NDT E Int.* **2000**, *33*, 1–6. [CrossRef]
16. Drobiec, L.; Jasiński, R.; Piekarczyk, A. *Diagnosis of Reinforced Concrete Structures. Methodology, Field Tests, Laboratory Tests of Concrete and Steel*; Wydawnictwo Naukowe PWN: Warsaw, Poland, 2010. (In Polish)
17. Soshiroda, T. Effects of bleeding and segregation on the internal structure of hardened concrete. In *Properties of Fresh Concrete*; Taylor & Francis: Mild Park, UK, 1990; pp. 225–232.
18. PN-EN 12390-4:2001. *Testing Hardened Concrete. Part 4: Compressive Strength. Specification for Testing Machines*; Polish Committee for Standardization: Warsaw, Poland, 2001.
19. PN-EN 12504-1:2011. *Testing Concrete in Structures. Part 1: Cored Specimens. Taking, Examining and Testing in Compression*; Polish Committee for Standardization: Warsaw, Poland, 2011.
20. Schabowicz, K. Ultrasonic tomography—The latest nondestructive technique for testing concrete members—Description, test methodology, application example. *Arch. Civ. Mech. Eng.* **2014**, *14*, 295–303. [CrossRef]
21. Schabowicz, K.; Suvorov, V.A. Nondestructive testing of a bottom surface and construction of its profile by ultrasonic tomography. *Russ. J. Nondestruct. Test.* **2014**, *50*, 109–119. [CrossRef]

11

Enhanced Singular Value Truncation Method for Non-Destructive Evaluation of Structural Damage using Natural Frequencies

Qiuwei Yang [1], Chaojun Wang [1], Na Li [1], Wei Wang [1,*] and Yong Liu [2]

[1] School of Civil Engineering, Shaoxing University, Shaoxing, Zhejiang 312000, China; yangqiuwei79@gmail.com (Q.Y.); qtwcj123@163.com (C.W.); lina@usx.edu.cn (N.L.)

[2] State Key Lab Water Resources & Hydropower Engineering School, Wuhan University, Wuhan 430072, China; liuy203@whu.edu.cn

* Correspondence: wellswang@usx.edu.cn

Abstract: As natural frequencies can be easily and accurately measured, structural damage evaluation by frequency changes is very common in engineering practice. However, this type of method is often limited by data, such as when the available natural frequencies are very few or contaminated. Although much progress has been made in frequency-based methods, there is still much room for improvement in calculation accuracy and efficiency. To this end, an enhanced singular value truncation method is proposed in this paper to evaluate structural damage more effectively by using a few lower order natural frequencies. The main innovations of the enhanced singular value truncation method lie in two aspects: The first is the normalization of linear systems of equations; the second is the multiple computations based on feedback evaluation. The proposed method is very concise in theory and simple to implement. Two numerical examples and an experimental example are employed to verify the proposed method. In the numerical examples, it was found that the proposed method can successively obtain more accurate damage evaluation results compared with the traditional singular value truncation method. In the experimental example, it was shown that the proposed method possesses more precise and fewer calculations compared with the existing optimization algorithms.

Keywords: non-destructive evaluation; structural damage; natural frequency; singular value truncation; multiple feedbacks; data noise

1. Introduction

Structural damage often leads to changes in the dynamic response parameters of a structure. By testing the vibration parameters and observing their changes, structural damages can be monitored in a timely manner to avoid disastrous consequences. In recent decades, structural damage evaluation has become a key issue in the field of civil engineering, mechanical engineering, aerospace engineering and so on. The method based on natural frequency changes [1–11] is one of the mainstream methods for structural damage evaluation, since the natural frequencies are most easily and accurately measured in comparison with other dynamic characteristics of a structure. Messina et al. [3] proposed a damage detection method termed the multiple damage location assurance criterion by using the natural frequency sensitivity analysis. Yu et al. [4] made use of natural frequency perturbation theory and artificial neural network to detect small structural damage. Yang and Liu [5] proposed a frequency-based method with added masses to identify damages of the symmetrical structures. Khiem and Toan [6] proposed a method to calculate the natural frequencies of a multiple-cracked beam and detect an unknown number of multiple cracks from the measured natural frequencies. Ding et al. [7] presented an improved artificial bee colony algorithm for crack identification in beam structure. Krishnanunni et al. [8] defined an objective function using the frequency sensitivity equation and minimized it using a cuckoo search algorithm to evaluate structural damage. Choi and Han [9] studied frequency-based damage detection in cantilever beam by using a vision-based monitoring system with a motion magnification technique. Pan et al. [10] proposed a novel concept of noise response rate (NRR) to evaluate the sensitivity of each mode of the frequency shift to noise. It was shown that selecting vibration modes with low NRR values improves the prediction accuracy of frequency-based damage detection. Ercolani et al. [11] studied the inverse method of damage detection from the measurement of the first three natural frequencies of vibration on two experimental beams.

Although much progress has been made in frequency-based methods, there is still much room for improvement in the calculation accuracy and efficiency since the available natural frequencies are very few and contaminated. For the damage identification problem, the damaged elements in the structure are often only a small minority because the actual damage usually occurs only in a few local areas. This particularity of damage identification has not been fully utilized in the previous frequency-based methods. In this paper, an enhanced singular value truncation (ESVT) method is proposed for structural damage evaluation by using only a few natural frequencies. Central to the proposed method is the normalization of linear systems of equations and the multiple computations based on feedback evaluation. The above particularity of damage detection is fully utilized in the proposed procedure by removing many undamaged elements in each computation according to the feedback evaluation. This operation can significantly reduce the computational complexity and obtain more accurate damage evaluation results. The presentation of this work is organized as follows. In Section 2, the natural frequency sensitivity theory is brief reviewed and then an enhanced singular value truncation method is proposed for structural damage evaluation. Two numerical examples and an experimental example are used to demonstrate the feasibility and superiority of the developed method in Sections 3 and 4, respectively. From the numerical results, it was found that the proposed method can successively obtain more accurate damage evaluation results compared with the traditional singular value truncation method. From the experimental results, it was shown that the proposed method possesses more precise and fewer calculations compared with the existing optimization algorithms. The conclusions of this work are summarized in Section 5.

2. Theoretical Development

2.1. Natural Frequency Sensitivity for Damage Detection

As is well known, the low-order natural frequencies of structural vibration can be easily and accurately measured in engineering practice. Thus the natural frequency is the most commonly used parameter in structural model updating or damage detection. In this section, the basis for the natural frequency sensitivity technique [1–5] is briefly reviewed. According to the vibration theory, the modes of structural free vibration can be obtained theoretically by solving the following generalized eigenvalue problem:

$$K\phi_j = \lambda_j M\phi_j \tag{1}$$

where M and K are the mass and stiffness matrices of the structure, and λ_j and ϕ_j are the jth eigenvalue and eigenvector, respectively. Note that the eigenvalue λ_j can be obtained from the corresponding natural frequency f_j by

$$\lambda_j = (2\pi \cdot f_j)^2 \tag{2}$$

Generally, the mass matrix $M$is assumed constant in model updating or damage detection. Then the first-order sensitivity of the jth eigenvalue λ_j can be computed by

$$\frac{\partial \lambda_j}{\partial x_i} = \phi_j^T K_i \phi_j \tag{3}$$

where x_i and K_i are the ith elemental stiffness perturbed parameter (also called as damage parameter) and stiffness matrix, respectively. The goal of model updating or damage detection is to obtain the values of these stiffness perturbed parameters by the changes between the measured eigenvalues and the theoretical eigenvalues. Assuming λ_j^* is the jth measured eigenvalue, the eigenvalue change $\Delta\lambda_j$ can be calculated as

$$\Delta\lambda_j = \lambda_j^* - \lambda_j \tag{4}$$

On the other hand, the eigenvalue change $\Delta\lambda_j$ can be approximated using Taylor's series expansion and linear superposition principle as

$$\Delta\lambda_j = \sum_{i=1}^{N} x_i \frac{\partial\lambda_j}{\partial x_i} \tag{5}$$

where N is the number of total elements in structural finite element model (FEM). For m measured eigenvalues, the first-order sensitivity equation of natural frequencies can be obtained as

$$A \cdot x = b \tag{6}$$

$$A = \begin{bmatrix} \frac{\partial\lambda_1}{\partial x_1} & \cdots & \frac{\partial\lambda_1}{\partial x_N} \\ \vdots & \ddots & \vdots \\ \frac{\partial\lambda_m}{\partial x_1} & \cdots & \frac{\partial\lambda_m}{\partial x_N} \end{bmatrix} \tag{7}$$

$$x = \begin{Bmatrix} x_1 \\ \vdots \\ x_N \end{Bmatrix} \tag{8}$$

$$b = \begin{Bmatrix} \Delta\lambda_1 \\ \vdots \\ \Delta\lambda_m \end{Bmatrix} \tag{9}$$

By solving the linear Equation (6), the unknown stiffness perturbed parameters α_i can be obtained, which will be used for model updating or damage evaluation. For example, the generalized inverse [12–15] is used in many cases to compute x in Equation (6), that is

$$x = A^+ b \tag{10}$$

where the superscript "+" denotes the Moore–Penrose generalized inverse [16].

2.2. Enhanced Singular Value Truncation Method

In engineering practice, only a few lower order natural frequencies with noise can be obtained through structural vibration testing [17–20]. Thus the results obtained by Equation (10) are often very unstable and inaccurate. This leads to the failure of model updating and damage detection. Therefore it is very necessary to develop a new computational method to compute the stiffness perturbed parameters more reliably. Traditionally, the singular value truncation (SVT) method [21–28] can be used to replace the generalized inverse to solve Equation (6) more effectively. However, as will be shown in the next example, the results obtained by the common SVT are still undesirable for many cases. In view of this, an ESVT method is proposed in this section to obtain more accurate x for structural damage evaluation. The proposed ESVT method is very concise in theory and very easy in calculation. The innovations of the ESVT method lie in two aspects: (1) Normalization of linear systems of equations; (2) multiple computations based on feedback evaluation. The ESVT method is illustrated in detail as follows.

Using the similar idea of the total least squares method [29–33], the linear systems of Equation (6) can be normalized by the division operation as

$$A^* x = 1_v \tag{11}$$

$$1_v = \begin{Bmatrix} 1 \\ \vdots \\ 1 \end{Bmatrix} \tag{12}$$

$$A^* = \begin{bmatrix} \frac{a_{11}}{b_1} & \cdots & \frac{a_{1N}}{b_1} \\ \vdots & \ddots & \vdots \\ \frac{a_{m1}}{b_m} & \cdots & \frac{a_{mN}}{b_m} \end{bmatrix} \tag{13}$$

where a_{ij} denotes the (i, j)th coefficient of A and b_i denotes the ith coefficient of b in Equation (6). The advantage of this normalization is that all errors, including measurement errors and model errors, are placed in the new coefficient matrix A^*. It will be found that this normalization process can improve the accuracy and robustness of the solution for the linear systems of equations. After the normalization process, Equation (11) can then be solved through the singular value truncation technique as follows. Performing the singular value decomposition on A^* in Equation (11), one has

$$U \Lambda V^T \cdot x = 1_v \tag{14}$$

$$U = [u_1, u_2, \cdots, u_n] \tag{15}$$

$$V = [v_1, v_2, \cdots, v_N] \tag{16}$$

$$\Lambda = \begin{bmatrix} Z & 0 \\ 0 & 0 \end{bmatrix}, Z = diag(\sigma_1, \sigma_2, \cdots, \sigma_t) \tag{17}$$

where U and V are the orthogonal matrices, and $\sigma_1, \sigma_2, \cdots, \sigma_t$ are the nonzero singular values of A^* with $\sigma_1 \geq \sigma_2 \geq \cdots \geq \sigma_t$. By ignoring some smaller singular values, the singular value truncation solution of x for the first time can be obtained from Equation (14) as

$$x = (\sum_{y=1}^{s} \sigma_y^{-1} v_y u_y^T) \cdot 1_v \tag{18}$$

where s is the number of remained singular values, $s \leq t$. The suitable value of s is determined by the L-curve method [34–36]. The main steps of the L-curve method are as follows: (1) Compute all possible solutions of x by Equation (18) when s is taken from 1 to t. (2) For each solution of x, calculate the 2-norm of x and $Ax - b$(or $A^*x - 1_v$). (3) Draw the scatter plot with $\|Ax - b\|_2$ as abscissa and $\|x\|_2$ as ordinate ($\| \cdot \|_2$ denotes the 2-norm). (4) Connect the resulting scatters with straight lines to form the L-curve. (5) Determine the suitable value of s according to the inflection point of the L-curve. The L-curve method will be further illustrated in Section 3.1.

For the damage evaluation problem, the perturbed elements in the FEM due to damages are often only a small minority. This particularity of damage detection problem has not been fully utilized in the published frequency-based algorithms. This particularity results in the existence of a large number of coefficients close to zero in the x obtained by Equation (18). Thus these coefficients close to zero in x should be seen as a product of data noise and set to zeros to simplify the equation (11) for the next recalculation. Generally, those values in x that satisfy $\frac{x_i}{\max(x)} \leq 0.05$ should be deemed to correspond to those undamaged elements in the structure. Then Equation (11) can be further simplified for the recalculation by removing some column vectors in A^* and coefficients in x corresponding to those undamaged elements. That is

$$A_2^* \cdot x' = 1_v \tag{19}$$

where A_2^* is the remained matrix of A^* after removing some column vectors related to those undamaged elements, x_2 is the remained vector of x after removing the corresponding coefficients. From Equation (19), the solution of x' can be obtained again using the similar singular value truncation progress between Equations (14) and (18) as

$$x' = (\sum_{y=1}^{s'} \sigma_y'^{-1} v_y' u_y'^T) \cdot 1_v \quad (20)$$

Note that the result obtained by Equation (20) is maybe still not the final solution. When x' is the same as the corresponding coefficients in x, the x' in Equation (20) is the final solution of the damage parameters. If not, the above recalculation process should be repeated and the new solution x'' of the stiffness perturbed parameters can be obtained. The above process should be repeated until the solutions of the two adjacent cases are exactly the same (for example, $x'' = x'$). At the last, structural damage evaluation can be carried out according to the final result. In the above process, it is important to note that the computational complexity of each computation in ESVT gradually decreases since the number of unknowns decreases gradually.

3. Numerical Examples

3.1. A Truss Structure

A cantilever truss structure as shown in Figure 1 was taken as the numerical example to demonstrate the effectiveness of the proposed method. The basic parameters of the structure were as follows: Young's modulus $E = 200$ GPa, density $\rho = 7.8 \times 10^3$ kg/m^3, and cross-sectional area $A = 3.14 \times 10^{-4}$ m^2. Two damage cases were studied in the example. The first one was a single damage case where element 10 has a 20% stiffness reduction. The second was a multiple damage case where elements 7 and 18 have 15% and 20% stiffness reductions, respectively. Only the first six natural frequencies (shown in Table 1) of the undamaged and damaged structures were used in the structural damage evaluation.

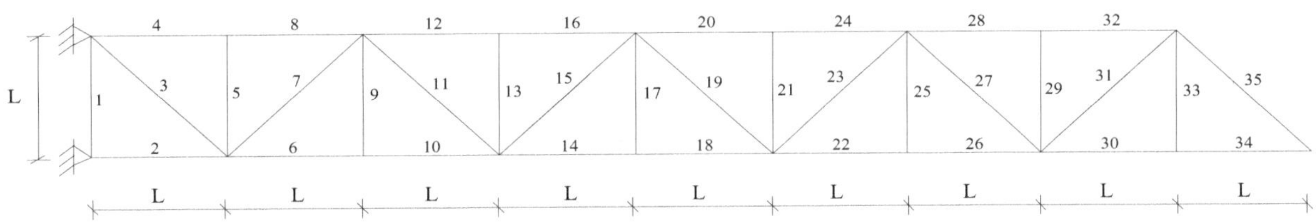

Figure 1. A cantilever truss structure (L = 0.5 m).

Table 1. The first six natural frequencies of the undamaged and damaged truss structures.

Natural Frequencies	Undamaged Structure	Damage Case 1	Damage Case 2
1	24.034	23.8588	23.9621
2	119.9908	119.3298	117.8529
3	195.9065	193.7641	194.9649
4	274.2121	273.7942	270.9942
5	436.9691	436.5461	436.3045
6	569.0272	568.812	565.0954

For each of damage cases, the evaluation results obtained by the SVT and ESVT are both given to illustrate the superiority of the ESVT method. For case 1, the SVT method was firstly employed to compute the damage parameters. As stated before, the suitable value of s in the computation process is determined by the L-curve as shown in Figure 2. Note that the scatters from right to left in Figure 2 correspond to s = 1, s = 2, etc. One can see from Figure 2 that the inflection point of the L-curve just corresponded to s = 5. Subsequently Figure 3 presents the damage evaluation result obtained by SVT method with s = 5. One can see from Figure 3 that the result was not satisfactory since element 10 cannot be uniquely determined as the damage element. Using the proposed ESVT method, Figures 4–8 give the damage evaluation results of the first to fifth calculations in ESVT. Apparently, the accuracy of damage evaluation result in Figures 4–8 was improving gradually and Figure 8 was the final result. It can be seen from Figure 8 that, after five operations, element 10 could be uniquely determined as the damage element. It was thus shown that the proposed ESVT method can achieve higher evaluation accuracy than the traditional SVT method.

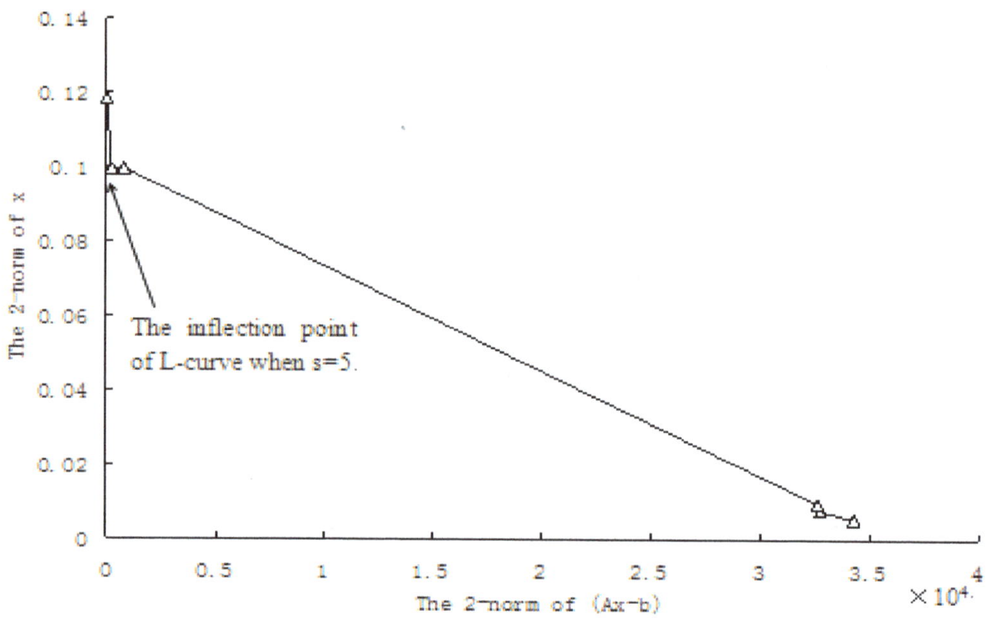

Figure 2. The L-curve used to determine the number of remained singular values in singular value truncation (SVT) for case 1.

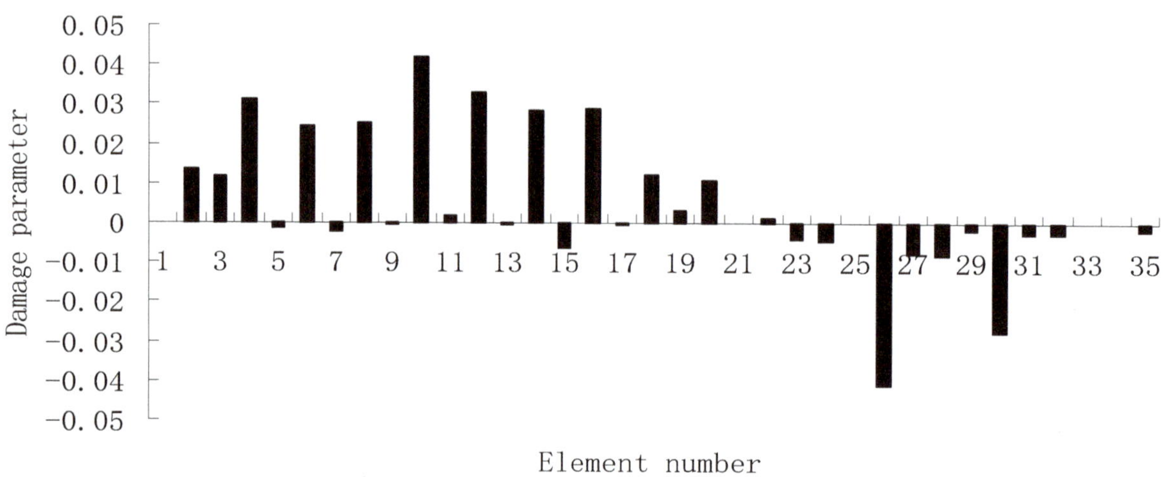

Figure 3. Damage evaluation result by the traditional SVT method for case 1 (element 10 had 20% stiffness reduction).

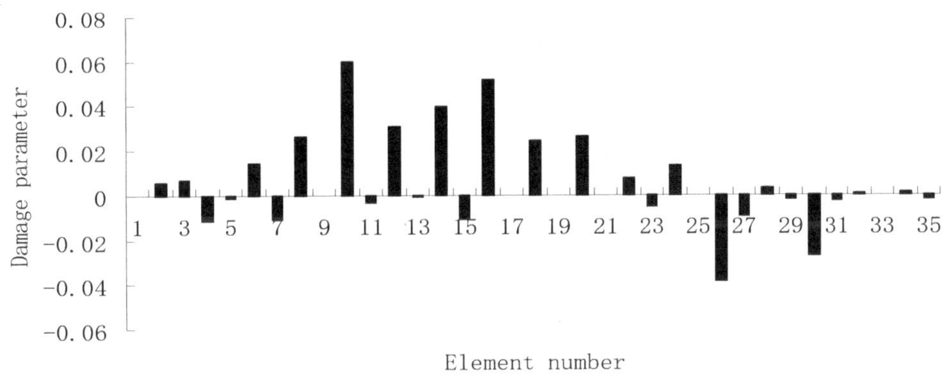

Figure 4. Damage evaluation result by the first computation of enhanced singular value truncation (ESVT) for case 1 (element 10 had 20% stiffness reduction).

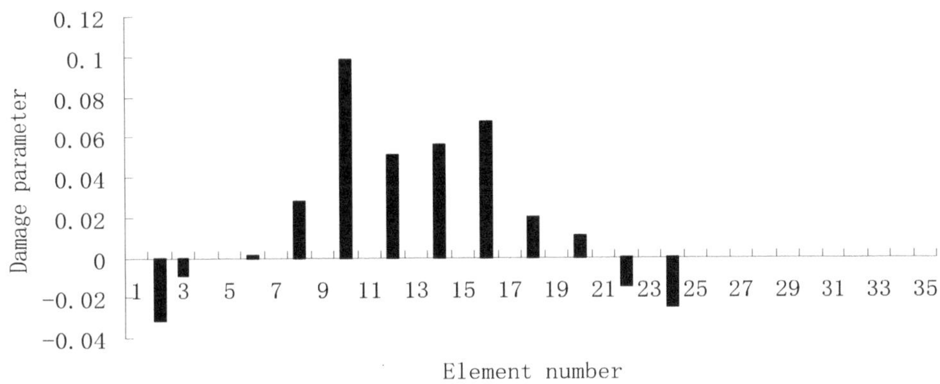

Figure 5. Damage evaluation result by the second computation of ESVT for case 1 (element 10 had 20% stiffness reduction).

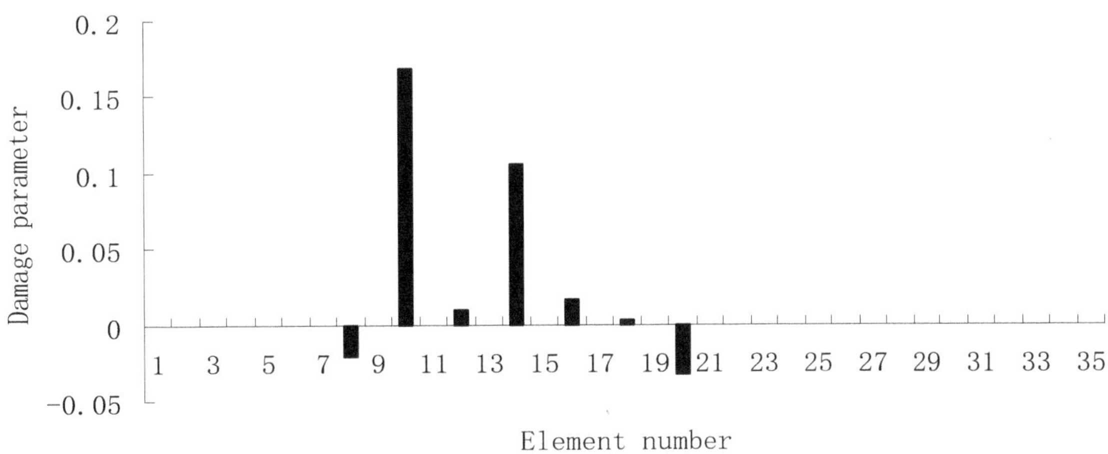

Figure 6. Damage evaluation result by the third computation of ESVT for case 1 (element 10 had 20% stiffness reduction).

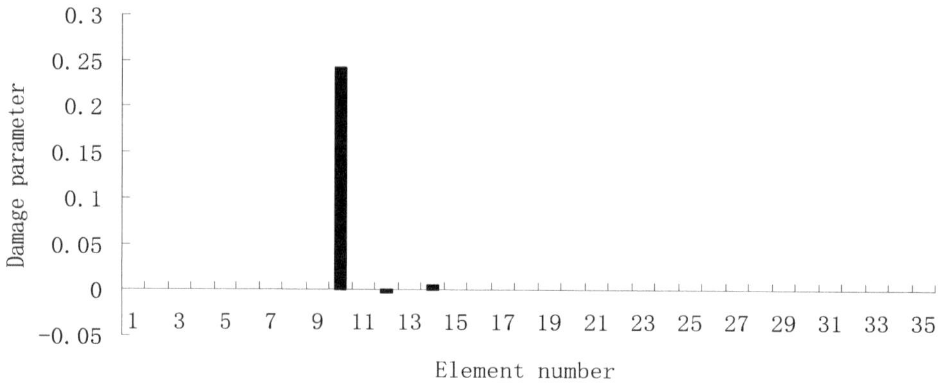

Figure 7. Damage evaluation result by the fourth computation of ESVT for case 1 (element 10 had 20% stiffness reduction).

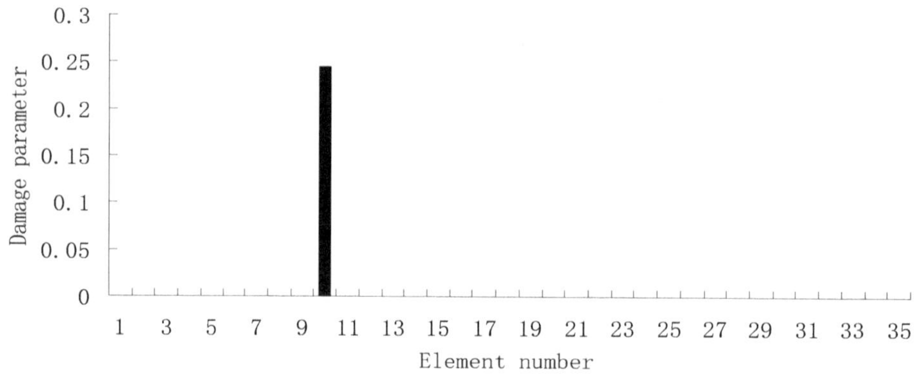

Figure 8. Damage evaluation result by the fifth computation of ESVT for case 1 (element 10 had 20% stiffness reduction).

For the second damage case, Figure 9 presents the damage evaluation result obtained by the traditional SVT method. From Figure 9, it was found that the result was not satisfactory since many elements besides 7 and 18 were determined as the damaged elements. Using the proposed ESVT method, Figures 10–14 provide the damage evaluation results of the first to fifth calculations in ESVT. It was clear that the accuracy of damage evaluation result in Figures 10–14 was improving gradually and Figure 14 was the final result for this case. The final result of Figure 14 clearly indicated that elements 7 and 18 were the true damaged elements. These results again show that the proposed ESVT method can achieve higher evaluation accuracy than the traditional SVT method.

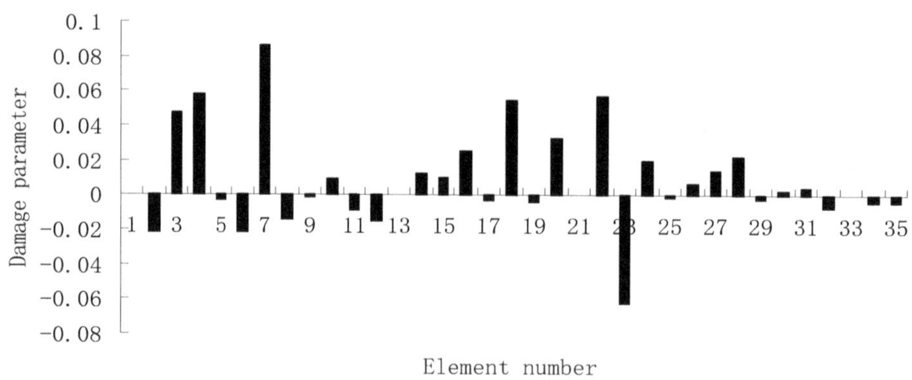

Figure 9. Damage evaluation result by the traditional SVT method for case 2 (elements 7 and 18 had 15% and 20% stiffness reductions).

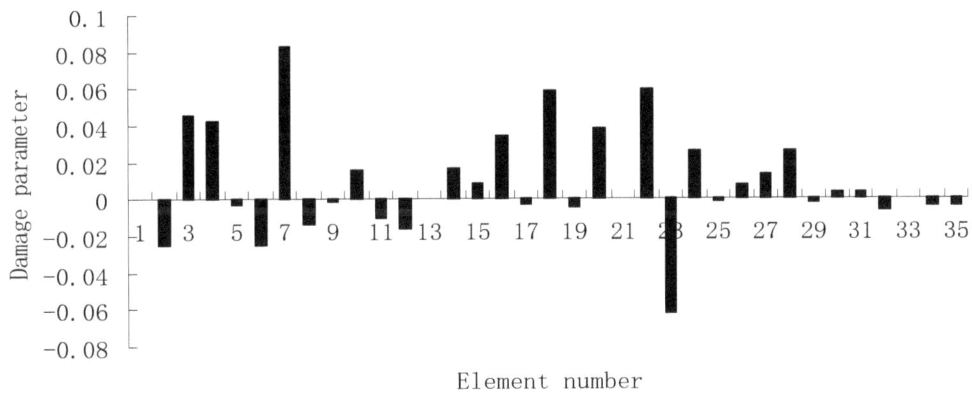

Figure 10. Damage evaluation result by the first computation of ESVT for case 2 (elements 7 and 18 had 15% and 20% stiffness reductions).

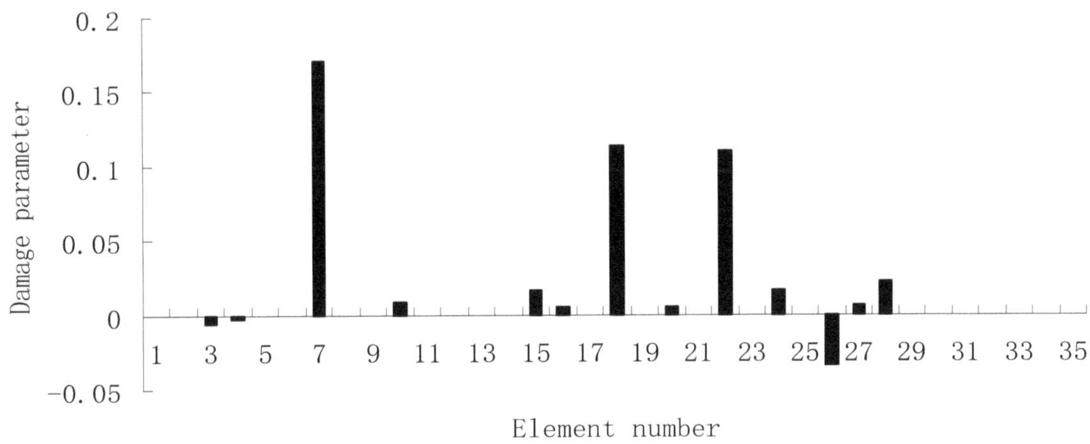

Figure 11. Damage evaluation result by the second computation of ESVT for case 2 (elements 7 and 18 had 15% and 20% stiffness reductions).

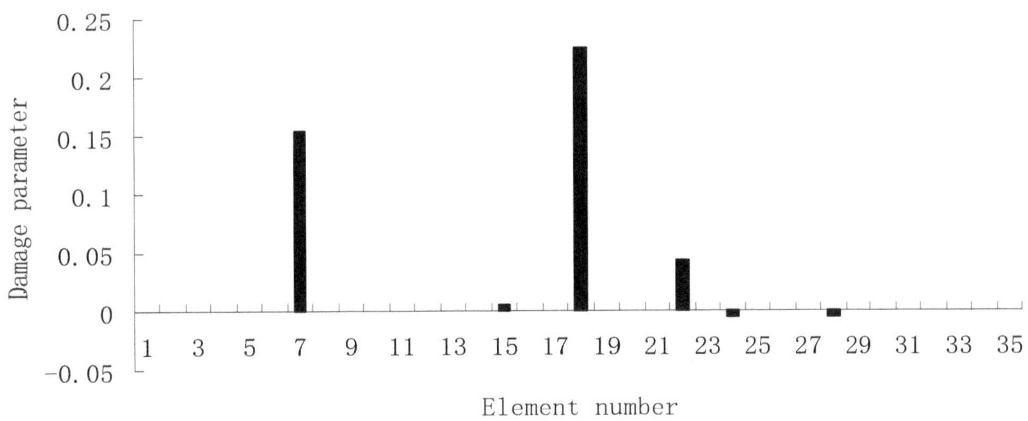

Figure 12. Damage evaluation result by the third computation of ESVT for case 2 (elements 7 and 18 had 15% and 20% stiffness reductions).

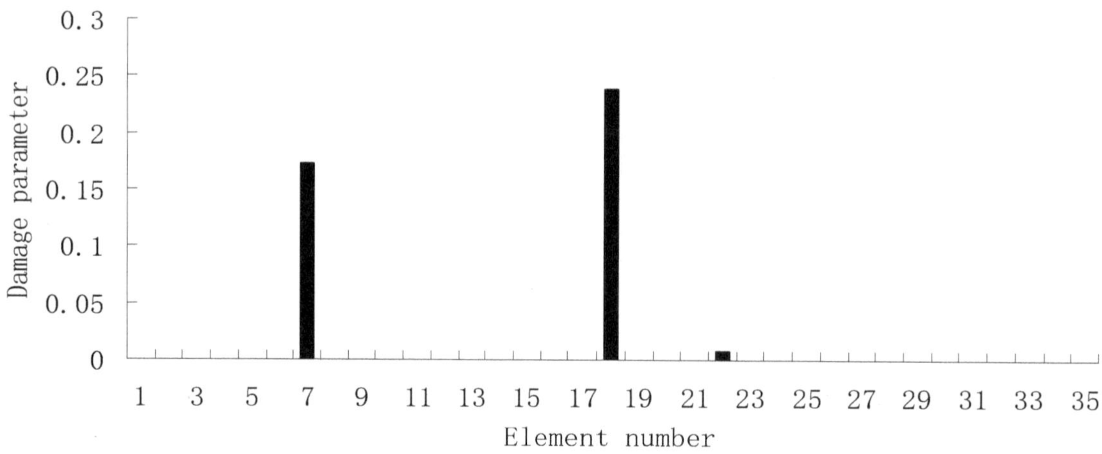

Figure 13. Damage evaluation result by the fourth computation of ESVT for case 2 (elements 7 and 18 had 15% and 20% stiffness reductions).

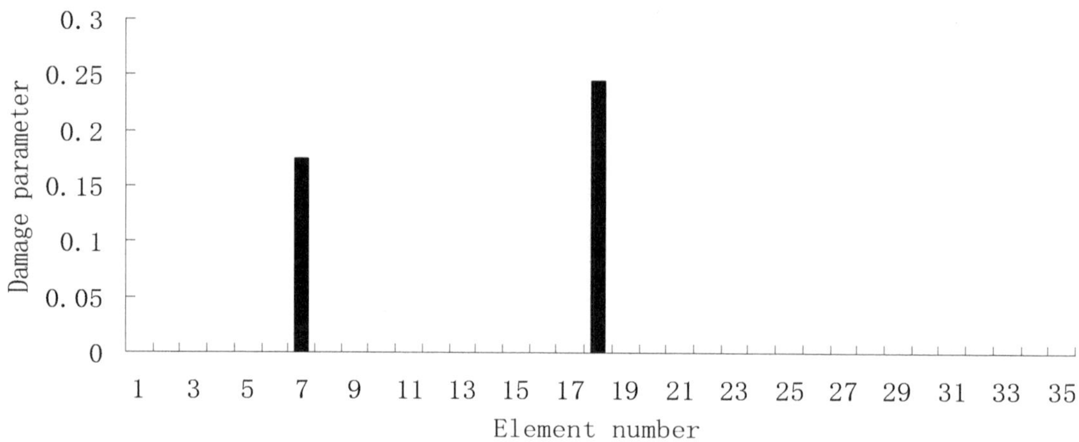

Figure 14. Damage evaluation result by the fifth computation of ESVT for case 2 (elements 7 and 18 had 15% and 20% stiffness reductions).

Next, Figures 15–17 present damage evaluation results using the first three, four and five frequencies to investigate the effect of the frequency number on the calculation results. From Figure 15, one can see that the result was not satisfactory when only three frequencies were used since element 10 cannot be uniquely determined as the damaged element in the final result of Figure 15e. From Figure 16, the result was also not satisfactory when only four frequencies were used since element 10 cannot be uniquely determined as the damage element in the final result of Figure 16f. When five frequencies were used, it can be seen from Figure 17 that the result was satisfactory since element 10 can be uniquely determined as the damaged element after six computations. It was thus shown that the results of damage evaluation become more accurate as the number of used frequencies increases. For this example, at least five frequencies were needed to obtain sufficient accurate damage evaluation results.

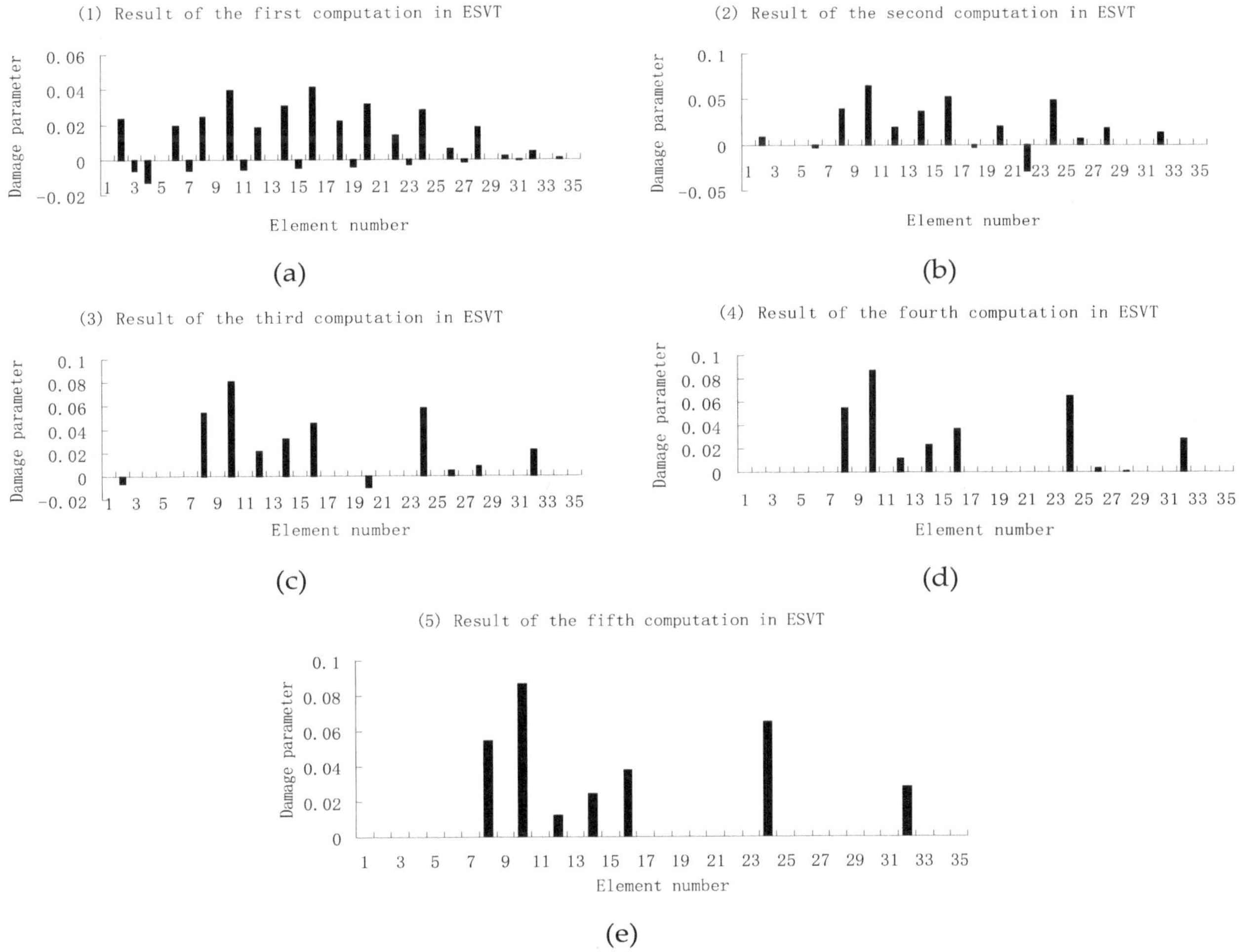

Figure 15. Damage evaluation results of the first to fifth computations in ESVT using the first three frequencies (element 10 had 20% stiffness reduction). (**a**) Result of the first computation in ESVT; (**b**) Result of the second computation in ESVT; (**c**) Result of the third computation in ESVT; (**d**) Result of the fourth computation in ESVT; (**e**) Result of the fifth computation in ESVT.

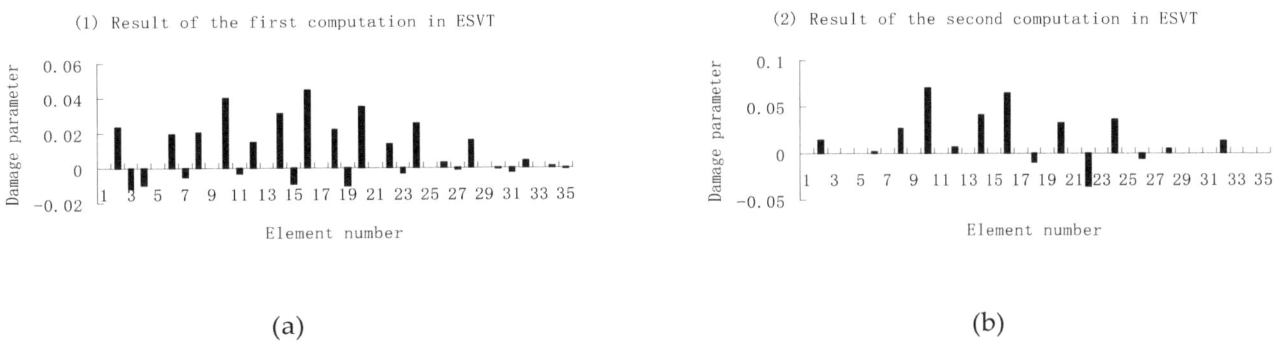

Figure 16. *Cont.*

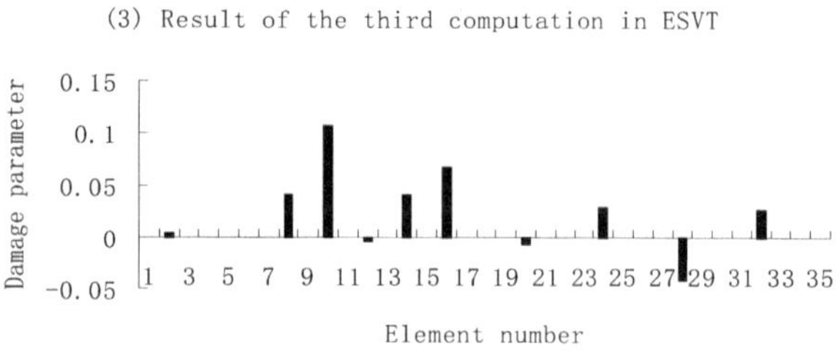

(c)

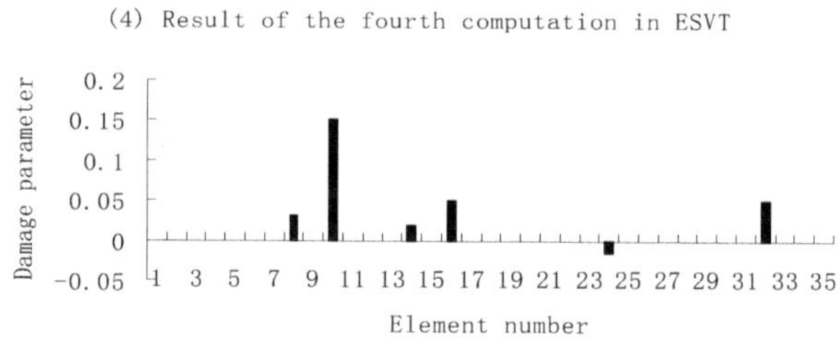

(d)

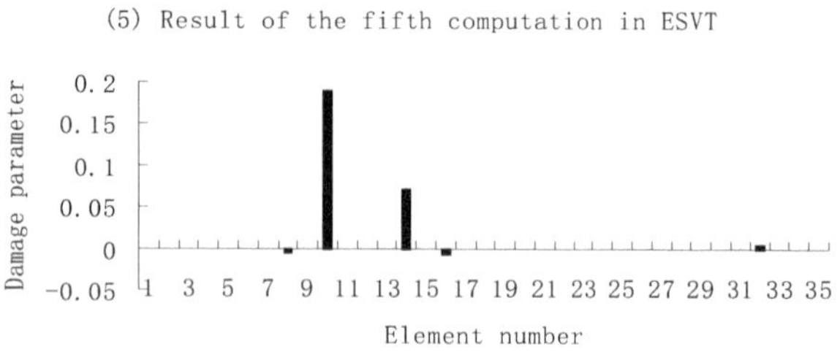

(e)

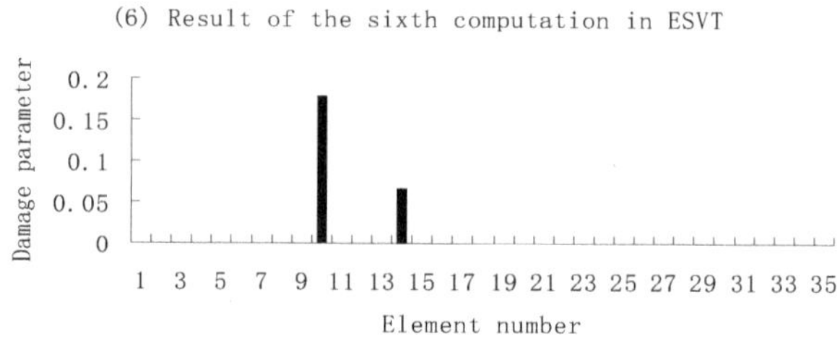

(f)

Figure 16. Damage evaluation results of the first to sixth computations in ESVT using the first four frequencies (element 10 had 20% stiffness reduction). (**a**) Result of the first computation in ESVT; (**b**) Result of the second computation in ESVT; (**c**) Result of the third computation in ESVT; (**d**) Result of the fourth computation in ESVT; (**e**) Result of the fifth computation in ESVT; (**f**) Result of the sixth computation in ESVT.

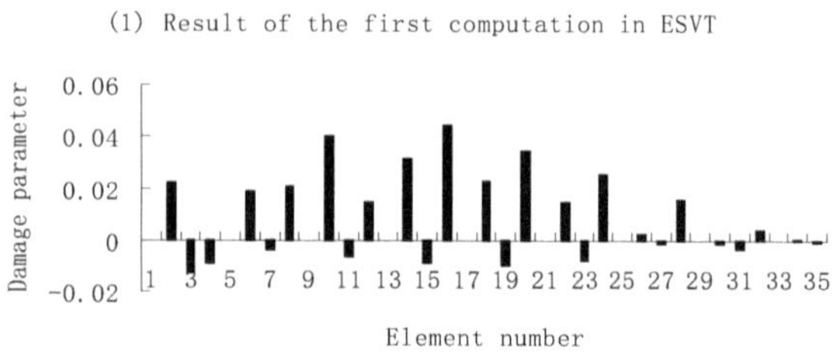

(a)

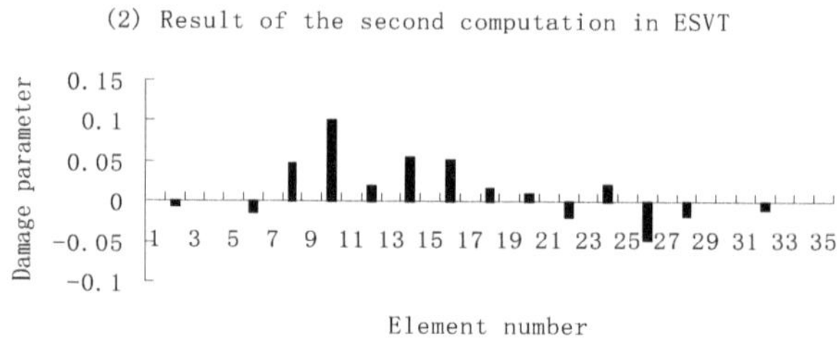

(b)

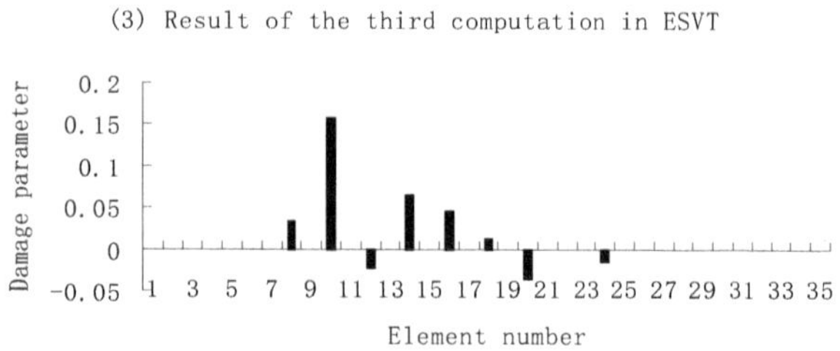

(c)

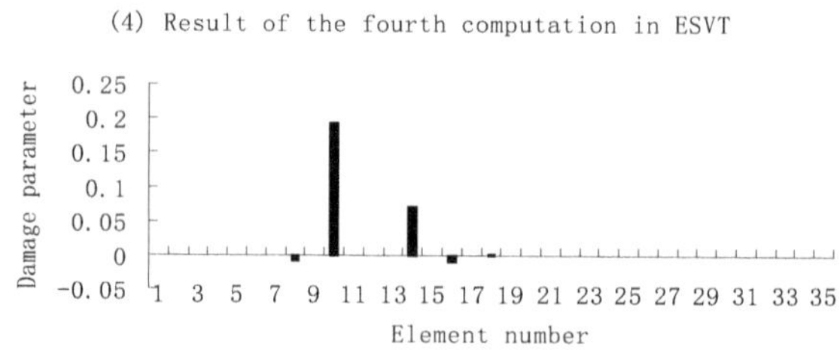

(d)

Figure 17. *Cont.*

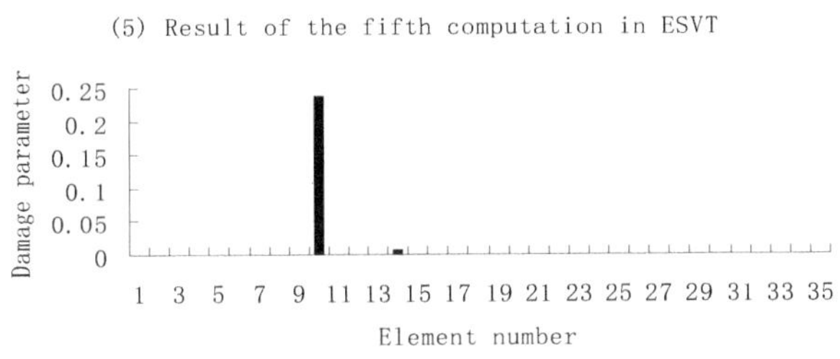

(e)

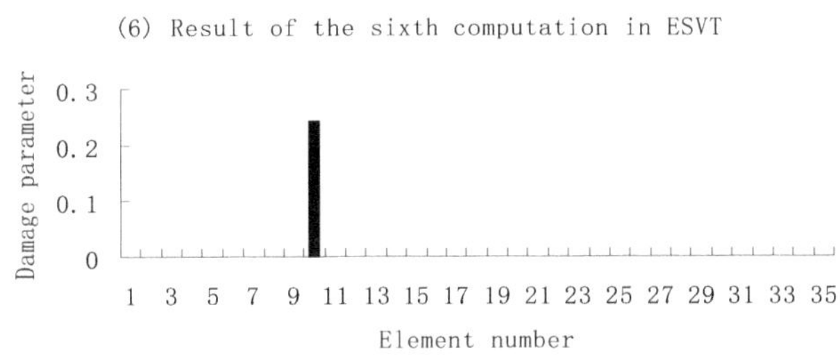

(f)

Figure 17. Damage evaluation results of the first to sixth computations in ESVT using the first five frequencies (element 10 had 20% stiffness reduction). (**a**) Result of the first computation in ESVT; (**b**) Result of the second computation in ESVT; (**c**) Result of the third computation in ESVT; (**d**) Result of the fourth computation in ESVT; (**e**) Result of the fifth computation in ESVT; (**f**) Result of the sixth computation in ESVT.

3.2. A Plate Structure

A plate structure as shown in Figure 18 was used as the second example to verify the proposed method. The main purpose of using this example was to verify the effectiveness of the proposed method for structures that require solid finite elements. The modulus of elasticity, mass density, and Poisson's ratio of this steel material were 200 GPa, 7800 kg/m^3, and 0.3, respectively. The plate was modeled using 50 identical solid elements as shown in Figure 18. In the following damage simulation, it was assumed that elements 12 and 20 had 20% and 15% stiffness reductions, respectively. Using the first eight frequencies, damage evaluation results obtained by the proposed ESVT method are shown in Figure 19. One can see from Figure 19 that the solution accuracy of the first to sixth computations was improving gradually and the sixth solution was the final result. The final result in Figure 19 clearly indicated that elements 12 and 20 were the true damaged elements. These results show that the proposed ESVT method can also be used successfully in structures that require solid finite elements.

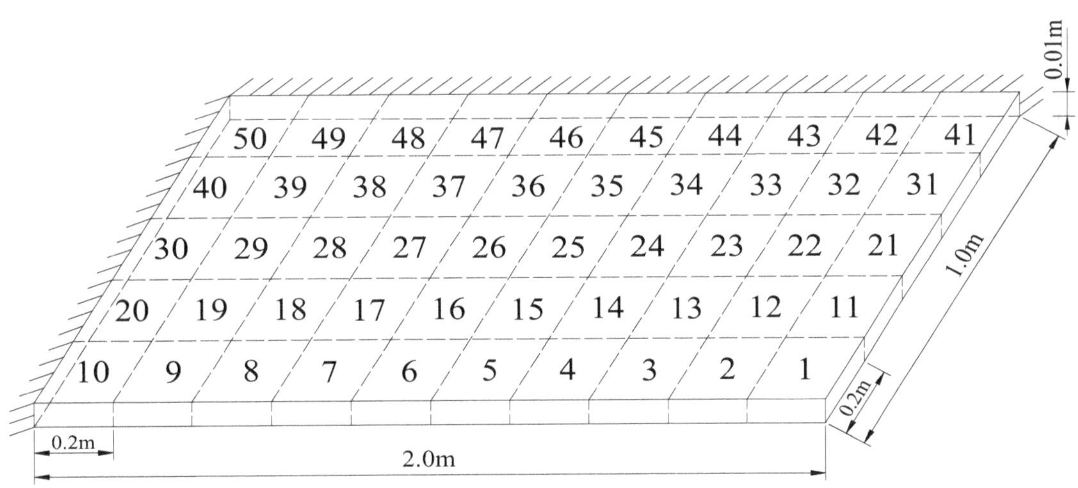

Figure 18. A plate structure.

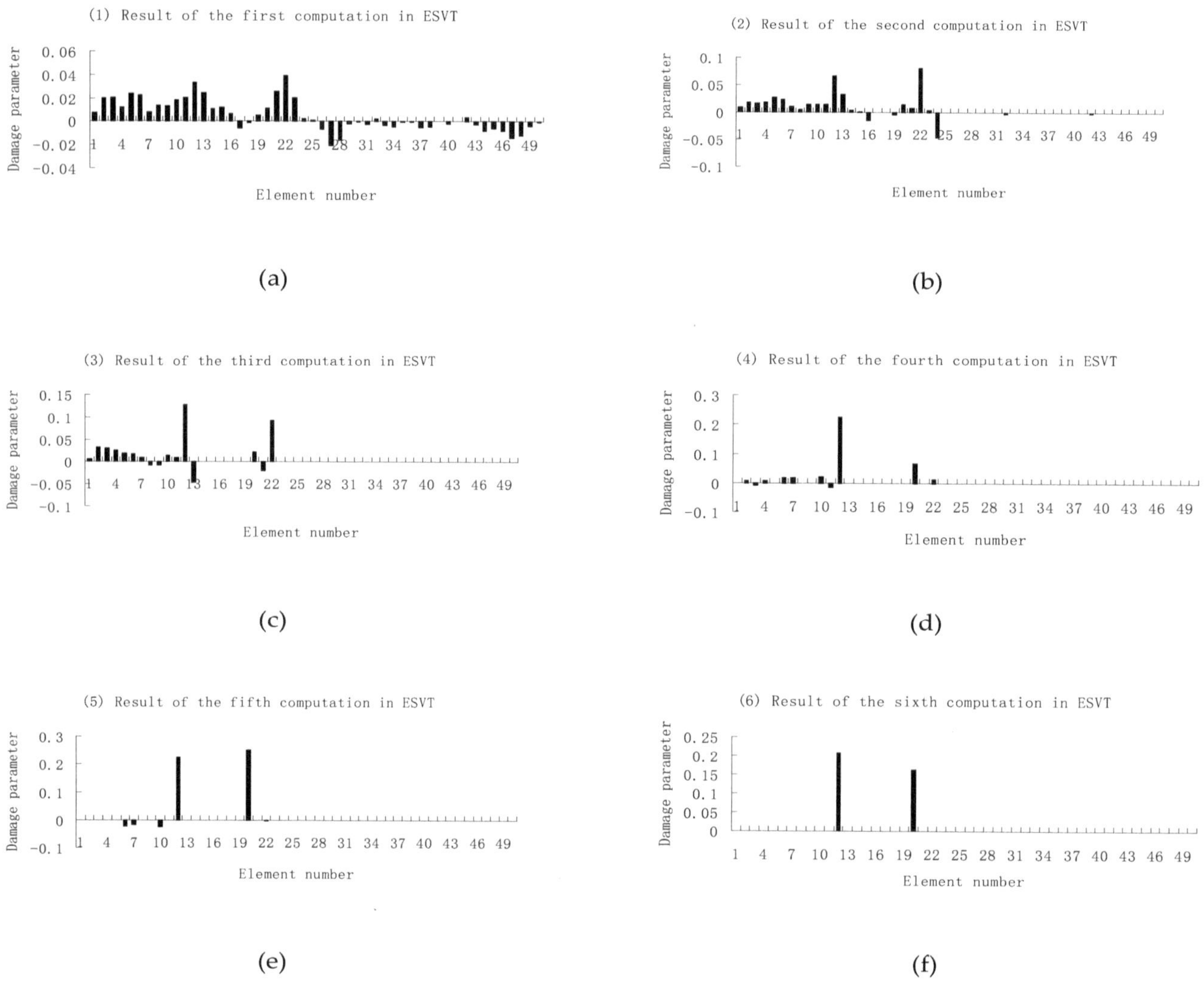

Figure 19. Damage evaluation results of the first to sixth computations in ESVT for the plate structure. (**a**) Result of the first computation in ESVT; (**b**) Result of the second computation in ESVT; (**c**) Result of the third computation in ESVT; (**d**) Result of the fourth computation in ESVT; (**e**) Result of the fifth computation in ESVT; (**f**) Result of the sixth computation in ESVT.

4. Experimental Validation

In this section, the experimental beam conducted by Yang et al. [37] was used as an example to verify the proposed method. As shown in Figure 20a, the length, width and height of the intact beam were 495.3 mm, 25.4 mm and 6.35 mm, respectively. The modulus of elasticity and mass density of this aluminium material were 71 GPa and 2210 kg/m^3, respectively. The beam was modeled using 20 equal-length elements and the damage was induced in the ninth element by a saw cut as shown in Figure 20b. The analytical and experimental values of the first six natural frequencies for the undamaged and damaged structures are all shown in Table 2.

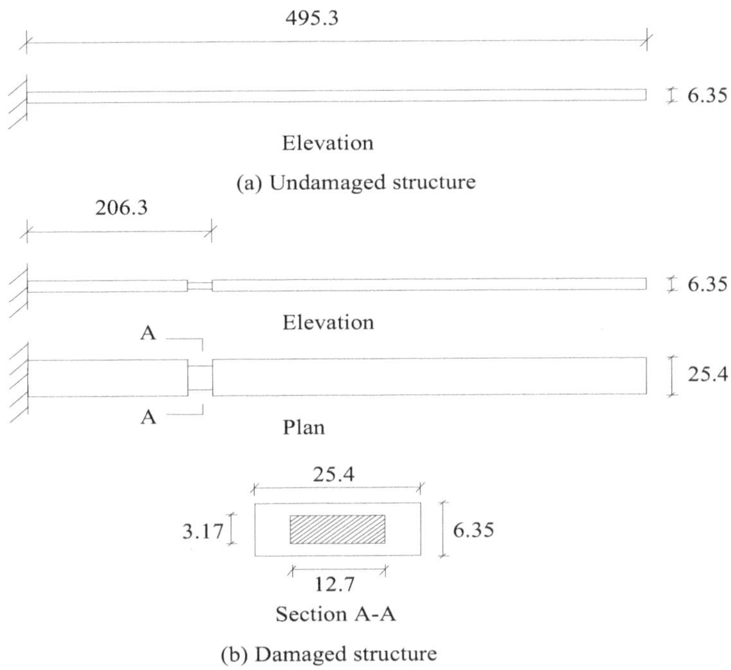

Figure 20. Configuration of the experimental beam [37]. (**a**) Undamaged structure; (**b**) Damage structure.

Table 2. The first six natural frequencies of the undamaged and damaged beams [37].

Natural Frequencies	Analytical Values (Hz)	Experimental Values (Undamaged)	Experimental Values (Damaged)
1	23.7	19.53	19.00
2	148.5	122.05	115.85
3	415.7	339.26	332.36
4	814.2	661.73	646.91
5	1345.3	1085.22	1037.46
6	2008.7	1594.59	1591.36

From columns 2 and 3 in Table 2, one can see that the differences between the analytical values obtained by FEM and the experimental values obtained by dynamic testing of the undamaged beam were very large. This means that the original FEM constructed by the software was not accurate enough to represent the undamaged beam. Thus the FEM of the undamaged beam was firstly corrected according to the natural frequency changes between the analytical values and the undamaged experimental values. Only the modified FEM could be used in the subsequent evaluation of structural damage. Note that the natural frequency sensitivity technique introduced in Section 2.1 can be used not only in damage evaluation but also in model updating. It should also be noted that the stiffness perturbed parameters of the modified FEM were computed only by one calculation process of the ESVT method in the model updating. This is the difference between the model updating problem and the damage identification problem. From the variations between column 2 and column 3 in Table 2, Figure 21 presents the stiffness perturbed parameters of the modified FEM and Table 3 gives the analytical values of the first six natural frequencies obtained by the modified FEM. From Table 3, one can see that the analytical values of the modified FEM were much closer to the undamaged experimental values than those of the original FEM. After model updating, structural damage evaluation can be subsequently carried out based on the modified FEM by using the gradual ESVT method. Figures 22–26 give the damage evaluation results of the first to fifth calculations in the ESVT. It was obvious that the accuracy of damage evaluation result in Figures 22–26 was improving gradually and Figure 26 was the final evaluation result. The final evaluation result of Figure 26 was very good since the true damage was correctly detected in element 9. For comparisons, the damage detection results reported by Krishnanunni et al. [8] and Hao et al. [38] are presented in Figure 27, obtained by Cuckoo Search algorithm (CSA) and Genetic algorithm (GA), respectively. Meanwhile, the result of Figure 26 is also shown in Figure 27 for easy comparison. From Figure 27, one can see that the damage evaluation result obtained by the proposed ESVT method had the highest accuracy among the three methods. Moreover, the computational complexity of the ESVT method was significantly lower compared to the other methods because both CSA and GA needed many iterations for good convergence. For example, the computation process using CSA reported by Krishnanunni et al. [8] was iterated 65,000 times for good convergence. Note that the proposed ESVT method only needed five calculations and the complexity of each calculation decreased gradually.

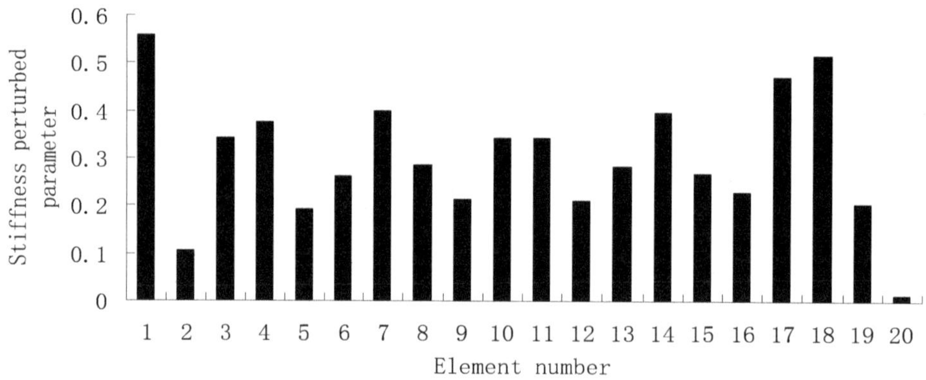

Figure 21. The stiffness perturbed parameters of the modified finite element model (FEM).

Table 3. Comparisons of natural frequencies obtained by the original FEM, the modified FEM and the experiment.

Natural Frequencies	Analytical Values of Original FEM (Hz)	Experimental Values (Undamaged Beam)	Analytical Values of Modified FEM (Hz)
1	23.7 (21.4%*)	19.53	19.0 (2.7%)
2	148.5 (21.7%)	122.05	119.8 (1.8%)
3	415.7 (22.5%)	339.26	333.7 (1.6%)
4	814.2 (23.0%)	661.73	651.2 (1.6%)
5	1345.3 (24.0%)	1085.22	1068.7 (1.5%)
6	2008.7 (26.0%)	1594.59	1582.6 (0.8%)

* The data in brackets denote the relative errors between analytical and experimental values.

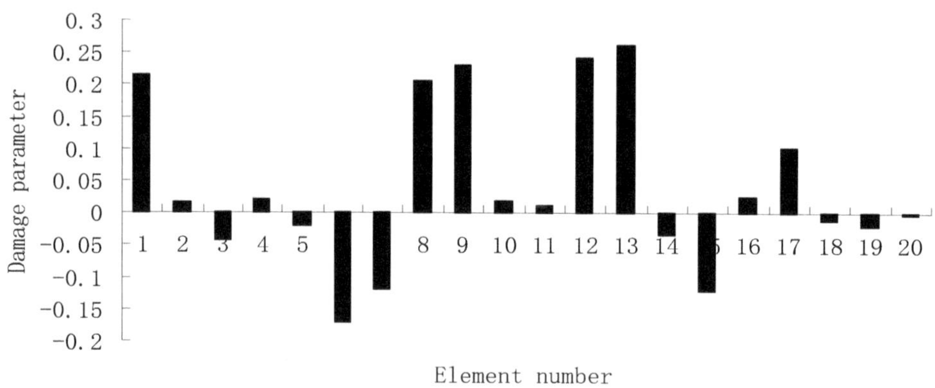

Figure 22. Damage evaluation result by the first computation of ESVT for the experimental beam.

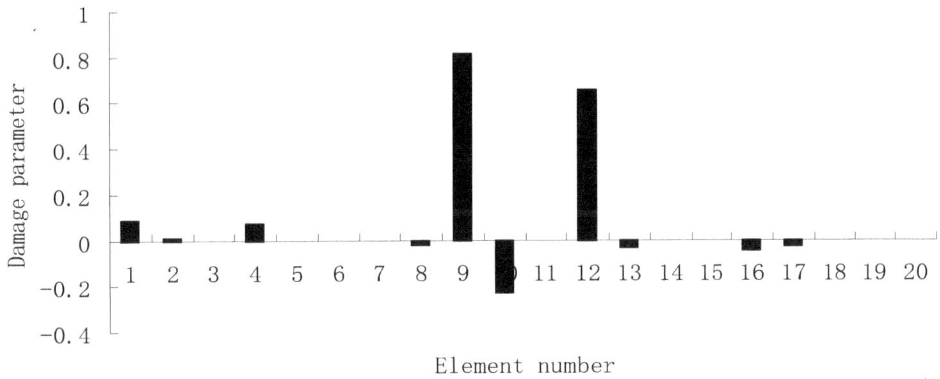

Figure 23. Damage evaluation result by the second computation of ESVT for the experimental beam.

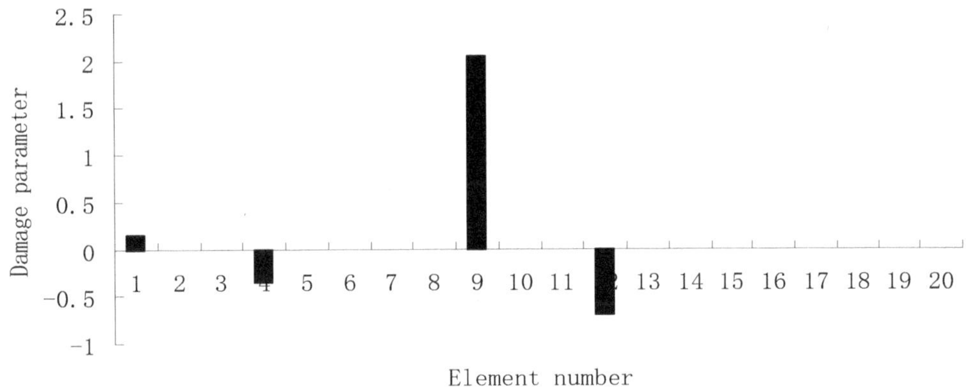

Figure 24. Damage evaluation result by the third computation of ESVT for the experimental beam.

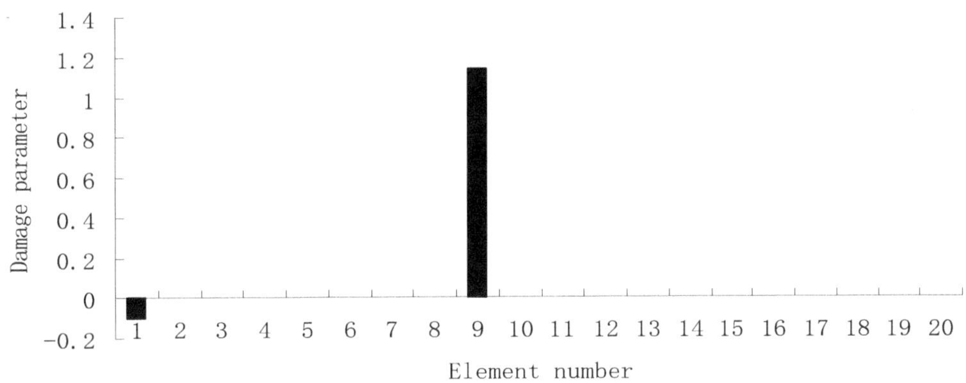

Figure 25. Damage evaluation result by the fourth computation of ESVT for the experimental beam.

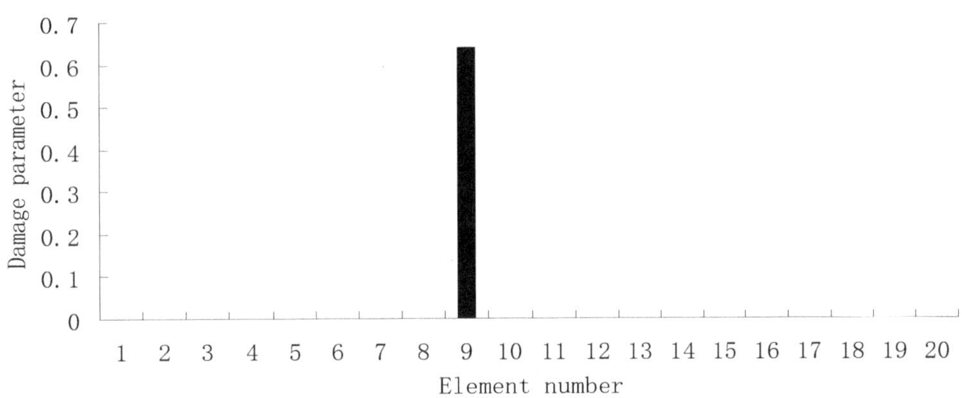

Figure 26. Damage evaluation result by the fifth computation of ESVT for the experimental beam.

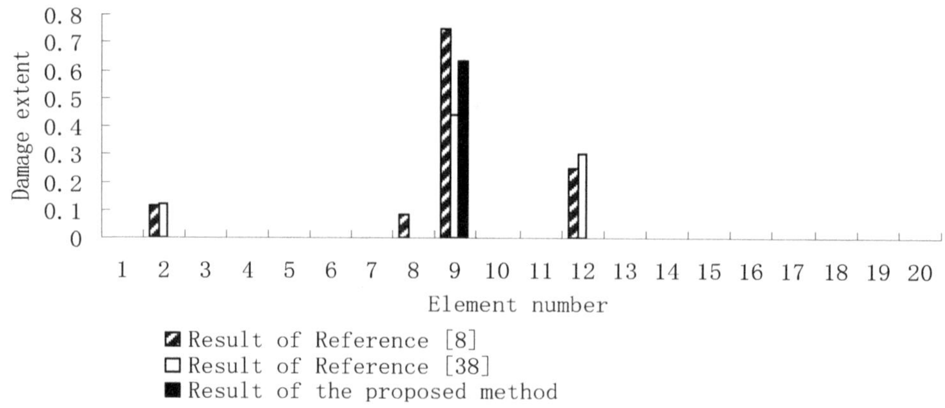

Figure 27. Comparison of damage evaluation results by the three methods.

5. Conclusions

For the damage evaluation problem, the damaged elements in the structure are often only a small minority because the actual damage usually occurs only in a few local areas. Using this particularity of damage evaluation, an ESVT method was proposed in this paper for structural damage detection using only a few lower order natural frequencies. Central to the ESVT method is the normalization of linear systems of equations and multiple computations based on feedback evaluation. In each computation of ESVT, many undamaged elements are removed according to the feedback evaluation to reduce the number of unknowns. This operation can significantly reduce the computational complexity and obtain more accurate damage evaluation results. The proposed method is very concise in theory and simple to implement. Two numerical examples and an experimental example were used to demonstrate the proposed method. From the numerical examples, it was found that the proposed method can successively obtain more accurate damage evaluation results compared with the traditional SVT method. From the experimental example, it was shown that the proposed method possesses more precise and fewer calculations compared with the existing optimization algorithms. It was shown that the proposed ESVT method may be a promising technique in non-destructive evaluation of structural damage. In practical applications, the proposed method can be applied to various types of structural damage such as reduction in elastic modulus and cracks, as long as these damages can cause observable frequency changes. Specific examples of crack detection by the proposed method will be further studied in the future.

Author Contributions: The authors confirm contribution to the paper as follow: Q.Y. and C.W. conducted the tests and analyzed the data; Q.Y. and W.W. proposed the idea and wrote the paper; N.L. revised the manuscript; Y.L. reviewed the results and approved the final version of the manuscript.

References

1. Salawu, O.S. Detection of structural damage through changes in frequency: a review. *Eng. Struct.* **1997**, *19*, 718–723. [CrossRef]
2. Yang, Q.W.; Liu, J.K. A coupled method for structural damage identification. *J. Sound Vib.* **2006**, *296*, 401–405. [CrossRef]
3. Messina, A.; Williams, J.E.; Contursi, T. Structural damage detection by a sensitivity and statistical-based method. *J. Sound Vib.* **1996**, *216*, 791–808. [CrossRef]
4. Yu, L.; Cheng, L.; Yam, L.H.; Yan, Y.J. Application of eigenvalue perturbation theory for detecting small structural damage using dynamic responses. *Compos. Struct.* **2007**, *78*, 402–409. [CrossRef]

5. Yang, Q.W.; Liu, J.K. Structural damage identification by adding given masses. *Eng. Mech.* **2009**, *26*, 159–163. (in Chinese).
6. Khiem, N.T.; Toan, L.K. A novel method for crack detection in beam-like structures by measurements of natural frequencies. *J. Sound Vib.* **2014**, *333*, 4084–4103. [CrossRef]
7. Ding, Z.; Lu, Z.; Huang, M.; Liu, J. Improved artificial bee colony algorithm for crack identification in beam using natural frequencies only. *Inverse Prob. Sci. Eng.* **2017**, *25*, 218–238. [CrossRef]
8. Krishnanunni, C.G.; Raj, R.S.; Nandan, D.; Midhun, C.K.; Sajith, A.S.; Ameen, M. Sensitivity-based damage detection algorithm for structures using vibration data. *J. Civ. Struct. Health Monit.* **2019**, *9*, 137–151. [CrossRef]
9. Choi, A.J.; Han, J.H. Frequency-based damage detection in cantilever beam using vision-based monitoring system with motion magnification technique. *J. Intell. Mater. Syst. Struct.* **2018**, *29*, 3923–3936. [CrossRef]
10. Pan, J.; Zhang, Z.; Wu, J.; Ramakrishnan, K.R.; Singh, H.K. A novel method of vibration modes selection for improving accuracy of frequency-based damage detection. *Composites Part B* **2019**, *159*, 437–446. [CrossRef]
11. Ercolani, G.D.; Felix, D.H.; Ortega, N.F. Crack detection in prestressed concrete structures by measuring their natural frequencies. *J. Civ. Struct. Health Monit.* **2018**, *8*, 661–671. [CrossRef]
12. Bicanic, N.; Chen, H.P. Damage identification in framed structures using natural frequencies. *Int. J. Numer. Methods Eng.* **1997**, *40*, 4451–4468. [CrossRef]
13. Xia, Y.; Hao, H.; Brownjohn, J.M.W.; Xia, P.Q. Damage identification of structures with uncertain frequency and mode shape data. *Earthquake Eng. Struct. Dyn.* **2002**, *31*, 1053–1066. [CrossRef]
14. Udwadia, F.E. Structural identification and damage detection from noisy modal data. *J. Aerosp. Eng.* **2005**, *18*, 179–187. [CrossRef]
15. Yang, J.; Li, P.; Yang, Y.; Xu, D. An improved EMD method for modal identification and a combined static-dynamic method for damage detection. *J. Sound Vib.* **2018**, *420*, 242–260. [CrossRef]
16. Rakha, M.A. On the Moore–Penrose generalized inverse matrix. *Appl. Math. Comput.* **2004**, *158*, 185–200. [CrossRef]
17. Farrar, C.R.; Doebling, S.W.; Nix, D.A. Vibration—Based structural damage identification. *Philos. Trans. R. Soc. Lond. Ser. A* **2001**, *359*, 131–149. [CrossRef]
18. Ruggiero, E.J.; Park, G.; Inman, D.J. Multi-input multi-output vibration testing of an inflatable torus. *Mech. Syst. Sig. Process.* **2004**, *18*, 1187–1201. [CrossRef]
19. Sodano, H.A.; Park, G.; Inman, D.J. An investigation into the performance of macro-fiber composites for sensing and structural vibration applications. *Mech. Syst. Sig. Process.* **2004**, *18*, 683–697. [CrossRef]
20. Foti, D.; Diaferio, M.; Giannoccaro, N.I.; Mongelli, M. Ambient vibration testing, dynamic identification and model updating of a historic tower. *NDT and E Int.* **2012**, *47*, 88–95. [CrossRef]
21. Neumaier, A. Solving ill-conditioned and singular linear systems: A tutorial on regularization. *SIAM Rev.* **1998**, *40*, 636–666. [CrossRef]
22. Basseville, M.; Mevel, L.; Goursat, M. Statistical model-based damage detection and localization: subspace-based residuals and damage-to-noise sensitivity ratios. *J. Sound Vib.* **2004**, *275*, 769–794. [CrossRef]
23. Chen, H.P. Application of regularization methods to damage detection in large scale plane frame structures using incomplete noisy modal data. *Eng. Struct.* **2008**, *30*, 3219–3227. [CrossRef]
24. Weber, B.; Paultre, P.; Proulx, J. Consistent regularization of nonlinear model updating for damage identification. *Mech. Syst. Sig. Process.* **2009**, *23*, 1965–1985. [CrossRef]
25. Li, X.Y.; Law, S.S. Adaptive Tikhonov regularization for damage detection based on nonlinear model updating. *Mech. Syst. Sig. Process.* **2010**, *24*, 1646–1664. [CrossRef]
26. Huang, Q.; Gardoni, P.; Hurlebaus, S.A. probabilistic damage detection approach using vibration-based nondestructive testing. *Struct. Saf.* **2012**, *38*, 11–21. [CrossRef]
27. Zhao, Z.; Lin, R.; Meng, Z.; He, G.; You, L.; Zhou, Y. A modified truncation singular value decomposition method for solving ill-posed problems. *J. Algorithms Comput. Technol.* **2018**, 1748301818813609. [CrossRef]
28. Arcucci, R.; Mottet, L.; Pain, C.; Guo, Y.K. Optimal reduced space for Variational Data Assimilation. *J. Comput. Phys.* **2019**, *379*, 51–69. [CrossRef]
29. Golub, G.H.; Van Loan, C.F. An analysis of the total least squares problem. *SIAM J. Numer. Anal.* **1980**, *17*, 883–893. [CrossRef]
30. Markovsky, I.; Van Huffel, S. Overview of total least-squares methods. *Signal Process.* **2007**, *87*, 2283–2302. [CrossRef]

31. Weng, Y.; Xiao, W.; Xie, L. Total least squares method for robust source localization in sensor networks using TDOA measurements. *Int. J. Distrib. Sens. Netw.* **2011**, *7*, 172902. [CrossRef]
32. Tong, X.; Jin, Y.; Li, L. An improved weighted total least squares method with applications in linear fitting and coordinate transformation. *J. Surv. Eng.* **2011**, *137*, 120–128. [CrossRef]
33. Li, S.; Liu, L.; Liu, Z.; Wang, G. A robust total Kalman filter algorithm with numerical evaluation. *Surv. Rev.* **2019**, 1–8. [CrossRef]
34. Hansen, P.C. Analysis of discrete ill-posed problems by means of the L-curve. *SIAM Rev.* **1992**, *34*, 561–580. [CrossRef]
35. Hansen, P.C.; O'Leary, D.P. The use of the L-curve in the regularization of discrete ill-posed problems. *SIAM J. Sci. Comput.* **1993**, *14*, 1487–1503. [CrossRef]
36. Reichel, L.; Sadok, H. A new L-curve for ill-posed problems. *J. Comput. Appl. Math.* **2008**, *219*, 493–508. [CrossRef]
37. Yang, J.C.S.; Tsai, T.; PEVLIN, V. Structural damage detection by the system identification technique. *Shock Vibr. Bull.* **1985**, *55*, 57–65.
38. Hao, H.; Xia, Y. Vibration-based damage detection of structures by genetic algorithm. *J. Comput. Civ. Eng.* **2002**, *16*, 222–229. [CrossRef]

12

Non-Destructive Assessment of Masonry Pillars using Ultrasonic Tomography

Monika Zielińska [1] and Magdalena Rucka [2,*]

1 Department of Technical Fundamentals of Architectural Design, Faculty of Architecture, Gdansk University of Technology, Narutowicza 11/12, 80-233 Gdansk, Poland; monika.zielinska@pg.edu.pl
2 Department of Mechanics of Materials and Structures, Faculty of Civil and Environmental Engineering, Gdansk University of Technology, Narutowicza 11/12, 80-233 Gdansk, Poland
* Correspondence: magdalena.rucka@pg.edu.pl or mrucka@pg.edu.pl

Abstract: In this paper, a condition assessment of masonry pillars is presented. Non-destructive tests were performed on an intact pillar as well as three pillars with internal inclusions in the form of a hole, a steel bar grouted by gypsum mortar, and a steel bar grouted by cement mortar. The inspection utilized ultrasonic stress waves and the reconstruction of the velocity distribution was performed by means of computed tomography. The results showed the possibilities of tomographic imaging in characterizing the internal structure of pillars. Particular attention was paid to the assessment of the adhesive connection between a steel reinforcing bar, embedded inside pillars, and the surrounding pillar body.

Keywords: non-destructive testing; masonry structures; strengthening; ultrasonic tomography; adhesion assessment

1. Introduction

A significant part of engineering structures consists of masonry objects. Historic buildings are usually made of bricks or stones [1,2] while to make contemporary objects, both ceramic elements and blocks of autoclaved aerated concrete are used [3]. The technical condition of masonry structures requires a careful quality assessment and often intervention, enabling their further proper functioning, due to the influence of atmospheric conditions, excessive loads and processes of the natural ageing. Properly carried out works, aimed at strengthening, repairing or maintaining masonry, should be preceded by a precise diagnosis process. In general, two main diagnostic approaches are possible: invasive and non-invasive [4]. In destructive testing (DT), material samples acquired from an object are destroyed to evaluate their mechanical properties. Particularly important in the diagnostics of engineering structures are non-destructive testing (NDT) methods, because they do not violate the integrity of tested objects. In recent years, various non-destructive techniques dedicated to masonry structures have been developed, including ultrasonic methods, ground penetrating radar, thermography or acoustic emissions (e.g., [1,5–10]).

The basic methods for strengthening masonry structures are total or partial brick replacement and the introduction of new reinforcing elements. The brick replacing technique relies on substituting old, degraded material with new material. This method significantly affects the structure of the object, and it is not recommended for works carried out on objects of cultural heritage, where any interference with the historic substance should be avoided. The structural efficiency of masonry can be also increased by introducing new strengthening elements (e.g., [11,12]). Reinforced concrete elements, introduced in the form of columns and beams [13] or jackets [14] constitute a substitute supporting structure. Additional steel (or composite) elements are usually used as grouted anchors or tie-rods (e.g., [15–18]) as well as composite jackets made of fabric-reinforced cementitious matrix [19] or fiber-reinforced polymers [20,21]. If a replacement technique is used, the adhesion of an old wall to a new part is not as important, because usually the new wall has only the role of filling, and in this case, it is completely lightened. However, the adhesive connection is very important when steel elements in the form of bars are used, because their main purpose is to transfer stresses to the masonry. Incorrect anchoring of the steel elements, or damage developing at the interface between two materials, i.e., steel and mortar, may be the reason why the strengthening method does not fulfil its assumed function. Therefore, the non-destructive evaluation of the adhesive connection is crucial for properly conducting the repairing process.

In this work, an evaluation of masonry pillars is presented. Particular attention was paid to the condition assessment of the adhesive connection between a steel reinforcing bar embedded inside pillars and the surrounding pillar body. The research was carried out using ultrasound waves and tomography imaging. Ultrasound tomography is one of the more developed imaging methods, utilizing the properties of elastic waves. Many previous works concern Lamb-wave tomography for damage detection in metal or composite thin plates. Rao et al. [22] performed a study on the online corrosion monitoring of a steel plate and thickness reconstruction of the corrosion damage. Zhao et al. [23] investigated damage imaging in aluminum plates with an artificial thinning area. Leonard et al. [24] studied Lamb-wave tomography in both aluminum and composite plates, with defects of various sizes and thicknesses. A structural health monitoring system of composite plates with through-thickness holes was presented by Prasad et al. [25]. Ultrasound tomography was also widely used for defect imaging in concrete structures (e.g., [26,27]). Martin et al. [26] examined post-tensioned concrete beams. They identified both the location of ducts and the voiding in ducts. Chai et al. [28] developed attenuation tomography for visualizing defects in a concrete slab. Aggelis et al. [29] applied numerical simulations of wave propagation to investigate the possibility of detecting different types of inhomogeneities in concrete. Schabowicz and Suvorov [30] described an ultrasonic tomogram, equipped with a multi-element antenna array, and its application for the estimation of the thickness of concrete and the localization of flawed zones. Schabowicz [31] presented the one-sided non-destructive testing of concrete cubic specimens and a foundation concrete slab using ultrasonic tomography. Haach and Ramirez [32] analyzed different arrangements of transducers in a study on the detection of cylindrical polystyrene blocks in concrete prismatic specimens. Choi and Popovics [33] compared one-sided imaging with through-thickness tomography using the example of highly reinforced concrete elements with internal defects. Integrated ultrasonic tomography and a 3D computer vision technique was developed by Choi et al. [34]. They obtained volumetric internal images to detect defects within concrete. Chai et al. [27] identified the tendon duct filling, as well as honeycomb defects, in concrete specimens. In the case of masonry structures, ultrasonic tomography was used to assess the general condition or to evaluate the effectiveness of the conducted repair. Schuller et al. [35] performed a velocity reconstruction in a wall specimen made of bricks. Binda, Saisi, and Zanzi [36] examined the stone pillars of the temple of S. Nicolò l'Arena using sonic tests. They obtained maps of the velocity distribution that enabled the presence of the different building techniques applied during the erection of the pillars to be recognized. The detection of voids in laboratory masonry specimens was studied by Paasche et al. [7]. They applied ultrasonic tomography and compared the results with those of ground penetrating radar tomography. Pérez-Gracia et al. [37] determined the velocity distribution in columns of the Mallorca Cathedral. Santos-Assunçao et al. [9] evaluated laboratory models of masonry columns. Changes in the wave velocity in tomography maps were identified as damages or changes in the bricks and mortar. The literature on the assessment of reinforced masonry structures is rather limited.

This study presents a comprehensive condition assessment of masonry pillars. Experimental and numerical investigations were performed on an intact pillar as well as three pillars with internal inclusions in the form of a hole, a steel bar grouted by gypsum mortar, and a steel bar grouted by cement mortar. The inspection was conducted using ultrasonic waves and computed tomography. The investigations focused on the characterization of the internal structure of the pillars by velocity reconstruction in the examined cross-section. The influence of different numbers of pixels and different configurations of paths on tomographic velocity reconstruction was studied.

2. Theoretical Background of Ultrasonic Tomography

Ultrasonic tomography imaging allows the internal structure of an investigated object to be reconstructed by a set of projections through the sample in many different directions. A schematic diagram of ultrasound tomography is shown in Figure 1. At first, the tested specimen must be divided into small, geometrical cells, called pixels (marked with dotted lines in Figure 1a). Each pixel in the image represents a discrete area in the sample, and it is associated with an intensity value. The conducted calculations enable a quantitative description of each pixel, based on its physical characteristics, such as the wave propagation velocity. The number of pixels can be changed, enabling the study of the effect of spatial resolution on the quality of the obtained image.

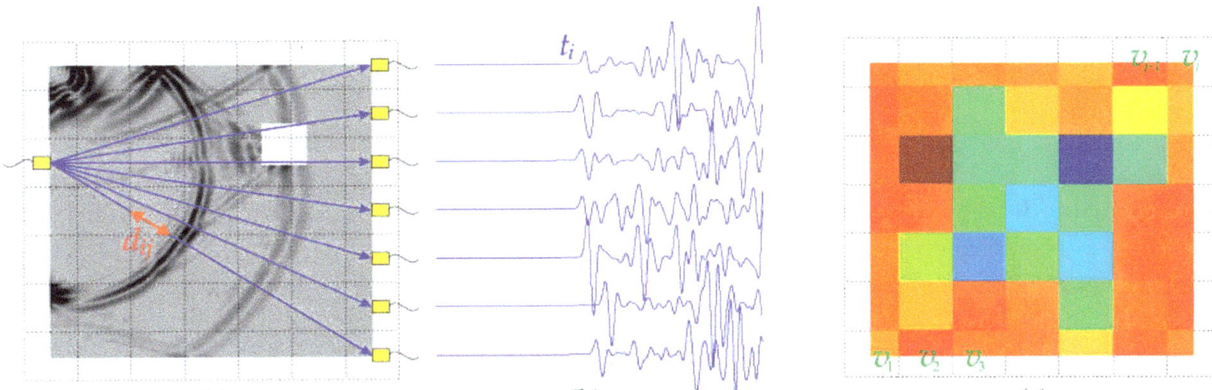

(a) (b) (c)

Figure 1. Schematic diagram of ultrasound tomography: (**a**) cross-section divided into pixels, with an indicated transmission point, receivers and a simulated wave field; (**b**) wave propagation signals; (**c**) tomographic image.

In this study, through-transmission tomography was applied (cf. [27,29,32–34]). The tomographic procedure is based on elastic waves passing through the tested specimen, from the transmitter (T) to receivers (R) (Figure 1b). The image reconstruction is performed based on the information obtained from the received pulses. Typically, this information is the time-of-flight (TOF), measured along many ray paths. Based on the known geometry of the specimen, the average velocity of wave propagation can be determined. This velocity depends on the mass density, modulus of elasticity and Poisson's ratio, characterizing the material inside the tested specimen. Any obstacle present in the path of the travelling wave results in a change of the propagation time. Defects, in the form of air voids or cracks on the wave path, cause a delay in the ray reaching the receiver. On the other hand, inclusions made of materials characterized by higher propagation velocities than the neighboring medium lead to an increase of the total velocity along all the rays passing through this inclusion.

The TOF between the transmitter and receiver can be represented by a line integral of the transition time distribution along the propagation way w:

$$t = \int_{T}^{R} dt = \int_{T}^{R} s\,dw = \int_{T}^{R} \frac{1}{v} dw, \tag{1}$$

where v denotes the average velocity, and s denotes the slowness, which is the inverse of the velocity. Therefore, by measuring the direct wave passing time, it is possible to determine the distribution of local velocities in the tested cross-section. The reconstitution of the velocity profile v_j for each cell in the plane of the transition of wave rays from the transmitter to the receiver can be performed based on the following formula:

$$t_i = \sum_{j=1}^{n} w_{ij} s_j, \quad i = 1,2,3,\ldots,m, \quad j = 1,2,3,\ldots,n, \tag{2}$$

where m is the number of measurement paths (rays), n is the number of cells (pixels), w_{ij} denotes the transition way of the i-ray through the j-pixel, t_i denotes the transition time of the P-wave, between the transmitter and receiver along the i-ray, and s_j is the slowness at pixel j, whose inverse is the velocity v_j, i.e., $v_j = 1/s_j$. It is assumed that the value of the velocity v_j in individual cells is constant.

Equation (2) can be expressed in matrix form, with known matrices t and w (representing the passage path through a given cell and the transition time along a given path, respectively) as well as the unknown matrix s (slowness):

$$\begin{bmatrix} t_1 \\ t_2 \\ t_3 \\ \vdots \\ t_i \end{bmatrix} = \begin{bmatrix} w_{11} & w_{12} & w_{13} & \cdots & w_{1j} \\ w_{21} & w_{22} & w_{23} & \cdots & w_{2j} \\ w_{31} & w_{32} & w_{33} & \cdots & w_{3j} \\ \vdots & \vdots & \vdots & \ddots & \vdots \\ w_{i1} & w_{i2} & w_{i3} & \cdots & w_{ij} \end{bmatrix} \begin{bmatrix} s_1 \\ s_2 \\ s_3 \\ \vdots \\ s_j \end{bmatrix}. \tag{3}$$

In computed tomography, the above system of equations is usually overdetermined or underdetermined, so the problem, described by Equation (3), is ill posed. It can be efficiently solved using the algebraic reconstruction technique (ART) [38]. In the first step of the calculations, each cell is assigned the same slowness (inverse of the velocity value, equal to the average velocity of the wave propagation in the examined material). In this way, an initial image is created. Next, the iteration process is started, and corrections are calculated according to the relation [39]:

$$s_j^{(k)} = s_j^{(k-1)} + \frac{w_{ij} \Delta t_i}{\sum_{j=1}^{n} w_{ij}^2}, \tag{4}$$

where Δt_i is the difference between the time of the original projection and the rebuilding time. The system of equations is iteratively solved until the reconstructed travel time reaches the measured travel time, with an established error.

3. Materials and Methods

3.1. Description of Specimens

A masonry pillar was used as a testing object. The pillar was manufactured using solid bricks, with dimensions of 25 cm × 12 cm × 6.5 cm. Joints between bricks were 1 cm thick and were filled with gauged (cement-lime) mortar. Nine brick layers were used to build the specimen. The model had a length of 66.5 cm and cross-section dimensions of 38 cm × 38 cm. The geometry of the pillar and a view of the even and odd layers are shown in Figure 2.

The test specimens included one pristine pillar and three pillars with inclusions (Figure 3). The first pillar (#1) was prepared as a reference model. Pillars #2, #3 and #4 had a square hole through the entire height of the specimen. The hole, with dimensions of 5.3 cm × 5.3 cm, was situated at a distance of 7.7 cm from the pillar edges (Figure 2b). The hole in pillar #2 remained empty, while in the case of pillars #3 and #4, an anchor was mounted in it. The threaded steel bar had a diameter of 32 cm, and was grouted using two types of mortar. In pillar #3, the anchor was mounted using gypsum mortar to represent a weak bond, while in the case of pillar #4, cement mortar was used to represent a strong bond.

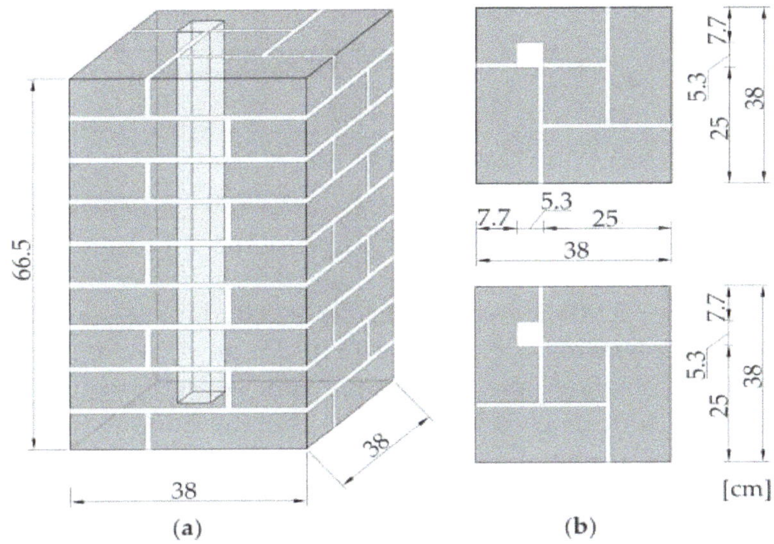

Figure 2. Geometry of the tested pillars: (**a**) 3D view; (**b**) plane view of even and odd layers.

Figure 3. Photographs of laboratory models of the brick pillars: (**a**) overall view; (**b**) plane views of specimens #1 to #4.

3.2. Identification of the Material Parameters

The tests were carried out on a single brick as well as on cubes made of cement mortar, gypsum mortar and gauged mortar, with dimensions of 10 × 10 × 10 cm^3. Each sample was measured and weighted, and then the mass density ρ was calculated. The Young's modulus E of brick and mortar was determined by a non-destructive approach using the ultrasonic pulse velocity (UPV) method ([40–42]). This approach was chosen because in this study a dynamic analysis was carried out in the ultrasonic frequency, which requires the identification of the dynamic modulus. In each sample, the transmission time of the P-wave was measured, and the P-wave velocity c_p was calculated. The Poisson's ratio ν was assumed, following Alberto et al. [43], as 0.15 for mortar and 0.2 for brick. Finally, the Young's modulus was calculated according to the following equation:

$$c_p = \sqrt{\frac{E(1-\nu)}{\rho(1+\nu)(1-2\nu)}}. \tag{5}$$

The results of the identified parameters are given in Table 1.

Table 1. Mechanical properties of the bricks and mortar used for the test specimens.

Material	Density [kg/m^3]	Young's Modulus [GPa]	Poisson's Ratio [-]
brick	1642.04	10.55	0.20
cement mortar	1719.10	8.53	0.15
gauged (cement lime) mortar	1799.90	8.20	0.15
gypsum mortar	1185.70	3.70	0.15

3.3. Experimental Setup

After 28 days, the masonry pillars were subjected to ultrasonic tests. The velocity of the stress waves was measured using the ultrasonic pulse velocity method. The PUNDIT PL-200 instrument (Proceq SA, Schwerzenbach, Switzerland) was utilized for measurements of the ultrasonic pulse velocity. Two exponential transducers of 54 kHz were used in the through transmission mode, and dry coupling was applied. The experimental setup is shown in Figure 4.

(a) (b)

Figure 4. Experimental setup: (**a**) testing in through transmission mode; (**b**) registration unit and exponential transducers.

Measurements were carried out on the middle (i.e., 5th) layer of bricks. Two experiments were conducted. In the first test (configuration #A), 7 measurement points were distributed at each pillar edge, while in the second test (configuration #B), measurements were taken at 13 points. The configurations of the measurement points are given in Figure 5. Transmitting points are denoted by T, and receiving points by R. The receiving transducer was set at a given point, and then the transmitting transducer was moved from the first to the last point on the opposite wall. There were 14 transmitting/receiving points for configuration #A and 23 transmitting/receiving points for configuration #B.

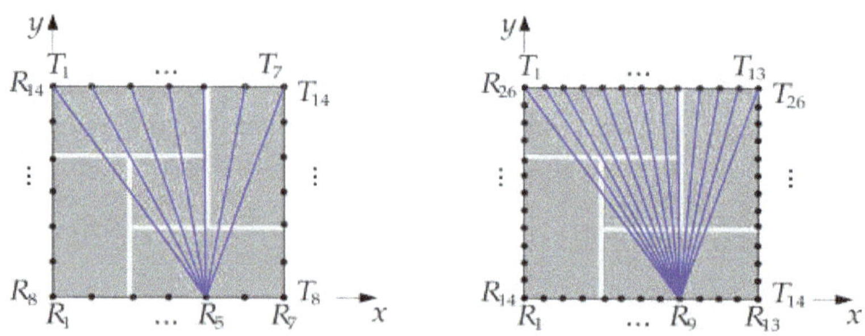

Figure 5. Configuration of the measurement points: (**a**) configuration #A (transmission of waves at points T_1-T_{14} and sensing of waves at points R_1-R_{14}), (**b**) configuration #B (transmission of waves at points T_1-T_{26} and sensing of waves at points R_1-R_{26}).

3.4. FEM Modeling

Numerical analyses of the stress wave propagation were carried out using the finite element method (FEM) in the Abaqus/Explicit. A two-dimensional model of the fifth layer of the pillar was developed using four-node plane strain elements, with reduced integration (CPE4R). The size of all elements in the model was 1 mm × 1 mm. The material of the pillar was assumed to have linear elastic behavior. The material properties of the bricks and mortar were determined experimentally (Table 1), and for the steel, the following parameters were adopted: E = 200 GPa, ρ = 7850 kg/m^3, v = 0.3. The connections between the bricks and mortar as well as between the mortar and steel were modelled as a tie constraint. The boundary conditions were assumed to be free on all edges. The excitation signal was a one-cycle sine wave of 54 kHz frequency, modulated by the Hann window. The size of the integration step was 10^{-7} s. The output acceleration signals were recorded at the same points as in the experimental investigations (Figure 5).

3.5. Configuration of Pixels and Paths

The acquired experimental and numerical signals were processed to obtain tomography images. At first, the cross-section of the pillar was divided into 49 pixels, in the case of configuration #A, and 169 pixels, in the case of configuration #B. Data analysis was conducted in three stages, with different arrangements of pats (Figure 6). In the first stage, paths were assumed only between opposite transmitting/receiving points. There were 14 paths in configuration #A.1 and 26 paths in configuration #B.1. In the second stage, from each transmitting point, two or three paths were considered, giving 38 paths in configuration #A.2 and 74 paths in configuration #B.2. The third stage took into account seven paths from each transmitting point. As a result, 98 paths were traced in configuration #A.3 and 338 paths in configuration #B.3.

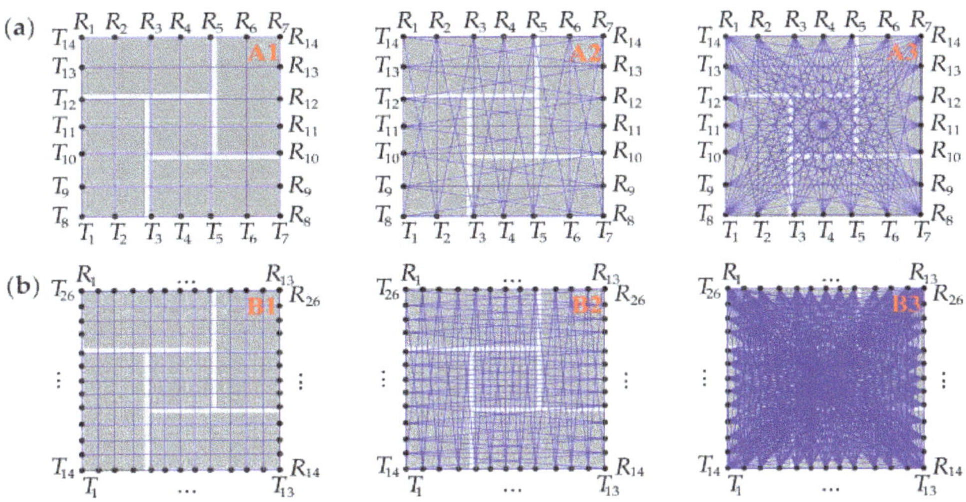

Figure 6. Configuration of paths: **(a)** configuration #A (#A1, #A2 and #A3); **(b)** configuration #B (#B1, #B2 and #B3).

4. Results and Discussion

Ultrasound tomography images, determined based on numerical FEM signals, are shown in Figures 7 and 8, with a division into 49 and 169 pixels, respectively (see Section 3.4). In general, the increase in the number of transmitters, as well as the increase in the number of paths, improved the resolution of the velocity maps and provided a more precise image of the internal structure of the pillars.

Tomograms for the intact pillar (#1) enabled the identification of the arrangement of joints (Figures 7a and 8a). Joints made of cement-lime mortar were characterized by a lower wave propagation velocity than the brick itself. The impact of the existence of joints on the tomographic images was visible, regardless of the number of transmitters used and the number of measurement paths. However, maps created based on the simplest configurations of paths (#A.1, #B.1) did not give satisfactory results regarding precise joint localization. The use of vertical and horizontal wave paths enabled only the identification of the directions of the mortar joints. More accurate tomographic images were obtained when a larger amount of measurement data was collected. Increasing the number of paths (configurations #A.2–A.3 and #B.2–B.3) allowed the position of joints in the considered cross-section and the exact brick arrangement pattern to be determined.

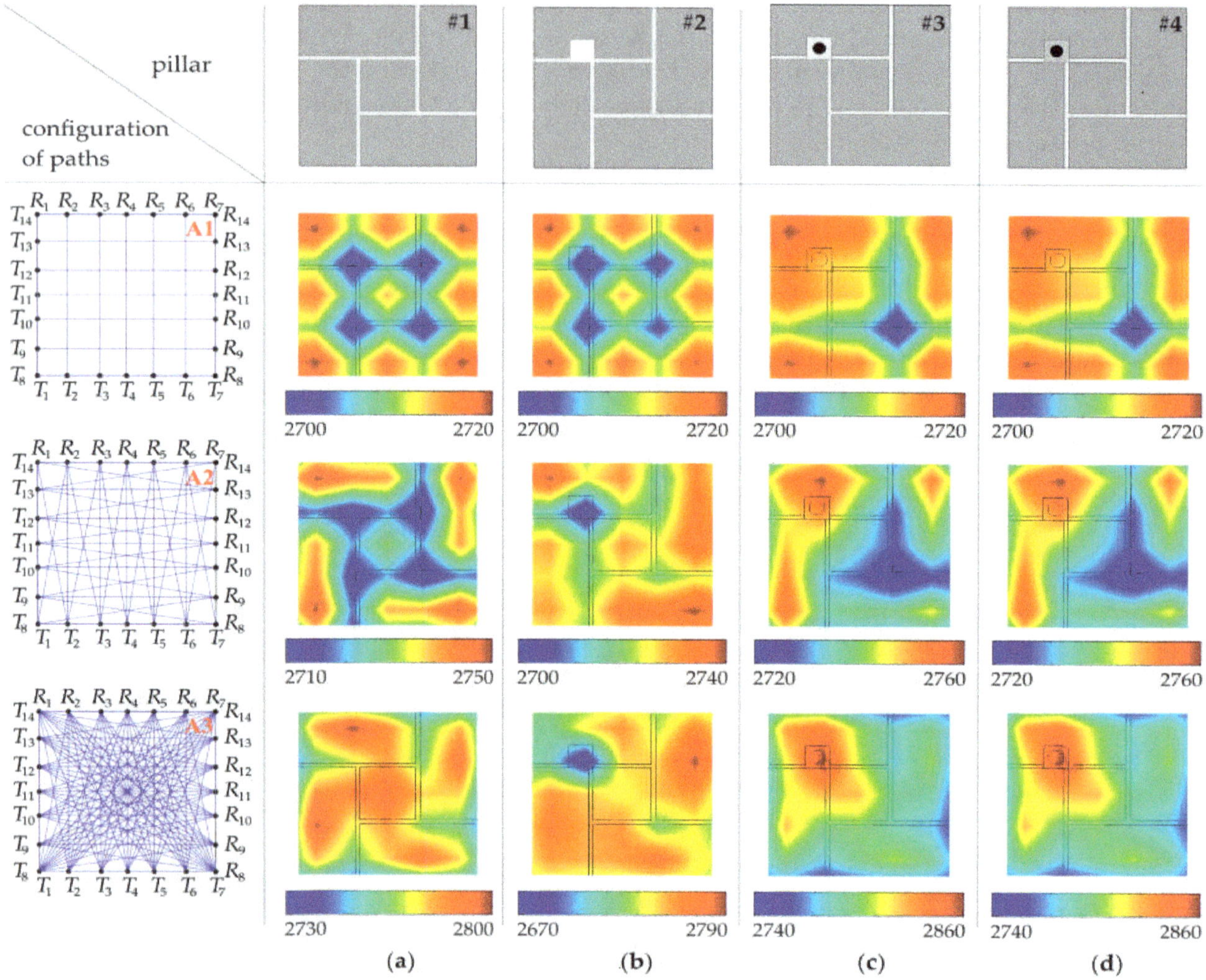

Figure 7. Tomographic velocity images (values in (m/s)), with a division into 49 pixels (configuration #A), obtained from FEM signals: (**a**) intact pillar; (**b**) pillar with the square hole; (**c**) pillar with the steel bar grouted by means of gypsum mortar; (**d**) pillar with the steel bar grouted by means of cement mortar.

Tests conducted for the pillar with the hole (pillar #2) allowed its position to be estimated, but only when a sufficiently large number of ray paths was used (Figures 7b and 8b). In the case of configurations of vertical and horizontal paths (configurations #A.1 and #B.1), the obtained tomographic images were very similar to those obtained for the intact pillar. To obtain the precise localization of the hole, the configuration, including all paths between the transmitting point and receiving points, were required. For such configurations (#A.3 and #B.3), a clear area, with a lower value of the velocity, can be observed. It was also possible to identify the arrangement of mortar joints, especially in the case of the image divided into 169 pixels.

The results for the pillars with the steel bar grouted by means of gypsum cement mortar are shown in Figure 7c,d (49 pixels) and Figure 8c,d (169 pixels). Tomographic velocity images of the simplest paths differed considerably between configuration #A.1 and #B.1. In the case of configuration #A.1, no path crossed the steel bar, therefore, it was not possible to localize the bar. However, since two paths passed directly along the joints, two straight lines with lower velocities appeared in the map, indicating the direction of the mortar joints. When 26 instead of 14 horizontal and vertical paths were used, two perpendicular paths crossed the steel bar. Therefore, for configuration #B.1, two longitudinal areas with a higher velocity appeared on the maps. These lines were the effect of blurring, which occurs in linear tomography. They indicated the presence of the bar but did not allow for its exact localization. The precise localization was achieved after considering more paths crossing the bar (configurations #A.3 and B.3). As a result, a distinct area with higher velocity values can be observed around the bar position. It should also be noted that the maps for pillars #3 and #4 are almost the same. This means that the main influence on the velocity distribution was the presence of the bar, and therefore, it was not possible to determine the type of mortar (gypsum or cement) used.

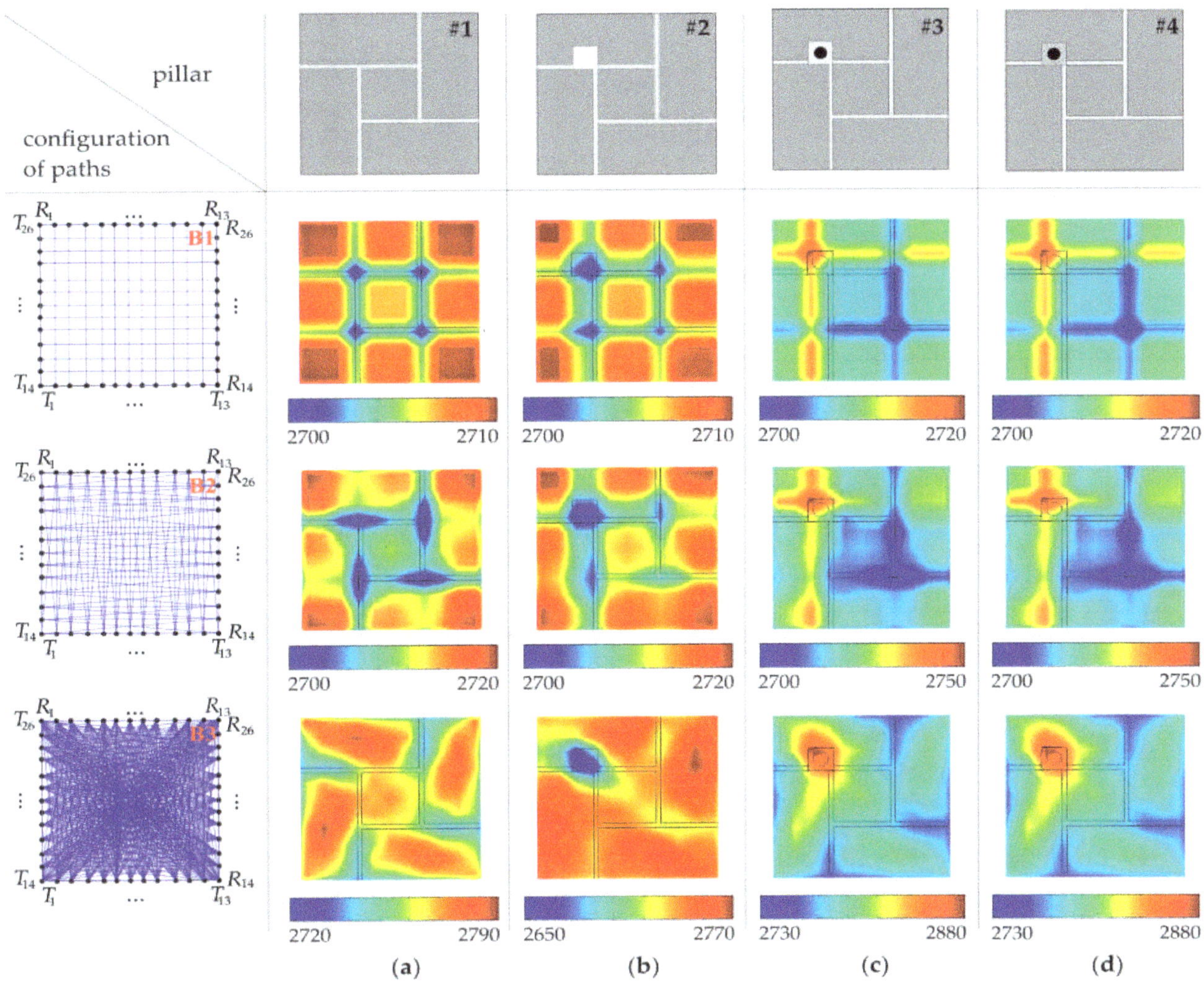

Figure 8. Tomographic velocity images (values in (m/s)), with a division into 169 pixels (configuration #B), obtained from FEM signals: (**a**) intact pillar; (**b**) pillar with the square hole; (**c**) pillar with the steel bar grouted by means of gypsum mortar; (**d**) pillar with the steel bar grouted by means of cement mortar.

The quantitative analysis of the wave propagation values through pillars along many different directions, is given in Tables 2 and 3. For each configuration of paths (#A.1–A.3 and #B.1–B.3), the velocity values were calculated along all rays, and then the minimum and maximum values as well as the mean value of the wave velocity were listed. Moreover, measures of variability, i.e., the standard deviation (SD) and coefficient of variation (CV), were calculated. The average values of the velocity of the particular arrangements were close to each other and ranged from 2719.18 m/s to 2745.48 m/s. The standard deviation ranged from 17.2 m/s to 29 m/s, and the coefficient of variation ranged from 0.63% to 1.07%. This indicates a small variation in the speed of wave propagation along the considered paths. Due to the idealized connections between mortar and bricks, implemented in FEM models, the velocity variations were caused only by internal inclusions.

Table 2. Numerical wave propagation velocities for configuration #A.

Pillar	Arrangement	v_{min} [m/s]	v_{max} [m/s]	$\Delta v = v_{max} - v_{min}$ [m/s]	v_{avg} [m/s]	SD [m/s]	CV [%]
#1	A1	2699.52	2750.93	51.41	2732.93	22.0	0.81
	A2	2699.52	2765.47	65.95	2736.24	21.0	0.77
	A3	2681.93	2767.02	85.09	2730.40	20.4	0.75
#2	A1	2692.63	2751.52	58.89	2733.03	23.2	0.85
	A2	2640.62	2766.05	125.44	2731.99	27.7	1.02
	A3	2640.62	2766.05	125.44	2721.84	28.8	1.06
#3	A1	2700.46	2753.27	52.81	2738.87	17.4	0.64
	A2	2700.46	2822.93	122.47	2742.47	22.0	0.80
	A3	2682.24	2825.21	142.97	2737.58	23.0	0.84
#4	A1	2700.46	2753.27	52.81	2738.87	17.4	0.64
	A2	2700.46	2822.93	122.47	2742.49	21.9	0.80
	A3	2682.24	2825.21	142.97	2737.58	23.0	0.84

Table 3. Numerical wave propagation velocities for configuration #B.

Pillar	Arrangement	v_{min} [m/s]	v_{max} [m/s]	$\Delta v = v_{max} - v_{min}$ [m/s]	v_{avg} [m/s]	SD [m/s]	CV [%]
#1	B1	2699.52	2750.93	51.41	2737.67	17.4	0.64
	B2	2693.96	2756.58	62.62	2737.61	17.2	0.63
	B3	2667.36	2771.10	103.74	2727.88	20.9	0.77
#2	B1	2692.63	2751.52	58.89	2735.22	18.8	0.69
	B2	2647.13	2757.36	110.23	2734.08	22.5	0.82
	B3	2618.33	2771.67	153.34	2719.18	29.0	1.07
#3	B1	2700.46	2810.41	109.95	2745.15	21.5	0.78
	B2	2696.75	2810.41	113.66	2745.48	21.4	0.78
	B3	2666.05	2825.21	159.15	2735.68	24.5	0.90
#4	B1	2700.46	2810.41	109.95	2745.15	21.5	0.78
	B2	2696.75	2810.41	113.66	2745.48	21.4	0.78
	B3	2666.05	2825.21	159.15	2735.68	24.5	0.90

The conducted research also allowed the impact of the number of joints along the wave path on the velocity of propagating ultrasonic waves to be determined. This effect was investigated for each of the four pillars (#1–#4), taking into account the results obtained for arrangement #B. While checking the influence of the number of joints on the wave velocities, paths crossing the hole or the bar as well as the paths along the pillar edges were omitted. The velocity of the wave passing through one joint was calculated as the mean obtained between points T_{10}-R_{10}, T_{11}-R_{11}, T_{12}-R_{12}, T_{15}-R_{15}, T_{16}-R_{16}, T_{17}-R_{17}. The velocity of the wave passing through two joints was calculated as the mean obtained between points T_6-R_6, T_7-R_7, T_8-R_8, T_{19}-R_{19}, T_{20}-R_{20}, T_{21}-R_{21}, and finally, for calculations concerning three joints, the data obtained between points T_1-R_{11} and T_3-R_{13} were used. The results of the obtained average propagation velocities are summarized in Table 4. The velocities obtained for each pillar are almost the same, because no considered path crossed internal inclusions. However, the difference between the velocity of the waves passing through one, two, and three joints was observed. The velocity value decreased with the increase of the number of joints on the wave path.

Table 4. The influence of the number of joints on the numerical wave propagation velocities.

Pillar	v_{avg} [m/s] 1 joint	v_{avg} [m/s] 2 joints	v_{avg} [m/s] 3 joints
#1	2747.70	2733.58	2726.77
#2	2747.19	2734.16	2727.26
#3	2747.45	2734.48	2727.47
#4	2747.45	2734.48	2727.47

The experimental tomographic images are shown in Figures 9 and 10, with a division into 49 and 169 pixels, respectively (see Section 3.4). Contrary to numerical maps, the quality of experimental images did not increase with the increase of the number of pixels or paths used. In real masonry elements, connections between mortar and bricks are not ideal. Air voids, pores and a lack of adhesion between a brick and mortar may occur. The heterogeneous nature of the tested specimens as well as the imperfections of the connections between bricks and mortar caused a strong dissipation of the energy of propagating waves and consequently affected the quality of the obtained tomography images.

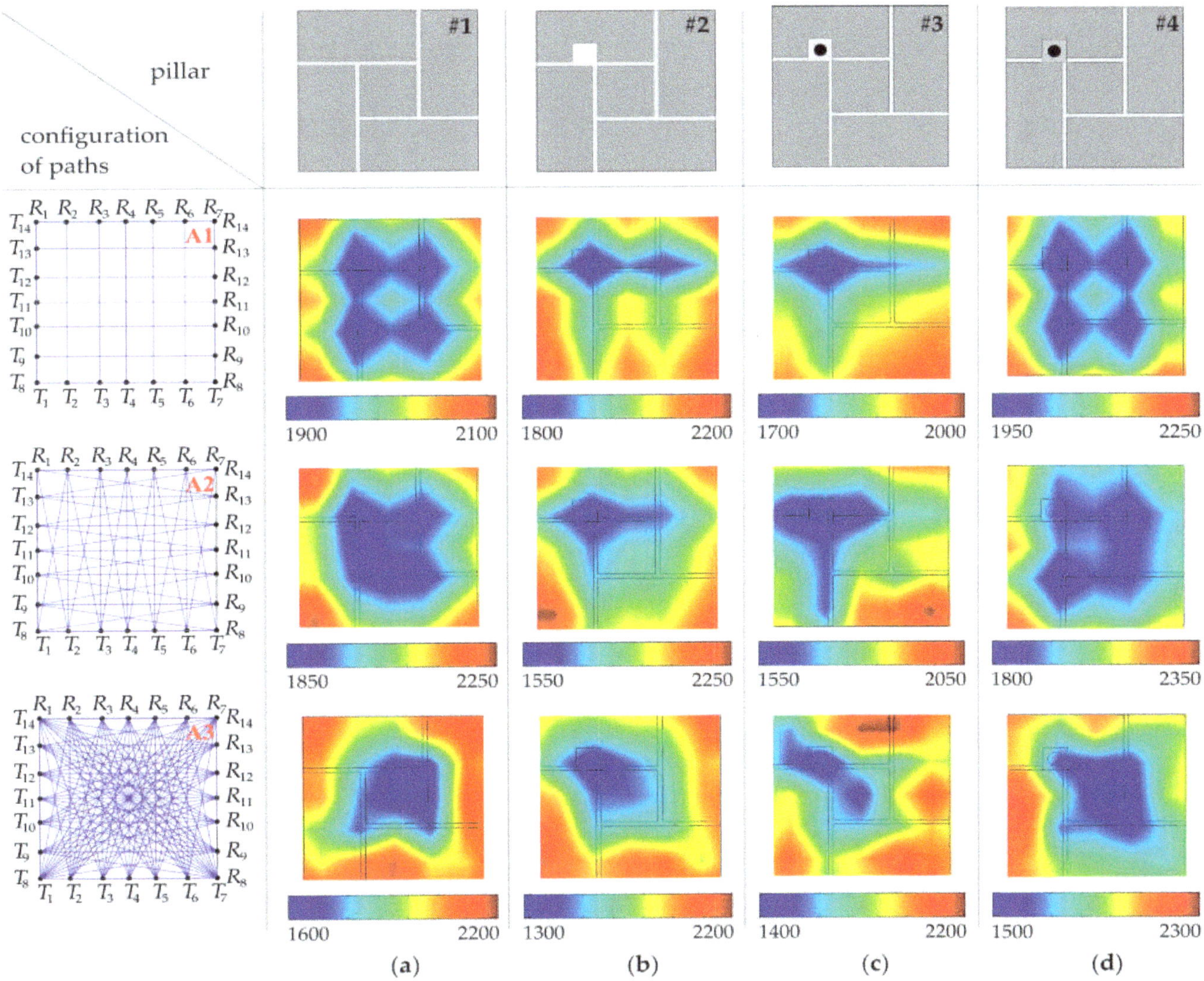

Figure 9. Tomographic velocity images (values in (m/s)), with a division into 49 pixels (configuration #A), obtained from experimental signals: (**a**) intact pillar; (**b**) pillar with the square hole; (**c**) pillar with the steel bar grouted by means of gypsum mortar; (**d**) pillar with the steel bar grouted by means of cement mortar.

Measurements made for the intact pillar (#1) using only vertical and horizontal paths revealed lower propagation velocities at joint intersections (Figures 9a and 10a). With the increase of the number of paths, for configurations #A.3 and #B.3 a zone with lower velocity values appeared in the central part of the pillar due to the accumulation of joints and paths crossing them.

Maps of the pillar with the hole are shown in Figures 9b and 10b. The hole unambiguously disturbed the wave transition. The location of the hole can be indicated as the area with lower velocity values. The identification of the position of the hole was clearest in the case of vertical and horizontal rays (#A.1 and #B.1), however, it was affected by the blurring effect. As in the case of the intact pillar, the accumulation of lower velocities can be seen for configurations #A.2, #A.3 and #B.2, #B.3.

The maps obtained for the pillar with the bar grouted by gypsum mortar (#3) revealed similar patterns as those for the pillar with the hole (Figures 9c and 10c). This may indicate poor adhesion between the bar and the gypsum mortar. On the other hand, the tomographic velocity images for the pillar with the bar grouted by cement mortar (#4) were very similar to those obtained for the intact pillar (Figures 9c and 10c). This means that the connection between the bar and gypsum mortar was of a good quality, and strong adhesion occurred.

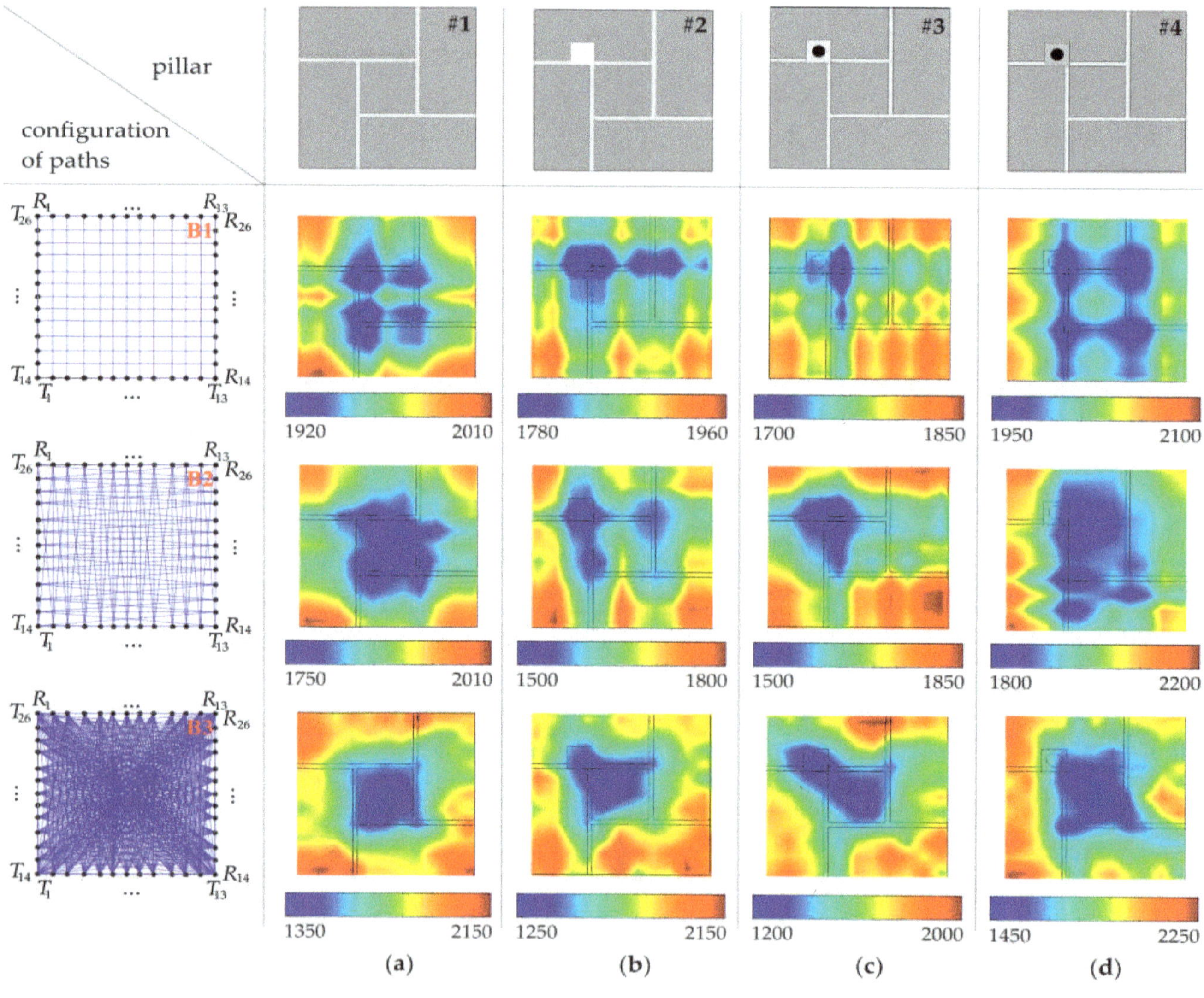

Figure 10. Tomographic velocity images (values in (m/s)), with a division into 169 pixels (configuration #B), obtained from experimental signals: (**a**) intact pillar; (**b**) pillar with the square hole; (**c**) pillar with the steel bar grouted by means of gypsum mortar; (**d**) pillar with the steel bar grouted by means of cement mortar.

The minimum, maximum and average velocities of ultrasonic waves, based on experimental tests for each configuration of paths (#A.1–A.3 and #B.1–B.3), are shown in Tables 5 and 6. The average values of the velocity for the particular pillars varied from 1635.34 m/s to 2163.11 m/s. The velocities for pillars #1 and #4 were higher than those for pillars #2 and #3, which proved weak adhesion in pillar #3 and strong adhesion in pillar #4. The standard deviation varied from 254.4 m/s to 465.7 m/s, and the coefficient of variation varied from 13.42% to 22.4%. These values were approximately 10–20 times larger than those obtained from the FEM simulations. At the same time, the experimental velocities appeared to be smaller than the numerical ones. This is the result of non-ideal connections between the joints and bricks in the real model of pillars.

Table 5. Experimental wave propagation velocities for configuration #A.

Pillar	Arrangement	v_{min} [m/s]	v_{max} [m/s]	$\Delta v = v_{max} - v_{min}$ [m/s]	v_{avg} [m/s]	SD [m/s]	CV [%]
#1	A1	1687.58	2651.26	963.68	2099.66	363.7	17.32
	A2	1687.58	2651.26	963.68	2104.60	296.1	14.07
	A3	1574.48	2651.26	1076.78	2030.73	258.4	12.72
#2	A1	1253.62	2908.28	1654.66	2100.92	463.0	22.04
	A2	1253.62	2908.28	1654.66	2042.58	395.5	19.36
	A3	1253.62	2908.28	1654.66	1927.77	322.5	16.73
#3	A1	1402.37	2615.69	1213.32	1936.08	304.9	15.75
	A2	1313.12	2615.69	1302.58	1876.67	310.2	16.53
	A3	1150.91	2615.69	1464.78	1731.45	291.4	16.83
#4	A1	1617.59	3068.45	1450.86	2163.11	465.7	21.53
	A2	1617.59	3068.45	1450.86	2107.37	365.2	17.33
	A3	1511.52	3068.45	1556.93	2009.67	316.0	15.72

Table 6. Experimental wave propagation velocities for configuration #B.

Pillar	Arrangement	v_{min} [m/s]	v_{max} [m/s]	$\Delta v = v_{max} - v_{min}$ [m/s]	v_{avg} [m/s]	SD [m/s]	CV [%]
#1	B1	1597.71	2651.26	1053.55	2014.90	330.1	16.38
	B2	1493.14	2651.26	1158.12	1969.88	283.3	14.38
	B3	1321.90	2651.26	1329.36	1896.30	254.4	13.42
#2	B1	1253.62	2908.28	1654.66	1887.48	422.8	22.40
	B2	1117.83	2908.28	1790.45	1807.67	362.3	20.04
	B3	1111.81	2908.28	1796.47	1722.46	311.5	18.08
#3	B1	1222.19	2615.69	1393.51	1814.67	314.8	17.35
	B2	1222.19	2615.69	1393.51	1748.82	314.9	18.01
	B3	1056.88	2615.69	1558.81	1635.34	277.0	16.94
#4	B1	1407.43	3068.45	1661.02	2055.87	424.4	20.64
	B2	1407.43	3068.45	1661.02	2001.88	378.7	18.92
	B3	1354.30	3068.45	1714.15	1921.07	315.8	16.44

The comparison of the propagation velocities, depending on the wave transition through one, two or three mortar joints, is shown in Table 7. The same wave paths as those in the numerical calculations were assumed. The velocities obtained for each pillar differed due to non-ideal connections between the mortar and bricks. Differences between the velocity of the waves passing through one, two, and three joints were also observed. As in the case of the FEM simulations, the velocity value decreased with the increase of the number of joints on the wave path.

Table 7. The influence of the number of joints on the experimental wave propagation velocities.

Pillar	v_{avg} [m/s] 1 joint	v_{avg} [m/s] 2 joints	v_{avg} [m/s] 3 joints
#1	2082.75	1727.25	1682.13
#2	1967.77	1657.43	1638.79
#3	1923.85	1578.54	1557.31
#4	2002.48	1800.36	1708.91

5. Conclusions

In this study, the ultrasonic tomography technique was applied to the non-destructive diagnostics of masonry pillars. Experimental and numerical investigations were performed on four laboratory specimens: one intact pillar and three pillars with inclusions. The conducted investigations focused on the tomographic velocity reconstruction and the assessment of the internal structure of the pillars.

The study of ultrasonic tomography applied to the assessment of masonry pillars led to the following conclusions:

1. The increase of the number of pixels and paths did not guarantee an improvement of the quality of the tomographic images. In numerical simulations made for the pillar models with idealized connections between the mortar and bricks, more accurate tomograms were obtained when a larger amount of measurement data was collected with a denser division in cells. This observation has not been confirmed in experimental tests conducted on real specimens, with connections between bricks and mortar influenced by air voids and non-ideal adhesion.
2. The change in the velocity value was observed depending on the number of joints through which the wave passed. The velocity decreased with the increase of the number of joints on the wave path.
3. Detection of the arrangement of joints in the cross-section was possible. The joints could be observed, based on numerical data, as line patterns with lower velocities. However, in experimental tests, the increase of the number of paths resulted in the appearance of a large zone, with lower velocity values in the central part of the pillar due to the accumulation of joints and paths crossing them.
4. The inclusion in the form of a hole was identified in both the numerical and experimental tests as an area with lower velocity values.
5. The inclusion in the form of an embedded bar was identified in the numerical data as an area with higher velocity values. This observation was not confirmed in experimental tests due to the existence of many factors that slowed the speed of the wave.
6. The experimental tests enabled the assessment of the quality of the adhesive connection between a steel reinforcing bar embedded inside pillars, and the surrounding pillar body. The tomograms obtained for the pillar with the bar grouted by gypsum mortar revealed similar patterns as those for the pillar with the hole, which indicated poor adhesion. The maps obtained for the pillar with the bar grouted by cement mortar were similar to those obtained for the intact pillar, which indicated strong adhesion.

To summarize, ultrasonic tomography appeared to be an effective technique for the reconstruction of the internal structure of brick pillars. The presented approach may be particularly useful in the diagnostics of applied strengthening and the assessment of the compatibility of reinforcement materials with brick structures, and it may become the basis for the strategic planning of repair procedures. Further investigations should consider the tomographic imaging of masonry structures, accessible from one or two sides, to develop efficient tomographic procedures based on a limited number of inputs and examined ray paths.

Author Contributions: Conceptualization, M.Z. and M.R.; Formal analysis, M.Z. and M.R.; Funding acquisition, M.Z.; Investigation, M.Z.; Methodology, M.Z. and M.R.; Software, M.Z.; Supervision, M.R.; Validation, M.R.; Visualization, M.Z.; Writing–original draft, M.Z.; Writing–review and editing, M.R.

Acknowledgments: Abaqus calculations were carried out at the Academic Computer Centre in Gdańsk. The help of Karol Grębowski during the preparation of the pillars is gratefully acknowledged.

References

1. Binda, L.; Saisi, A.; Tiraboschi, C. Investigation procedures for the diagnosis of historic masonries. *Constr. Build. Mater.* **2000**, *14*, 199–233. [CrossRef]
2. Piroglu, F.; Ozakgul, K. Site investigation of masonry buildings damaged during the 23 October and 9 November 2011 Van Earthquakes in Turkey. *Nat. Hazards Earth Syst. Sci.* **2013**, *13*, 689–708. [CrossRef]
3. Jasiński, R.; Drobiec, Ł. Comparison Research of Bed Joints Construction and Bed Joints Reinforcement on Shear Parameters of AAC Masonry Walls. *J. Civ. Eng. Archit.* **2016**, *10*, 1329–1343. [CrossRef]
4. Vasanelli, E.; Sileo, M.; Leucci, G.; Calia, A.; Aiello, M.A.; Micelli, F. Mechanical characterization of building stones through DT and NDT tests: Research of correlations for the in situ analysis of ancient masonry. *Key Eng. Mater.* **2015**, *628*, 85–89. [CrossRef]
5. McCann, D.M.; Forde, M.C. Review of NDT methods in the assessment of concrete and masonry structures. *NDT E Int.* **2001**, *34*, 71–84. [CrossRef]
6. Schuller, M.P. Nondestructive testing and damage assessment of masonry structures. *Prog. Struct. Eng. Mater.* **2003**, 239–251. [CrossRef]
7. Paasche, H.; Wendrich, A.; Tronicke, J.; Trela, C. Detecting voids in masonry by cooperatively inverting P-wave and georadar traveltimes. *J. Geophys. Eng.* **2008**, *5*, 256–267. [CrossRef]
8. Bosiljkov, V.; Uranjek, M.; Žarnić, R.; Bokan-Bosiljkov, V. An integrated diagnostic approach for the assessment of historic masonry structures. *J. Cult. Herit.* **2010**, *11*, 239–249. [CrossRef]
9. Santos-Assunçao, S.; Perez-Gracia, V.; Caselles, O.; Clapes, J.; Salinas, V. Assessment of complex masonry structures with GPR compared to other non-destructive testing studies. *Remote Sens.* **2014**, *6*, 8220–8237. [CrossRef]
10. Khan, F.; Rajaram, S.; Vanniamparambil, P.A.; Bolhassani, M.; Hamid, A.; Kontsos, A.; Bartoli, I. Multi-sensing NDT for damage assessment of concrete masonry walls. *Struct. Control Heal. Monit.* **2015**, *22*, 449–462. [CrossRef]
11. Micelli, F.; Cascardi, A.; Marsano, M. Seismic strengthening of a theatre masonry building by using active FRP wires. In *Brick and Block Masonry: Proceedings of the 16th International Brick and Block Masonry Conference*; CRC Press: Padova, Italy, 2016; pp. 753–761.
12. La Mendola, L.; Lo Giudice, E.; Minafò, G. Experimental calibration of flat jacks for in-situ testing of masonry. *Int. J. Archit. Herit.* **2018**. [CrossRef]
13. Rucka, M.; Lachowicz, J.; Zielińska, M. GPR investigation of the strengthening system of a historic masonry tower. *J. Appl. Geophys.* **2016**, *131*, 94–102. [CrossRef]
14. Lachowicz, J.; Rucka, M. Diagnostics of pillars in St. Mary's Church (Gdańsk, Poland) using the GPR method. *Int. J. Archit. Herit.* **2018**, 1–11. [CrossRef]
15. Paganoni, S.; D'Ayala, D. Testing and design procedure for corner connections of masonry heritage buildings strengthened by metallic grouted anchors. *Eng. Struct.* **2014**, *70*, 278–293. [CrossRef]
16. Collini, L.; Fagiani, R.; Garziera, R.; Riabova, K.; Vanali, M. Load and effectiveness of the tie-rods of an ancient Dome: Technical and historical aspects. *J. Cult. Herit.* **2015**, *16*, 597–601. [CrossRef]
17. Ural, A.; Firat, F.K.; Tuğrulelçi, S.; Kara, M.E. Experimental and numerical study on effectiveness of various tie-rod systems in brick arches. *Eng. Struct.* **2016**, *110*, 209–221. [CrossRef]
18. Pisani, M.A. Theoretical approach to the evaluation of the load-carrying capacity of the tie rod anchor system in a masonry wall. *Eng. Struct.* **2016**, *124*, 85–95. [CrossRef]
19. Ombres, L.; Verre, S. Masonry columns strengthened with Steel Fabric Reinforced Cementitious Matrix (S-FRCM) jackets: Experimental and numerical analysis. *Measurement* **2018**, *127*, 238–245. [CrossRef]
20. Micelli, F.; Cascardi, A.; Aiello, M.A. A Study on FRP-Confined Concrete in Presence of Different Preload Levels. In Proceedings of the 9th International Conference on Fibre-Reinforced Polymer (FRP) Composites in Civil Engineering—CICE, Paris, France, 17–19 July 2018; pp. 493–499.

21. Ferrotto, M.F.; Fischer, O.; Cavaleri, L. A strategy for the finite element modeling of FRP-confined concrete columns subjected to preload. *Eng. Struct.* **2018**, *173*, 1054–1067. [CrossRef]
22. Rao, J.; Ratassepp, M.; Lisevych, D.; Hamzah Caffoor, M.; Fan, Z. On-Line Corrosion Monitoring of Plate Structures Based on Guided Wave Tomography Using Piezoelectric Sensors. *Sensors* **2017**, *17*, 2882. [CrossRef]
23. Zhao, X.; Royer, R.L.; Owens, S.E.; Rose, J.L. Ultrasonic Lamb wave tomography in structural health monitoring. *Smart Mater. Struct.* **2011**, *20*, 105002. [CrossRef]
24. Leonard, K.R.; Malyarenko, E.V.; Hinders, M.K. Ultrasonic Lamb wave tomography. *Inverse Probl.* **2002**, *18*, 1795–1808. [CrossRef]
25. Prasad, S.M.; Balasubramaniam, K.; Krishnamurthy, C.V. Structural health monitoring of composite structures using Lamb wave tomography. *Smart Mater. Struct.* **2004**, *13*, N73–N79. [CrossRef]
26. Martin, J.; Broughton, K.J.; Giannopolous, A.; Hardy, M.S.A.; Forde, M.C. Ultrasonic tomography of grouted duct post-tensioned reinforced concrete bridge beams. *NDT E Int.* **2001**, *34*, 107–113. [CrossRef]
27. Chai, H.K.; Liu, K.F.; Behnia, A.; Yoshikazu, K.; Shiotani, T. Development of a tomography technique for assessment of the material condition of concrete using optimized elastic wave parameters. *Materials* **2016**, *9*, 291. [CrossRef] [PubMed]
28. Chai, H.K.; Momoki, S.; Kobayashi, Y.; Aggelis, D.G.; Shiotani, T. Tomographic reconstruction for concrete using attenuation of ultrasound. *NDT E Int.* **2011**, *44*, 206–215. [CrossRef]
29. Aggelis, D.G.; Tsimpris, N.; Chai, H.K.; Shiotani, T.; Kobayashi, Y. Numerical simulation of elastic waves for visualization of defects. *Constr. Build. Mater.* **2011**, *25*, 1503–1512. [CrossRef]
30. Schabowicz, K.; Suvorov, V.A. Nondestructive testing of a bottom surface and construction of its profile by ultrasonic tomography. *Russ. J. Nondestruct. Test.* **2014**, *50*, 109–119. [CrossRef]
31. Schabowicz, K. Ultrasonic tomography—The latest nondestructive technique for testing concrete members—Description, test methodology, application example. *Arch. Civ. Mech. Eng.* **2014**, *14*, 295–303. [CrossRef]
32. Haach, V.G.; Ramirez, F.C. Qualitative assessment of concrete by ultrasound tomography. *Constr. Build. Mater.* **2016**, *119*, 61–70. [CrossRef]
33. Choi, H.; Popovics, J.S. NDE application of ultrasonic tomography to a full-scale concrete structure. *IEEE Trans. Ultrason. Ferroelectr. Freq. Control* **2015**, *62*, 1076–1085. [CrossRef] [PubMed]
34. Choi, H.; Ham, Y.; Popovics, J.S. Integrated visualization for reinforced concrete using ultrasonic tomography and image-based 3-D reconstruction. *Constr. Build. Mater.* **2016**, *123*, 384–393. [CrossRef]
35. Schullerl, M.; Berra, M.; Atkinson, R.; Binda, L. Acoustic tomography for evaluation of unreinforced masonry. *Constr. Build. Makrials* **1997**, *11*, 199–204. [CrossRef]
36. Binda, L.; Saisi, A.; Zanzi, L. Sonic tomography and flat-jack tests as complementary investigation procedures for the stone pillars of the temple of S. Nicolo 1'Arena (Italy). *NDT E Int.* **2003**, *36*, 215–227. [CrossRef]
37. Pérez-Gracia, V.; Caselles, J.O.; Clapés, J.; Martinez, G.; Osorio, R. Non-destructive analysis in cultural heritage buildings: Evaluating the Mallorca cathedral supporting structures. *NDT E Int.* **2013**, *59*, 40–47. [CrossRef]
38. Kak, A.C.; Slaney, M. *Principles of Computerized Tomographic Imaging*; The Institiute of Electrical and Electronics Engineers, Inc.: New York, NY, USA, 1988.
39. Oliveira, E.F.; Dantas, C.C.; Vasconcelos, D.A.A.; Cadiz, F. Comparison Among Tomographic Reconstruction Algorithms With a Limited Data. In Proceedings of the International Nuclear Atlantic Conference-INAC 2011, Belo Horizonte, Brazil, 24–28 October 2011.
40. Lu, X.; Sun, Q.; Feng, W.; Tian, J. Evaluation of dynamic modulus of elasticity of concrete using impact-echo method. *Constr. Build. Mater.* **2013**, *47*, 231–239. [CrossRef]
41. Węglewski, W.; Bochenek, K.; Basista, M.; Schubert, T.; Jehring, U.; Litniewski, J.; Mackiewicz, S. Comparative assessment of Young's modulus measurements of metal-ceramic composites using mechanical and non-destructive tests and micro-CT based computational modeling. *Comput. Mater. Sci.* **2013**, *77*, 19–30. [CrossRef]
42. Wolfs, R.J.M.; Bos, F.P.; Salet, T.A.M. Correlation between destructive compression tests and non-destructive ultrasonic measurements on early age 3D printed concrete. *Constr. Build. Mater.* **2018**, *181*, 447–454. [CrossRef]
43. Alberto, A.; Antonaci, P.; Valente, S. Damage analysis of brick-to-mortar interfaces. *Procedia Eng.* **2011**, *10*, 1151–1156. [CrossRef]

13

Characteristic Curve and its use in Determining the Compressive Strength of Concrete by the Rebound Hammer Test

Dalibor Kocáb *, Petr Misák and Petr Cikrle

Faculty of Civil Engineering, Brno University of Technology, Veveří 331/95, 602 00 Brno, Czech Republic
* Correspondence: dalibor.kocab@vutbr.cz

Abstract: During the construction of concrete structures, it is often useful to know compressive strength at an early age. This is an amount of strength required for the safe removal of formwork, also known as stripping strength. It is certainly helpful to determine this strength non-destructively, i.e., without any invasive steps that would damage the structure. Second only to the ultrasonic pulse velocity test, the rebound hammer test is the most common NDT method currently used for this purpose. However, estimating compressive strength using general regression models can often yield inaccurate results. The experiment results show that the compressive strength of any concrete can be estimated using one's own newly created regression model. A traditionally constructed regression model can predict the strength value with 50% reliability, or when two-sided confidence bands are used, with 95% reliability. However, civil engineers usually work with the so-called characteristic value defined as a 5% quantile. Therefore, it appears suitable to adjust conventional methods in order to achieve a regression model with 95% one-sided reliability. This paper describes a simple construction of such a characteristic curve. The results show that the characteristic curve created for the concrete in question could be a useful tool even outside of practical applications.

Keywords: rebound hammer; SilverSchmidt; concrete; compressive strength; non-destructive testing

---

1. Introduction

Concrete structures have always been built for a long service life, safety, durability, load-bearing capacity, stability, and working and functional reliability. During construction or usage, however, there may arise a need to determine or verify the properties of the concrete. This is why accurate diagnostics of concrete elements or whole structures are essential. Non-destructive testing methods (NDT) are often used for this purpose [1]. Besides the ultrasonic pulse velocity test, the rebound hammer test is a popular method, because it is easy to use and is practically non-destructive [2–5]. There are two basic ways of using it to test concrete. The first is using a rebound hammer to diagnose older structures with the primary purpose of classing the concrete according to strength or uniformity [6,7]. Testing older structures often requires removing both plaster and the top layer of concrete, as well as combining the measurement with destructive compressive strength tests performed on cores [8–10]. Another use of the rebound test is assessing the quality of new concrete structures, especially those with a smooth surface.

The rebound hammer test is currently the most common method of testing the hardness of concrete and was first developed in Switzerland in the 1950s as the Schmidt rebound hammer (sometimes known as the Swiss hammer) [11]. In 1950 Ernst Schmidt created the first rebound hammer, which proved to be superior to indentation methods used until that time (the most common one involved pressing a steel ball into the surface) [12]. Testing hardness with the Schmidt hammer was gradually accepted as the best, and since then concrete hardness has been determined by measuring the rebound number instead of examining the indentation produced by the ball. Today, the Schmidt hammer is the most common concrete sclerometer worldwide [13]. One of the strong points of the Schmidt hammer is its ability to test materials other than concrete (see [14–19]).

Its benefits notwithstanding, the rebound hammer test has some weak points, which should be kept in mind during testing and evaluation. Its primary purpose has always been to determine the quality of new concrete elements or structures. The general relationships between hardness and compressive strength were adjusted for this purpose, having been created by the manufacturers and then carried over to technical standards; they apply mostly to concrete of 14 to 56 days of age. The measurement range is then designed for structural concrete used during the second half of the 20th century; i.e., compressive strength of approximately 10 to 70 N/mm^2 [20], or, more realistically, 15 to 60 N/mm^2 [21]. When testing older concretes, it is necessary to take into account the severity of carbonation or surface damage. Kim et al. [22] concerned themselves with the influence of concrete carbonation on the rebound value and compressive strength. They discovered the depth of carbonation affects compressive strength very differently than the rebound value, which implies that the same dependencies will not apply to differently carbonated concrete. This is why they used regression analysis to create an equation that includes a factor which corrects for a reduction in compressive strength due to age, and should only be a function of the rebound value and compressive strength. The influence of concrete age and carbonation was studied in greater detail by Szilágyi et al. [11], who say that even though the rebound hammer test has been available for over 60 years, available literature lacks models that understand surface hardness as an age-dependent property. Their paper [11] presents a SBZ model; a phenomenological constructive model, which operates with concrete hardness measured by the rebound hammer test in relation to time. The SBZ model is based on the development of the capillary pore system in hardened cement paste, which, for reasons of simplicity, is replaced by the w/c ratio. The model includes the relationship between w/c ratio and 28-day compressive strength, development of compressive strength over time, relationship between compressive strength and rebound number at the age of 28 days, and the progress of carbonation depth over time and its influence on the rebound number. As with carbonation, high temperatures also affect surface hardness and compressive strength. Panedpojaman and Tonnayopas [23] found that high temperatures are detrimental to compressive strength. If concrete is exposed to fire, up to about 420 °C the rebound number does not change in any major way, but compressive strength is lost. This is because calcium carbonate crystals form in the pore structure, causing hardness to decrease at a much slower rate and thus rendering the rebound hammer test unusable for determining compressive strength.

Correct statistical analysis of the measured values is critical for evaluating Schmidt hammer tests. Alwash et al. [24] analysed the influence of several factors on the reliability of rebound hammer test evaluation, such as within-test variability, variability of true compressive strength, number of test locations and cores used to determine the relationship between strength and the rebound number, way of choosing the test locations (random or conditional) and the model identification programme (regression or bi-objective). El Mir and Nehme [1] tested several hundred specimens focusing on the coefficient of variance in rebound values. They found that greater carbonation depth, higher w/c ratio, or higher porosity of the concrete cause the coefficient of variance to increase and reduce repeatability. However, concrete containing additions, such as metakaolin or silica fume, high-strength and high-performance self-compacting concrete, exhibit a lower coefficient of variance in rebound values. Szilágyi et al. [25] conducted an extensive statistical analysis of the variability of concrete hardness using a database of both laboratory and in situ measurements spanning over the past 60 years. The study covers several thousand tests (over eighty thousand individual rebound numbers) and shows that current sources and standards leave much to be desired in terms of the evaluation of concrete using a Schmidt impact hammer. The normality tests (the Shapiro–Wilk normality test) yielded rather inconsistent results: "the hypothesis of normality can only be accepted at very low levels of probability for individual test locations." Given the density function of the coefficient of variance of the rebound values, a strong positive skewness can be seen. Szilágyi et al. [25] speak against directly correlating the average rebound number with compressive strength as univariate functions, and argue for a "series of multivariate functions with independent variables of the degree of hydration, type and amount of cement and aggregate, environmental conditions, and testing conditions."

Besides assessing the quality of mature concrete or measuring its compressive strength, which, of course, carries certain associated problems, more modern types of impact hammers are capable of estimating the very early strength (also known as stripping strength) of concrete of only a few hours or days of age. This paper focuses primarily on creating a conversion formula to assess this early compressive strength, which is closely related to e.g., determining a time when it is safe to remove formwork. In fact, this is close to the original purpose of Schmidt hammers–testing concrete which is only several months old.

Many types of sclerometers have been developed over the past 70 years; however, the most common type used presently is the Schmidt type. There are several types available, differing from one another in the impact energy, shape and size of the plunger, or mechanical construction. In the past, and possibly even today, the most common Schmidt hammer is the Original Schmidt: a traditional rebound hammer, which became the basis for all the major rebound tests worldwide [26]. The basic type is the Schmidt N with impact energy of 2.207 Nm, but there is also the L type with energy of 0.735 Nm. A little over ten years ago a new model was introduced: the SilverSchmidt, which uses optical sensors to measure the impact and rebound velocity immediately before and after impact. It does not return the rebound number, but the Q-value. The different construction of the SilverSchmidt gives it the ability to measure concretes of lower as well as higher strength, and is supplied in the N and L configuration. A SilverSchmidt L can be fitted with a mushroom plunger accessory that enables measuring compressive strength as low as 5 N/mm^2. It is designed primarily for determining concrete strength uniformity and identifying inferior areas [27]. Upon introduction to the market, the manufacturer promised better quality of measurement, but this may not be entirely true: Viles et al. [17] consider the SilverSchmidt better than its predecessor, while Szilágyi et al. [25] rank it worse than the Original Schmidt. This may be why the newest model of the Schmidt hammers can only measure the rebound number and not the Q-value: it is the Original Schmidt Live supplied with a complex application (Apple iOS and Android) for measurement, reporting, and analysis [28].

The experiment described below was performed with an Original Schmidt N, SilverSchmidt N, SilverSchmidt L, and SilverSchmidt L with a mushroom plunger (MP) accessory.

2. Models of Relationship between Hardness Tests and Compressive Strength

2.1. Common Dependence Models

This section discusses some models of dependence commonly used in contemporary civil engineering for estimating compressive strength based on rebound hammer tests. They will be further compared with models created using the experimental data.

The standard [21] defines two models of dependence between test results obtained by an Original Schmidt N and compressive strength. These are two lines for which two different ranges apply.

Line A:

$$f_c = 1.75a - 29, \tag{1}$$

where $a = 25 - 40$ [-] is the rebound value and
Line B:

$$f_c = 1.786a - 30.44, \tag{2}$$

for rebound number range $a = 41 - 54$ [-].

The document [29] presents dependence models for the SilverSchmidt. The relationship for test results obtained by the SilverSchmidt N is discussed in two ways: First is the median relationship with 50% reliability:

$$f_c = 1.8943e^{0.064Q} \tag{3}$$

for $Q = 20 - 62$ [-]. This curve was created based on test results obtained by the BAM institute (Federal Institute for Materials Research and Testing in Berlin, Germany) with three different kinds of concrete, which differed in the w/c ratio and type of cement, covering a strength range of $f_c = 10 - 100$ N/mm^2 [29].

Based on the results obtained by the BAM institute, The Shaanxi Province Construction Science Research Institute, China, and Hunan University, China, the document [29] defines a curve with 90% reliability

$$f_c = 2.77e^{0.048Q} \tag{4}$$

for a range of $Q = 22 - 75$ [-]. The curve was created on the basis of recommendations in EN 13791 [30], ASTM C805 [31] and ACI 228.1 [32]. They state that the dependence model should be created so that 90% of the experimental data would lie above the curve.

The [29] also describes a model for results obtained by the SilverSchmidt L for a range of $Q = 20 - 62$ [-] as

$$f_c = 1.9368e^{0.0637Q}. \tag{5}$$

Measurements by the SilverSchmidt L can also be made with the mushroom attachment, which should enable measurements of young concrete with low compressive strength. The relationship

$$f_c = 0.0108Q^2 + 0.2236Q \tag{6}$$

is defined by [27] for a range of $Q = 13 - 44$ [-] and $f_c = 5 - 30$ N/mm^2.

2.2. Characteristic Curve

Drawing a general relationship between the rebound number and compressive strength can be a complex task. The aim of this part of the article is to show a new approach to the evaluation of NDT test results by designing the so-called characteristic curve, which ensures 95% reliability.

Several papers were published saying that formulating strength as a single parameter can be misleading [11]. Single-parameter formulation means that determining compressive strength requires knowing only the rebound number. Several publications [11,13] warn that the model of the relationship should also include e.g., the type of cement, aggregate, or w/c ratio. These and other parameters have an undeniable influence on the strength of concrete and therefore the rebound number measured by a Schmidt hammer as well.

The actual method of seeking the optimal model of relationship also requires attention. The most common method, known as regression analysis, or the least squares method, involves several assumptions, some of which may be violated during data evaluation. A determination of an optimal relationship between the rebound test and compressive strength often violates the assumption of homoscedasticity, i.e., homogeneity of variance [11]. A violation of the homogeneity of variance means that changes in the rebound number (x axis) change variation in compressive strength (y axis) as well. In other words, the higher the rebound number measured, the higher the variance in compressive strength. If the violation of homoscedasticity is ignored, the resulting model of relationship can underrepresent the value of compressive strength, especially in higher-strength concrete.

However, the experimental data presented in this paper shows that under certain circumstances the above-mentioned issues can be legitimately dismissed. These are cases where the goal is not to find a general model for every concrete, but a specific model for only one. Moreover, such a model is not intended for measuring compressive strength across the whole spectrum, but for estimating the stripping rebound number, which should correspond to stripping compressive strength (see below).

Most previously published models of the relationship between the rebound number and compressive strength are designed to estimate the median value of compressive strength. The model is therefore designed as a median curve plotted through the experimental data. In theory, a model thus designed is 50% reliable: the measured value of compressive strength is 50% likely to be higher or lower.

In the civil engineering practice, however, most cases do not use 50% reliability, but 95%. Such a value of a material property is called a characteristic value. Concerning compressive strength, there is the term characteristic strength [30]. It is essentially always a 5% quantile, meaning that 95% of test results should be higher than this value. It is, therefore, a sort of one-sided interval estimate of the parameter value. Even when assessing concrete strength based on NDT results, it would often be useful to have such a one-sided estimate in the form of a curve. Points on this curve, which could be called a characteristic curve, would determine 95% one-sided interval estimates of compressive strength; i.e., characteristic strength. This section presents one of the possible ways how to construct such a curve using experimental data.

This is done using the above-mentioned method of least squares in its simplest form of a linear regression model [33]:

$$y = b_1 + b_2 \cdot x, \tag{7}$$

where y is the conventional dependent variable (compressive strength), x is the independent variable (rebound number) and b_1 and b_2 are the regression coefficients being determined. As will be shown later, this simple model appears suitable for all the experimental data examined here. For a better understanding of how the characteristic curve is constructed, we shall demonstrate the principle of the least squares method.

The experimental data is essentially pairs (x_i, y_i), where x_i are the rebound number values and y_i are the corresponding values of compressive strength. Further, $i = 1, \ldots, n$, where n is the number of measurements, i.e., value pairs. An important role in the least squares method is played by the matrix

$$\mathbf{H} = \begin{bmatrix} n & \sum x_i \\ \sum x_i & \sum x_i^2 \end{bmatrix}, \tag{8}$$

where $\sum$ signifies $\sum_{i=1}^{n}$, and its determinant, which can be expressed as

$$\det \mathbf{H} = n \sum x_i^2 - \left(\sum x_i\right)^2. \tag{9}$$

The median values of regression coefficients can then be expressed using the following formulas:

$$b_2 = \frac{n \sum x_i y_i - \sum x_i \sum y_i}{\det \mathbf{H}}, \tag{10}$$

$$b_1 = \bar{y} - b_2 \bar{x}, \tag{11}$$

where $\bar{y}$ and $\bar{x}$ are the mean values of the properties being measured [34].

At this point the model has a curve passing through the experimental data. It is generally recommended to supplement this curve with so-called confidence bands. This means a confidence band for the median curve, which is normally determined by 95 % interval estimates of b_1 and b_2, and a prediction band, which determines the 95% interval estimate of compressive strength for the given rebound number. In order to construct such bands it is necessary to express the minimum value of the sum of square errors [33,34]

$$S^*_{min} = \sum_{i=1}^{n} (y_i - b_1 - b_2 x_i)^2 \tag{12}$$

and point estimator of variance

$$s^2 = \frac{S^*_{min}}{n-2}. \tag{13}$$

Then, for any fixed x, it is necessary to denote the value of h^*:

$$h^* = \frac{1}{n} + \frac{n(x-\bar{x})^2}{\det \mathbf{H}}. \tag{14}$$

The confidence band for the median value is then determined using the formula

$$\left\langle (b_1 + b_2x) - t_{(1-\alpha/2)}s\sqrt{h^*}; (b_1 + b_2x) + t_{(1-\alpha/2)}s\sqrt{h^*} \right\rangle, \tag{15}$$

where $t_{(1-\alpha/2)}$ is $(1-\alpha/2)$ quantile of Student's t-distribution with $n-2$ degrees of freedom [33,34]. Next, the confidence band for the individual values (prediction band) is determined by the following formula

$$\left\langle (b_1 + b_2x) - t_{(1-\alpha/2)}s\sqrt{1+h^*}; (b_1 + b_2x) + t_{(1-\alpha/2)}s\sqrt{1+h^*} \right\rangle. \tag{16}$$

Both regression bands are shown in Figure 5 through Figure 11. It is useful to expand this traditional regression analysis method with regression coefficient testing and assessing the overall aptitude of the model with the multiple correlation coefficient, which in our case equals the correlation coefficient r. The aptitude of the model is most commonly denoted by the coefficient of determination r^2. The number $r^2 \times 100\%$ (conventionally) signifies the percentage of y_i, which is explained by the regression model.

It is important to remember that the confidence bands are constructed as two-sided interval estimates of y for every fixed x. This method of denotation works for most applications of regression analysis. However, seeing as civil engineering usually works with the characteristic value (5% quantile), this method is not quite ideal.

We construct the characteristic curve using the relationship for determining the prediction band. The breadth of the band is determined by variance s^2, regression coefficients b_1 and b_2, the h^* value for every fixed x, and Student's t-distribution. This quantile determines, among others, the aptitude of the confidence bands, i.e., the probability with which one could expect that the true y-value is indeed located within this band. Adjusting the $t_{(1-\alpha/2)}$ quantile to the $t_{(1-\alpha)}$ quantile with the same degree of freedom $n-2$ at $\alpha = 0.05$ makes it possible to obtain the desired one-sided interval estimate. The characteristic curve $y_{0.05}$ can then be written as

$$y_{0.05} = (b_1 + b_2x) - t_{(1-\alpha)}s\sqrt{1+h^*}. \tag{17}$$

This curve essentially determines the value of compressive strength for every fixed x, i.e., for every fixed result of a rebound test. It is, therefore, a half-plane

$$\langle +\infty; y_{0.05} \rangle. \tag{18}$$

The following sections show the use of this characteristic curve for determining the characteristic stripping hardness with real-world data.

3. Experiment

The goal of the measurements and their evaluation was to create a new conversion relationship for determining the compressive strength of concrete by rebound hammer tests. These relationships concern mainly early-age strength: the so-called stripping strength, i.e., in the range of 5 to 10 N/mm^2. The experiment took place in three stages.

3.1. Determining the Conversion Relationship for Compressive Strength of Concrete Used in Bridge Construction

The goal of the first stage was two-fold; to determine a conversion relationship for the rebound number to compressive strength for the selected concrete and to ascertain whether the Original Schmidt N and SilverSchmidt rebound hammers are suitable for this purpose: whether the statistical evaluation would confirm that the more advanced SilverSchmidt suffers from greater measurement variability [25].

The experiment required making 18 cubic specimens of 150 mm in size. The specimens were cast on-site, during the construction of an arch bridge on Svitavská street in Brno, see Figure 1. The concrete in question was C 30/37 XF4; its composition is shown in Table 1. The properties of used cement are shown in Tables 2 and 3. Basic properties of all used aggregates (including aggregates used in Sections 3.2 and 3.3) are shown in Table 4. The admixtures are described in Table 1 by name and further information can be found in the manufacturer's technical data [35,36].

The specimens were made from concrete taken from three concrete mixer trucks. Six cube specimens were made from each concrete sample, see Figure 2a. The fresh-state properties, determined according to EN 12350 [37–39], are in Table 5, and the slump test is pictured in Figure 2b. After pouring, the specimens were covered with a PE sheet and left at the construction site for 24 h. At the age of 48 h they were transported to a laboratory at the Faculty of Civil Engineering, BUT, where they were removed from moulds and divided into 6 groups of 3 so that each group would contain a cube from truck 1, truck 2, and truck 3. The first group of 3 specimens was tested (compressive strength according to EN 12390-3 [40]) and the remaining 15 were placed under water at a temperature of (20 ± 2) °C. The other groups were tested at the age of 3, 7, 14, 28, and 90 days. The reason the concrete was tested at different ages was to obtain data that show how the concrete's properties develop over time. Because the concrete's composition is known, it is possible to use the single-parameter formulation of compressive strength using the rebound number.

Figure 1. Bridge construction during which concrete C 30/37 XF4 was sampled.

Table 1. Composition of concrete C 30/37 XF4.

Component	Amount (kg Per 1 m^3)
Cement CEM I 42.5 R (Mokrá cement plant)	400
Aggregate 0-4 mm—Ledce (see Table 4)	700
Aggregate 8-16 mm—Olbramovice (see Table 4)	669
Aggregate 11-22 mm—Lomnička (see Table 4)	284
Water	172
Air-entraining admixture—Sika® Aer 200	0.60
Plasticiser—Sika® ViscoCrete®-5-800 Multimix (AT)	2.40

Table 2. Chemical properties of Cement CEM I 42.5 R (Mokrá cement plant) according to EN 196-2 [41].

Component/Property	Average Value (%)
CaO	63.7
SiO_2	19.6
Al_2O_3	4.8
Fe_2O_3	3.3
MgO	1.4
SO_3	3.1
Cl^-	0.040
K_2O	0.75
Na_2O	0.19
Na_2O Equivalent	0.69
Insoluble residue	0.7
Loss of ignition	3.4

Table 3. Basic physical and mechanical properties of Cement CEM I 42.5 R (Mokrá cement plant) according to EN 196-6 [42] and EN 196-8 [43].

Parameter	Average Value
Blain (m^2/kg)	408
Density (kg/m^3)	3110
Heat of hydration (7 days) (J/g)	310

Table 4. Basic properties of used aggregates.

Aggregate	Type	Rock	Specific Density (Mg/m^3)	Loose Bulk Density (Mg/m^3)
Ledce	Natural quarried	Gravel sand	2.553	1.408
Lípa	Natural quarried	Gravel sand	2.583	1.508
Bratčice	Natural quarried	Gravel sand	2.610	1.537
Olbramovice	Natural crushed	Granodiorite	2.640	1.450
Lomnička	Natural crushed	Gneiss	2.690	1.530
Litice	Natural crushed	Granodiorite	2.700	1.600

Table 5. Fresh-state properties of concrete C 30/37 XF4.

Mix Truck	Slump Test (mm)	Air Content (%)	Bulk Density (kg/m^3)
1	240	4.2	2320
2	180	3.8	2370
3	220	3.9	2360

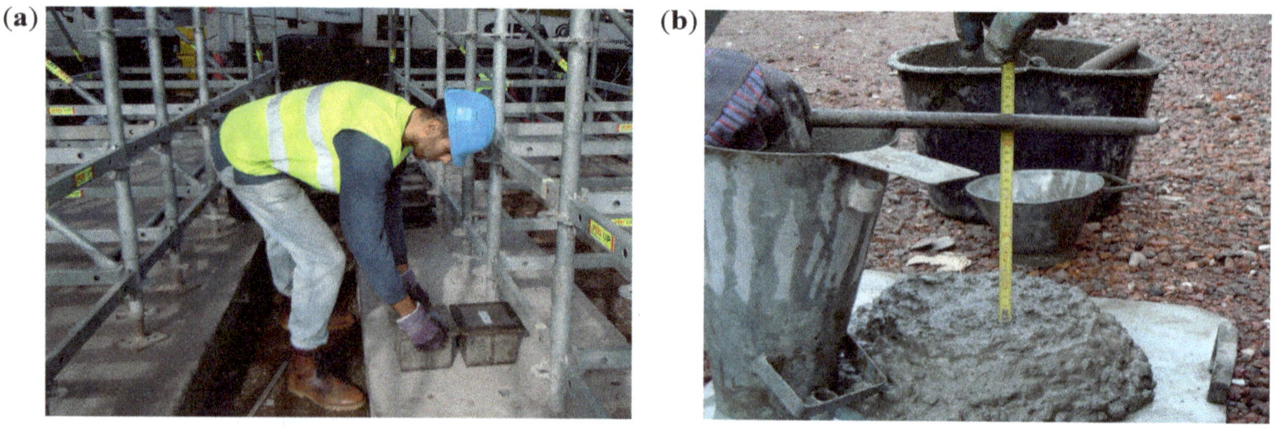

Figure 2. (**a**) cube specimens being made; (**b**) consistency test-slump.

Each cube was measured and weighed, and afterwards mounted in a testing press and compressed with a force of 50 kN. Two opposing sides were tested for hardness with a rebound hammer. A total of 5 readings of the rebound number R were taken using an Original Schmidt N and 5 readings of of the Q-value using a SilverSchmidt N, see Figure 3. Finally, compressive strength was determined according to [40].

(a) (b)

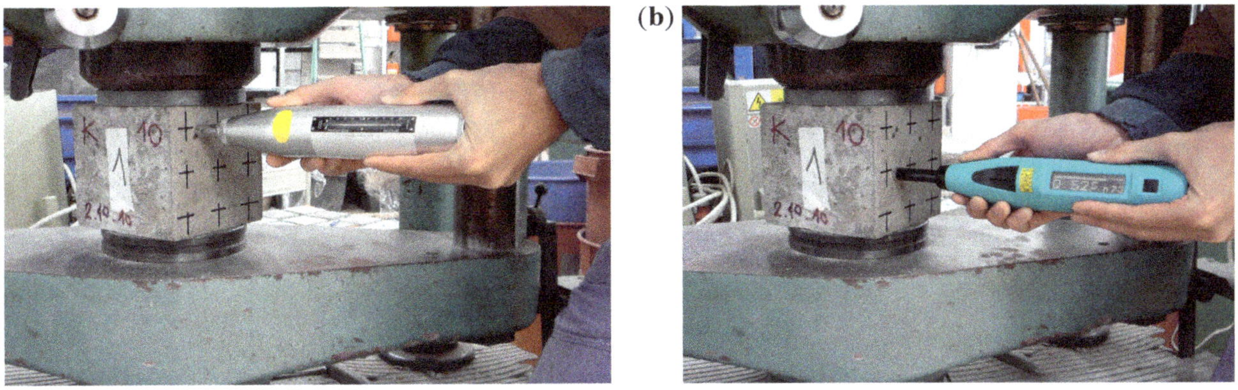

Figure 3. (**a**) testing a cube with an Original Schmidt N; (**b**) with a SilverSchmidt N.

3.2. Determining the Conversion Relationship for the Stripping Strength of Precast Concrete

The goal of the second part of the experiment was to determine the relationship between the rebound number and compressive strength for concrete C 50/60, which is used in precast prestressed girders. Effort was made to capture low (stripping) strength. The formula used for making the V03 girders is detailed in Table 6. V03 girders have an asymmetrical I cross-section, length of 25.346 m, height of 1.5 m and were used in the construction of a shopping centre.

Table 6. Composition of concrete C 50/60.

Component	Amount (kg Per 1 m^3)
Cement CEM I 42.5 R (Mokrá cement plant)	450
Aggregate 0-4 mm—Lípa (see Table 4)	690
Aggregate 4-8 mm—Litice (see Table 4)	215
Aggregate 8-16 mm—Litice (see Table 4)	845
Water	180
Plasticiser—Stachement 2180	4.50

While the girders were being made, concrete C 50/60 was sampled and made into 18 cube-shaped specimens with the size of 150 mm. While still freshly poured in steel moulds, they were covered with a PE sheet and left near the girders in the manufacturing hall, where the ambient temperature did not exceed 15 °C. Because the experiment aimed to test compressive strength at a very young age, the first three cubes were tested 27 h after cement was mixed with water. The other specimens were unmoulded at this age, placed under water and tested at the age of 42, 48, 68, 144, and 656 h. A total of 15 cubes were tested during the first 6 days and the last three at 28 days of age.

The testing was conducted similarly to the first stage of the experiment with the sole difference that besides the Original Schmidt N and SilverSchmidt N, the SilverSchmidt L both with and without the mushroom attachment was used. The SilverSchmidt L with the attachment is primarily used to test very young concrete, which is why it was used only during the first 2 days (i.e., first three measurement times), but not beyond. Figure 4 shows the manufacturing and testing of the cube specimens.

Figure 4. (**a**) making of the specimens; (**b**) manufacture of the V03 girders.

3.3. Determining the Conversion Relationship for the Stripping Strength of Concretes of Similar Composition

While the previous parts of the experiment were carried out with only one kind of concrete, the third involved making cube specimens from six different concrete mixtures. They only differed in the amount of cement, water, and admixtures; in essence, the w/c ratio. The actual components of the six concretes (marked I through VI) were identical. The amount of water was always balanced against the amount of admixtures so as to achieve the same workability in all the mixtures. Table 7 shows the composition of the concretes.

Table 7. Composition of the concretes.

Component	Amount (% Per 1 m^3)					
	I	II	III	IV	V	VI
Cement CEM I 42.5 R (Mokrá cement plant)	87.50	100.00	87.50	100.00	87.50	100.00
Aggregate 0/4 mm—Bratčice (see Table 4)	218.75	206.25	218.75	206.25	218.75	206.25
Aggregate 4/8 mm—Olbramovice (see Table 4)	46.25	46.25	46.25	46.25	46.25	46.25
Aggregate 8/16 mm—Olbramovice(see Table 4)	173.75	173.75	173.75	173.75	173.75	173.75
Water	44.75	44.75	43.50	43.50	41.00	41.00
Plasticiser—Sika® ViscoCrete® 4035	0.220	0.250	0.438	0.500	0.438	0.500
Air-entraining admixture—Sika® LPS A 94	0.000	0.000	0.000	0.000	0.188	0.188

The basic fresh-state properties determined in compliance with [37–39] are detailed in Table 8.

Table 8. Composition of the concretes.

Property	Concrete					
	I	II	III	IV	V	VI
Bulk Density (kg/m^3)	2300	2300	2270	2300	2190	2260
Slump test (mm)	50	60	70	50	60	50
Air content (%)	2.8	3.2	3.5	3.0	6.2	5.7

Each concrete was used to make 6 cube specimens sized 150 mm. This made a total of 36 test specimens. While still in polyurethane moulds, the specimens were covered with a PE sheet and stored at standard laboratory conditions (ambient temperature of (20 ± 2) °C and humidity of (60 ± 10)%), where they aged until testing. All the specimens were tested during the next day after being cast; the age ranged between 16 and 36 h. Immediately after unmoulding, the specimens were measured and weighed, and then mounted in a testing press and tested for the Q-value ten times using a SilverSchmidt PC L with the mushroom accessory attached; five measurements on two opposing sides. Finally, the compressive strength test was performed according to [40]. The goal was to determine whether it is possible to create one conversion relationship for the stripping compressive strength of several concretes of similar composition.

4. Results and Discussion

Test results presented in Section 3.1 (concrete bridge) and Section 3.2 (precast concrete) each time represent one concrete of known composition. The hardness and strength tests were designed so that the experimental data described the development of compressive strength over time. This justifies the single-parameter formulation of compressive strength based on the rebound number using linear regression model. The same type of model was used by the authors of the article [22], while most authors (see [3,7,10,19,44]) use rather polynomial or exponential function curves. In addition, authors of [11] refer to dozens of other publications, all of which use non-linear functions. The linear model was used in Section 3.3 as well, where there were several different kinds of concrete, but of very similar composition. These concretes varied in the amount of admixtures, aggregate, and w/c ratio, but the raw materials were the same.

The data evaluation was focused on the stripping rebound number, which should correspond to stripping compressive strength with 95% reliability; i.e., the value read from the characteristic curve (cf. Section 2.2). Stripping compressive strength, meaning strength at which formwork can be safely removed, is set here at 5 N/mm^2.

The evaluation of results of the bridge concrete (Section 3.1) focused on the aptitude of the linear regression model for tests performed with the Original Schmidt N and SilverSchmidt N impact hammers (Figures 5 and 6). The values of the coefficient of determination r^2 indicate that both these models show high aptitude, which, compared to Szilágyi's et al. conclusions [25] is slightly higher in the case of the SilverSchmidt N. The results also show that the linear regression model used herein is more suitable for the evaluation of measurements made with a SilverSchmidt N than the exponential model proposed in [29] (see Figure 6). In both cases, a characteristic curve was created and the stripping rebound numbers evaluated. Given the range of the data obtained, approx. 30 to 80 N/mm^2, these values are strongly extrapolated and thus are for illustration only. Table 7 summarises the results.

The results of tests described in Section 3.2 (precast concrete) are plotted in Figures 7–10 including linear regressions. In this case the linear regression model also showed aptitude, exhibiting high values of the coefficient of determination r^2 (see Table 9). The results of tests performed with an Original Schmidt N on this particular concrete show that relationships described in [21] may undervalue compressive strength (see Figure 7). The non-linear character of the dependence of rebound tests performed with a SilverSchmidt N, L, and L-MP [27,29] was not confirmed.

The greatest aptitude of the linear model was demonstrated by test results for the similar concretes described in Section 3.3 ($r^2 = 0.963$). These tests were performed with a SilverSchmidt L with the mushroom plunger accessory. As with the previous cases, these results indicated no need to use an exponential or other model.

Table 9. Summary.

Rebound Hammer	Concrete Type	$Q_{0.50}$ ($a_{0.50}$)	$Q_{0.05}$ ($a_{0.05}$)	r^2	Figure
Original Schmidt N	Bridge concrete	14.3	17.9	0.878	Figure 5
SilverSchmidt N	Bridge concrete	24.3	28.2	0.922	Figure 6
Original Schmidt N	Precast concrete	11.9	17.2	0.949	Figure 7
SilverSchmidt N	Precast concrete	17.6	26.7	0.922	Figure 8
SilverSchmidt L	Precast concrete	18.2	24.6	0.957	Figure 9
SilverSchmidt L–MP	Precast concrete	17.4	23.6	0.953	Figure 10
SilverSchmidt L–MP	Concretes of similar composition	13.5	15.9	0.963	Figure 11

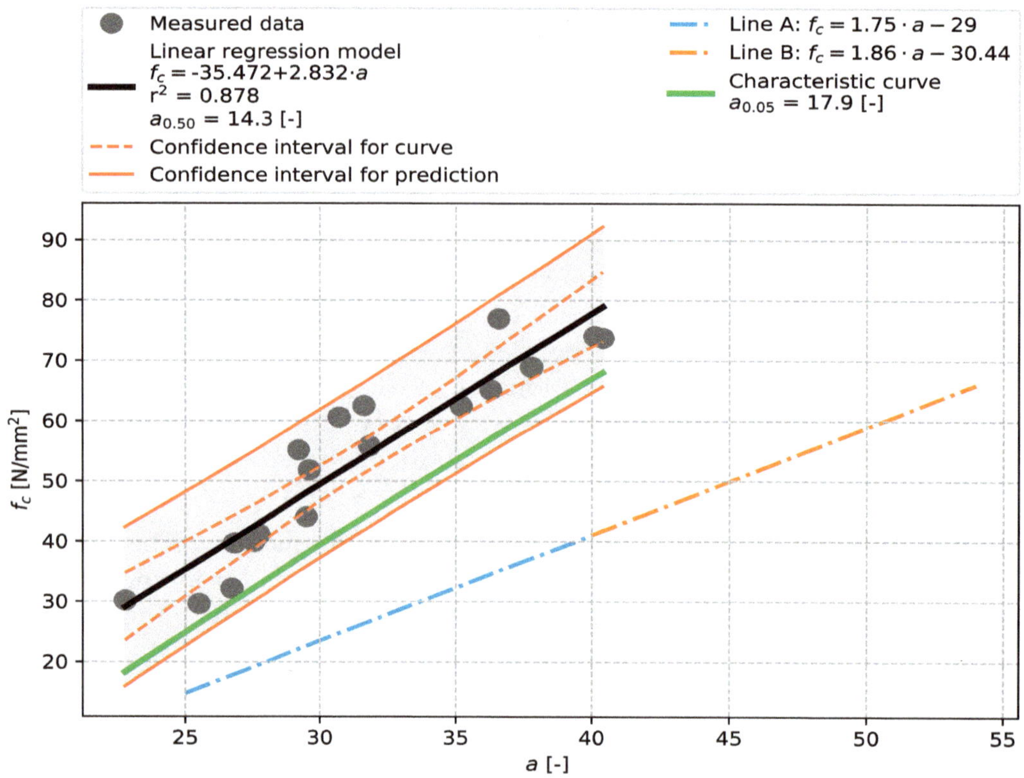

Figure 5. Bridge concrete described in Section 3.1—model of the dependence of the rebound number [-] vs. compressive strength f_c [N/mm^2]—Original Schmidt N.

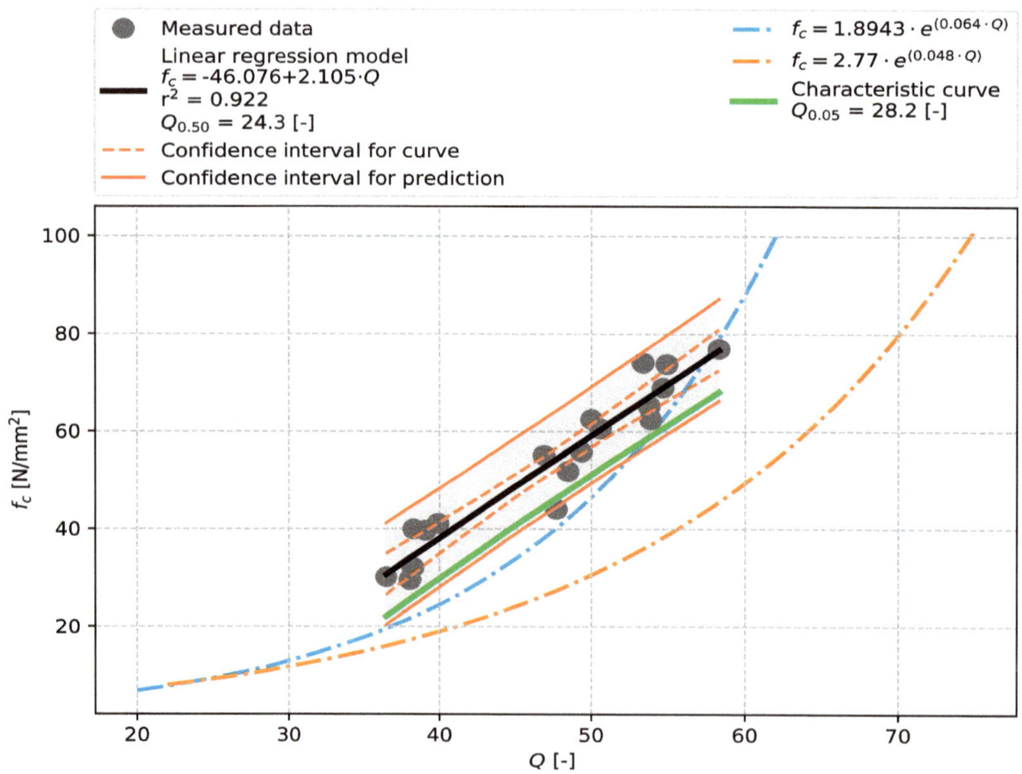

Figure 6. Bridge concrete described in Section 3.1—model of the dependence of the Q-value [-] vs. compressive strength f_c [N/mm^2]—SilverSchmidt N.

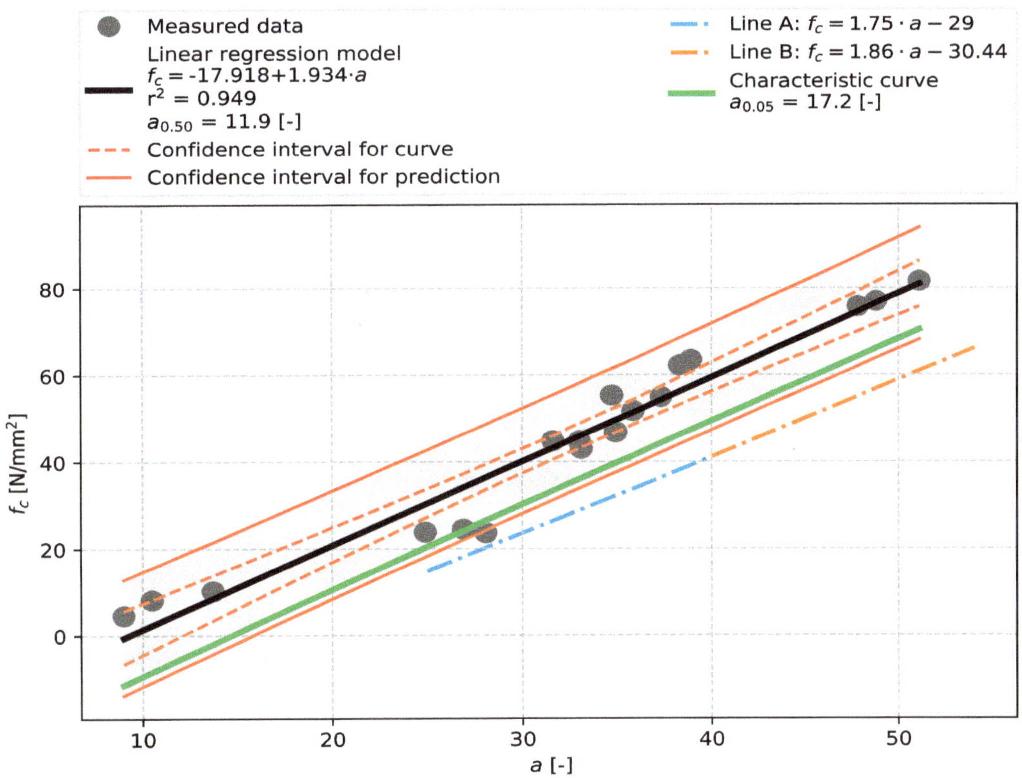

Figure 7. Precast concrete described in Section 3.2—model of the dependence of a [-] vs. compressive strength f_c [N/mm^2]—Original Schmidt N.

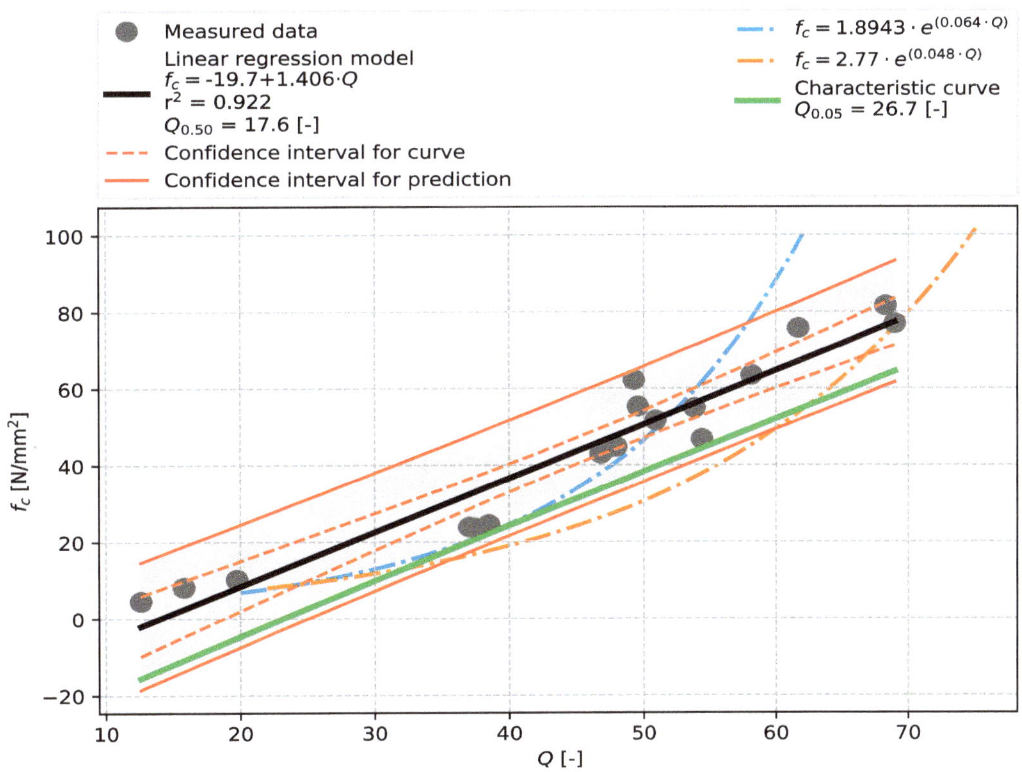

Figure 8. Precast concrete described in Section 3.2—model of the dependence of Q-value [-] vs. compressive strength f_c [N/mm^2]—SilverSchmidt N.

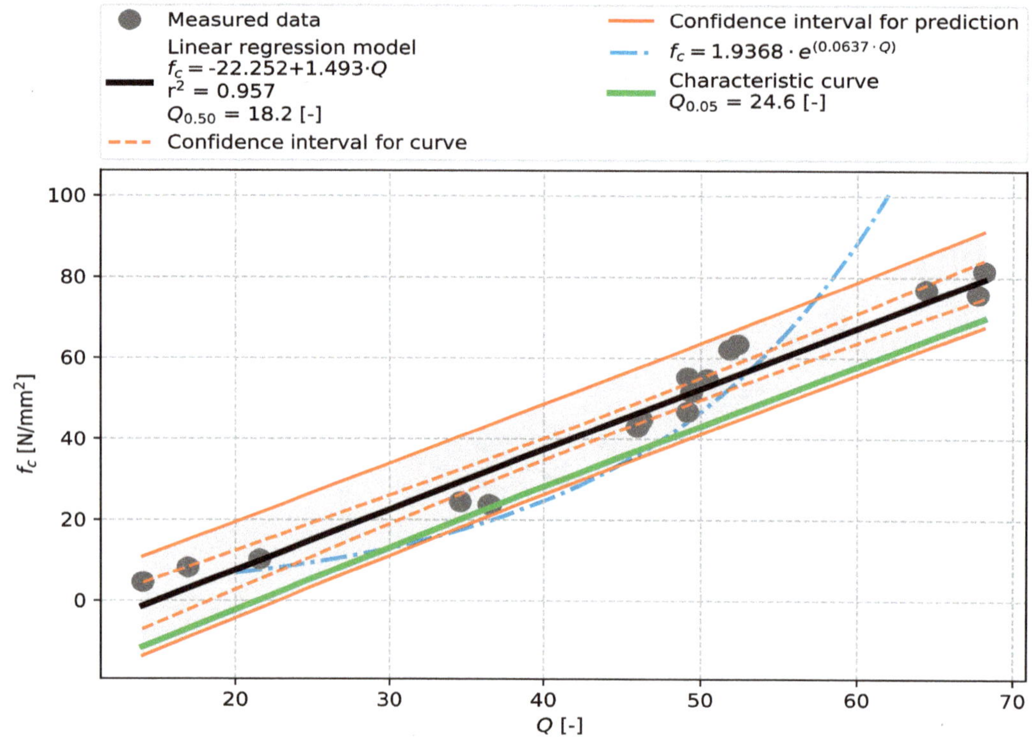

Figure 9. Precast concrete described in Section 3.2—model of the dependence of Q-value [-] vs. compressive strength f_c [N/mm^2]—SilverSchmidt L.

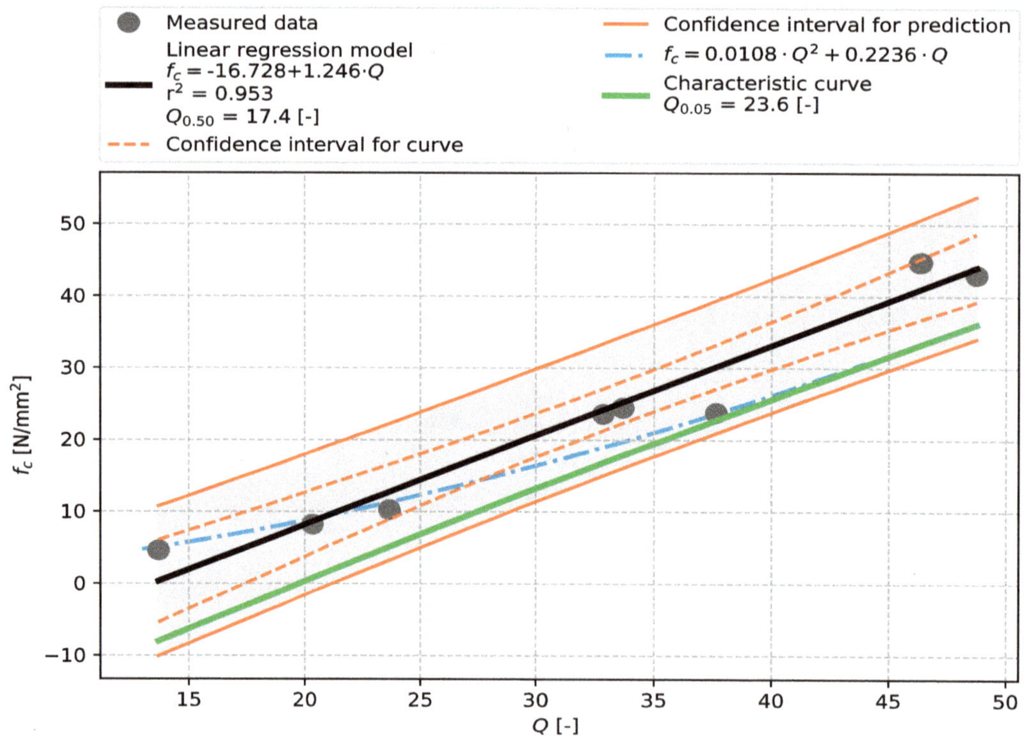

Figure 10. Precast concrete described in Section 3.2—model of the dependence of Q-value [-] vs. compressive strength f_c [N/mm^2]—SilverSchmidt L—MP.

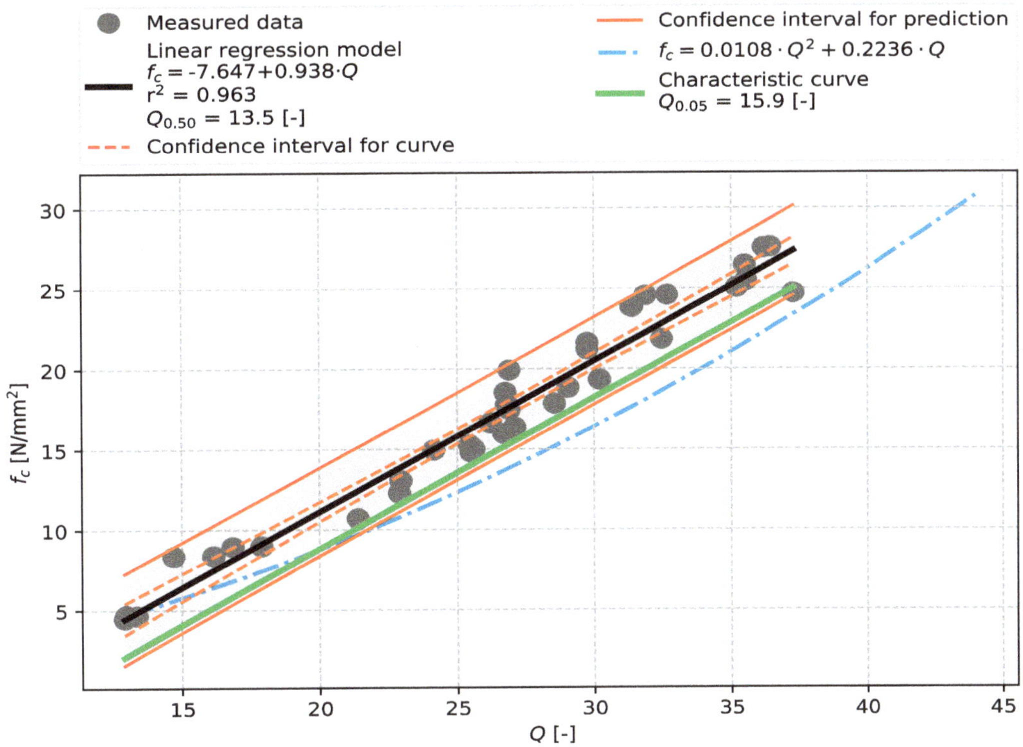

Figure 11. Precast concrete described in Section 3.3—model of the dependence of Q-value [-] vs. compressive strength f_c [N/mm^2]—SilverSchmidt L – MP.

5. Conclusions

The experiments indicate the following:

- The paper describes an innovative design of constructing a characteristic curve, which can be used to find the stripping rebound number obtained by different types of rebound hammers. Using the characteristic curve appears suitable for finding a specific rebound number which would indicate, with 95% reliability, that the concrete has the required stripping strength (in this paper it is 5 N/mm^2).
- The data shows that the concretes tested herein (including concretes with different formulas but the same raw materials) do not require a regression model more complex than a simple line. Moreover, this simplest shape of regression model enables a relatively simple formulation of the characteristic curve.
- We certainly do not question the validity of relationships described elsewhere (Section 2.1), or their ability to represent the relationship between the rebound number and compressive strength. It should also be remembered that these general models were created for concretes of virtually any composition, and a simple line would not be enough.
- Our regression models and rebound numbers representing stripping strength always apply to the one specific concrete they were designed for and do not work with others. The goal of the paper was to show how such regression models may be created for any concrete.

Author Contributions: Conceptualization, D.K. and P.M.; methodology, D.K., P.M. and P.C.; software, P.M.; validation, D.K. and P.M.; formal analysis, D.K. and P.M.; investigation, D.K. and P.C.; resources, D.K.; data curation, P.M.; writing–original draft preparation, D.K. and P.M.; writing–review and editing, D.K. and P.M.; visualization, P.M.; supervision, P.C.; project administration, D.K.; funding acquisition, D.K.

Acknowledgments: Authors would like to thank to Zahrada, Bílek and Žalud for support.

References

1. Mir, A.E.; Nehme, S.G. Repeatability of the rebound surface hardness of concrete with alteration of concrete parameters. *Constr. Build. Mater.* **2017**, *131*, 317–326. [CrossRef]
2. Bungey, J.H.; Millard, S.G.; Grantham, M. *Testing of Concrete in Structures*, 4th ed.; CRC Press: London, UK, 2006.
3. Ali-Benyahia, K.; Sbartaï, Z.M.; Breysse, D.; Kenai, S.; Ghrici, M. Analysis of the single and combined non-destructive test approaches for on-site concrete strength assessment. *Case Stud. Constr. Mater.* **2017**, *6*, 109–119. [CrossRef]
4. *Handbook on Nondestructive Testing of Concrete*, 2nd ed.; CRC Press: Boca Raton, FL, USA, 2004.
5. Xu, T.; Li, J. Assessing the spatial variability of the concrete by the rebound hammer test and compression test of drilled cores. *Constr. Build. Mater.* **2018**, *188*, 820–832. [CrossRef]
6. Review of the Rebound Hammer Method Estimating Concrete Compressive Strength on Site. In Proceedings of the International Conference on Architecture And Civil Engineering (ICAACE'14), Dubai, UAE, 25–26 December 2014; pp. 118–127. [CrossRef]
7. Breysse, D. Nondestructive evaluation of concrete strength. *Constr. Build. Mater.* **2012**, *33*, 139–163. [CrossRef]
8. Balayssac, J.P.; Garnier, V. *Non-Destructive Testing and Evaluation of Civil Engineering Structures*; Elsevier: Kidlington, UK, 2018.
9. Breysse, D.; Martínez-Fernández, J.L. Assessing concrete strength with rebound hammer. *Mater. Struct.* **2014**, *47*, 1589–1604. [CrossRef]
10. Jain, A.; Kathuria, A.; Kumar, A.; Verma, Y.; Murari, K. Combined Use of Non-Destructive Tests for Assessment of Strength of Concrete in Structure. *Procedia Eng.* **2013**, *54*, 241–251. [CrossRef]
11. Szilágyi, K.; Borosnyói, A.; Zsigovics, I. Rebound surface hardness of concrete. *Constr. Build. Mater.* **2011**, *25*, 2480–2487. [CrossRef]
12. Schmidt, E. *Rebound Hammer for Concrete Testing*; Schweiz Bauztg: Berlin, Germany, 1950.
13. Szilágyi, K.; Borosnyói, A.; Dobó, K. Static indentation hardness testing of concrete: A long established method revived. *Epa.-J. Silic. Based Compos. Mater.* **2011**, *63*, 2–8. [CrossRef]
14. Basu, A.; Aydin, A. A method for normalization of Schmidt hammer rebound values. *Int. J. Rock Mech. Min. Sci.* **2004**, *41*, 1211–1214. [CrossRef]
15. Bilgin, N.; Dincer, T.; Copur, H. The performance prediction of impact hammers from Schmidt hammer rebound values in Istanbul metro tunnel drivages. *Tunn. Undergr. Space Technol.* **2002**, *17*, 237–247. [CrossRef]
16. Liang, R.; Hota, G.; Lei, Y.; Li, Y.; Stanislawski, D.; Jiang, Y. Nondestructive Evaluation of Historic Hakka Rammed Earth Structures. *Sustainability* **2013**, *5*, 298–315. [CrossRef]
17. Viles, H.; Goudie, A.; Grab, S.; Lalley, J. The use of the Schmidt Hammer and Equotip for rock hardness assessment in geomorphology and heritage science. *Earth Surf. Process. Landforms* **2011**, *36*, 320–333. [CrossRef]
18. Bui, Q.B. Assessing the Rebound Hammer Test for Rammed Earth Material. *Sustainability* **2017**, *9*, 1904. [CrossRef]
19. Mohammed, B.S.; Azmi, N.J.; Abdullahi, M. Evaluation of rubbercrete based on ultrasonic pulse velocity and rebound hammer tests. *Constr. Build. Mater.* **2011**, *25*, 1388–1397. [CrossRef]
20. Original Schmidt Live Cue Cards. Available online: https://www.proceq.com/compare/schmidt-hammers/ (accessed on 25 June 2019).
21. ÚNMZ, Prague. *ČSN 73 1373: Non-Destructive Testing of Concrete—Determination of Compressive Strength by Hardness Testing Methods*, 1st ed.; ÚNMZ: Prague, Czech Republic, 2011.
22. Kim, J.K.; Kim, C.Y.; Yi, S.T.; Lee, Y. Effect of carbonation on the rebound number and compressive strength of concrete. *Cem. Concr. Compos.* **2009**, *31*, 139–144. [CrossRef]
23. Panedpojaman, P.; Tonnayopas, D. Rebound hammer test to estimate compressive strength of heat exposed concrete. *Constr. Build. Mater.* **2018**, *172*, 387–395. [CrossRef]
24. Alwash, M.; Breysse, D.; Sbartaï, Z.M.; Szilágyi, K.; Borosnyói, A. Factors affecting the reliability of assessing the concrete strength by rebound hammer and cores. *Constr. Build. Mater.* **2017**, *140*, 354–363. [CrossRef]
25. Szilágyi, K.; Borosnyói, A.; Zsigovics, I. Extensive statistical analysis of the variability of concrete rebound hardness based on a large database of 60years experience. *Constr. Build. Mater.* **2014** *53*, 333–347. [CrossRef]

26. Schmidt Rebound Hammers. Available online: https://www.proceq.com/compare/schmidt-hammers/ (accessed on 21 June 2019).
27. SilverSchmidt. Available online: https://www.proceq.com/uploads/tx_proceqproductcms/import_data/files/SilverSchmidt_Operating%20Instructions_English_high.pdf (accessed on 21 June 2019).
28. Original Schmidt Live. Available online: https://www.proceq.com/product/original-schmidt-live/ (accessed on 21 June 2019).
29. The SilverSchmidt Reference Curve. Available online: https://www.pcte.com.au/silver-schmidt-rebound-hammer (accessed on 21 June 2019).
30. CEN. *EN 13791: Assessment of In-Situ Compressive Strength In Structures and Precast Concrete Components*; CEN: Brussels, Belgium, 2006.
31. ASTM International. *ASTM C805/C805M-18: Standard Test Method for Rebound Number of Hardened Concrete*; ASTM International: West Conshohocken, PA, USA, 2018.
32. American Concrete Institute. *ACI 228.1R-03: In-Place Methods to Estimate Concrete Strength*; American Concrete Institute: Farmington Hills, MI, USA, 2003.
33. Chatterjee, S.; Simonoff, J.S. *Handbook of Regression Analysis*; Wiley: Hoboken, NJ, USA, 2013.
34. Pedhazur, E.J.; Kerlinger, F.N. *Multiple Regression in Behavioral Research*, 2nd ed.; Holt, Rinehart and Winston: New York, NY, USA, 1982.
35. Sika Company. Available online: https://cze.sika.com/ (accessed on 21 June 2019).
36. Stachema. Available online: https://prisadydobetonu.stachema.cz/ (accessed on 21 June 2019).
37. CEN. *EN 12350-2: Testing Fresh Concrete—Part 2: Slump-Test*, 1st ed.; CEN: Brussels, Belgium, 2009.
38. CEN. *EN 12350-6: Testing Fresh Concrete—Part 6: Density*; CEN: Brussels, Belgium, 2009.
39. CEN. *EN 12350-7: Testing Fresh Concrete—Part 7: Air Content—Pressure Methods*; CEN: Brussels, Belgium, 2009.
40. CEN. *EN 12390-3: Testing Hardened Concrete—Part 3: Compressive Strength of Test Specimens*; CEN: Brussels, Belgium, 2009.
41. CEN. *EN 196-2: Method of Testing Cement—Part 2: Chemical Analysis of Cement*; CEN: Brussels, Belgium, 2013.
42. CEN. *EN 196-6: Methods of Testing Cement—Part 6: Determination Of Fineness*; CEN: Brussels, Belgium, 2018.
43. CEN. *EN 196-8: Methods of Testing Cement—Part 8: Heat of Hydration—Solution Method*; CEN: Brussels, Belgium, 2010.
44. Pucinotti, R. Reinforced concrete structure. *Constr. Build. Mater.* **2015**, *75*, 331–341. [CrossRef]

14

Inverse Contrast in Non-Destructive Materials Research by using Active Thermography

Paweł Noszczyk * and Henryk Nowak

Department of Building Physics and Computer Design Methods, Faculty of Civil Engineering, Wroclaw University of Science and Technology, 50-370 Wrocław, Poland; henryk.nowak@pwr.edu.pl

* Correspondence: pawel.noszczyk@pwr.edu.pl

Abstract: Background: it is undesirable for defects to occur in building partitions and units. There is a need to develop and improve research techniques for locating such defects, especially non-destructive techniques for active thermography. The aim of the experiment was to explore the possibility of using active thermography for testing large-sized building units (with high heat capacity) in order to locate material inclusions. Methods: as part of the experiment, two building partition models—one made of gypsum board (GB) and another made of oriented strand board (OSB)—were built. Three material inclusions (styrofoam, granite, and steel), considerably differing in their thermal parameters, were placed in each of the partitions. A 7.2 kW infrared radiator was used for thermally exciting (heating) the investigated element for 30 min. The distribution of the temperature field was studied on both sides of the partition for a few hours. Results: using the proposed investigative method, one can detect defects in building partitions under at least 22 mm of thick cladding. At a later cooling down phase, inverse temperature contrasts were found to occur—the defects, which at the beginning of cooling down were visible as warmer areas, at a later phase of cooling down are perceived as cooler areas, and vice versa (on the same front surface). In the transmission mode, the defects are always visible as areas warmer than defect-free areas. Moreover, a quantitative (defect location depth) analysis with an accuracy of up to 10% was carried out using the Echo Defect Shape method. Conclusions: active thermography can be used in construction for non-destructive materials testing. When the recording of thermograms is conducted for an appropriate length of time, inverse contrasts can be observed (on the same front surface).

Keywords: thermovision; active thermography; thermal contrast; defect detection; location of inclusions; non-destructive testing; materials research; building partition

---

1. Introduction

The material structure of a building unit determines its physical parameters (e.g., its resistance to various external impacts). In the case of existing building structures, one can use either destructive, semi-destructive, or non-destructive tests to assess their structural components. If it is technically possible, it is desirable to carry out non-destructive tests, which do not adversely affect the tested member. A wide range of non-destructive tests is used in construction to investigate the material structure of individual structural components. The applications of most tests are described in the extensive literature on this subject [1–4]. Thermography, which is also referred to as thermal (infrared) imaging, has been widely used in construction for non-destructive testing purposes for many years [5–9]. It is anticipated that the number of its applications and the quality of thermal imaging surveys will increase in the nearest future [10,11]. A special kind of thermography is active thermography, which consists of thermally stimulation of the tested element and registering its thermal response to the set controlled excitation over time [12]. This method has been successfully applied in materials research to investigate thin elements (up to a few millimeters thick) [13–15]. The most commonly used active thermography methods are: lock in thermography (LT), stepped thermography (ST), and phase thermography (PT) [16–18]. Stepped thermography consists of thermally exciting (using an external heat source) the tested element with a continuous heat pulse and recording thermograms as the element cools down. There occur two main phases in this method: heating up (or cooling down) and cooling down (or heating up) of the investigated surface. According to the relevant European standard [19] in step of thermography, an energy source (e.g., a halogen or induction lamp) is switched on or/and off at a particular time for excitation. Contrary to pulse thermography, the thermal signature of the defects or of the rear side of the layer already appears during excitation. The image sequence may be analyzed in the time domain or in the frequency domain. Since it is more difficult to detect inclusions in large-sized and massive members, they are less often investigated using this method [20–48]. No experiment in which the time of recording thermograms is considerably extended until a thermal equilibrium between the tested element and the environment is reached. This was reported in the literature on the subject. As part of the present research, elements with a large thermal mass were tested and thermograms were recorded simultaneously in the transmission mode and the reflection mode for three defects made of different materials. A difference in defect detection depending on the distance from which the tested surface was heated up was observed. In most cases, the occurrence of material inclusions in building units is undesirable since it adversely affects the durability of the whole structure. It is vital to locate such material inclusions in order to ensure the safe serviceability of civil structures and their durability. For this reason, an attempt was made to employ active thermography to investigate building partition models and find out if modelled material inclusions in large-sized (a few centimeters thick) members could be located in this way. In the course of the experiment, a new kind of thermal contrast, which has not been described in the literature, was noticed. Building partition models made of oriented strand board (OSB) and gypsum board (GB) were investigated. Material inclusions made of steel, granite, and XPS (extruded polystyrene), i.e., materials markedly differing in their thermal parameters (Table 1), were modelled in each of the two partition models. The tested element was excited by a long-duration heat pulse. The distribution of the temperature field on the surface was registered simultaneously in the reflection mode (the thermal imaging camera situated on the thermal excitation side) and in the transmission mode (the camera situated on the opposite—relative to the thermal excitation—side of the partition). In this way, the material inclusions inside the investigated building partition model were located and identified. Thanks to the recording of the thermograms of the investigated surface over a longer time, an inverse contrast phenomenon (described further in this paper) was observed. In the earlier publications [49,50], the inverse contrast phenomenon was mentioned in connection with the testing of thin (a few mm) elements of a small size (up to 10–20 cm). In addition, in the paper [51], an inverse contrast occurring on different sides of the tested element (the front and rear side of the material sample) was mentioned. In the present research, the inverse contrast phenomenon occurs on one (the same) side of the tested building partition after a long observation time. The principal aim of this research was to examine the effect of the type of partition materials and the type of material inclusions for the possibility of detecting the latter in massive building partitions, i.e., elements with a large thermal capacity.

2. Materials and Methods

Two building partition models, consisting of four 1250 $\times$ 1250 mm (oriented strand and gypsum) boards bolted together, were investigated. The surface (outer) boards were 22 mm thick, while the inner boards were 10 mm thick. The overall thickness of the partition model amounted to 64 mm. Materials markedly differing in their thermal parameters, i.e., steel, granite, and styrofoam (XPS), were placed inside the model (inserted into the 10 mm thick boards). The inclusions were 200 $\times$ 100 $\times$ 20 mm in size and spaced from one another at a distance greater than their largest dimension in order to eliminate the influence of one inclusion on another during heat conduction in the course of the experiment. This distance was estimated on the basis of previously carried out numerical studies of heat conduction in the partition model. A precise geometrical model of the tested element, which shows the arrangement of the material inclusions inside the partition, is presented in Figure 1.

Two thermal imaging cameras are placed on the two sides of the partition model. A thermal excitation source and an air temperature and humidity sensor were used in the experiment. A Flir P65 camera with an infrared (IR) detector resolution of 320 × 240 pixels (the camera on the unheated side of the partition—the transmission mode) and an Optris PI400 camera with an IR detector resolution of 382 × 288 pixels (the camera on the side heated by the infrared radiator—the reflection mode) were used in the experiment. The thermal sensitivity of the two infrared cameras was below 80 mK. A Fobo infrared radiator with a total power of 7.2 kW (six lamps, each with a power of 1.2 kW) was used for thermal excitation. In order to prevent the tested element's edges from cooling (which would disturb the temperature field there), a Styrofoam band was placed around the partition model. A schematic of the test stand is shown in Figure 2.

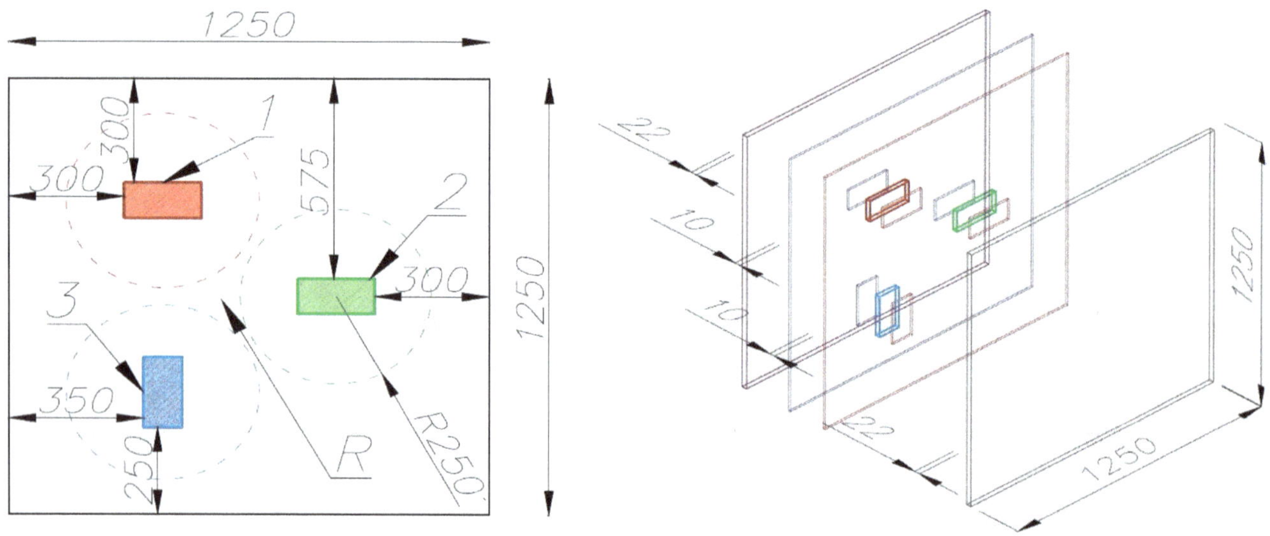

(a) (b)

Figure 1. Geometrical structure of the partition model (given units in mm): (**a**) arrangement of material inclusions, internal structure. (**b**) View of the 3D partition model where inner boards with holes and inclusions are marked blue and red, while the outer boards are marked violet and grey. Designations in figures: 1—XPS inclusion, 2—granite inclusion, 3—steel inclusion, and R—homogenous area without inclusion.

(**a**)

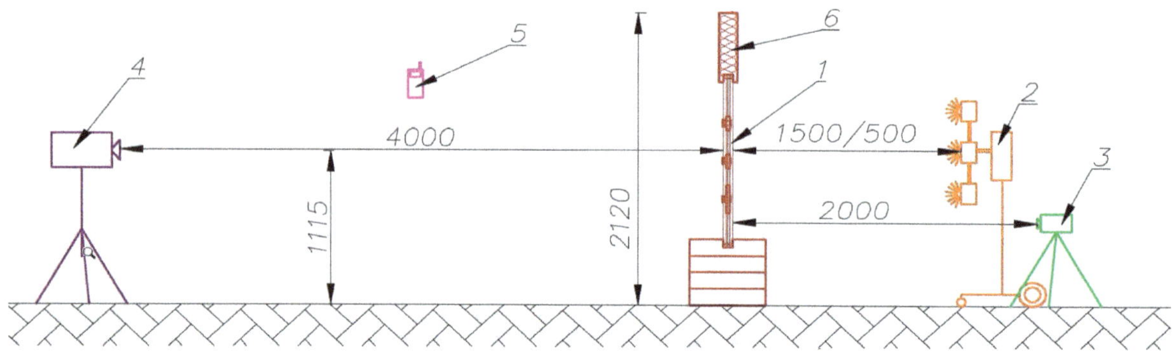

Figure 2. *Cont.*

(**b**)

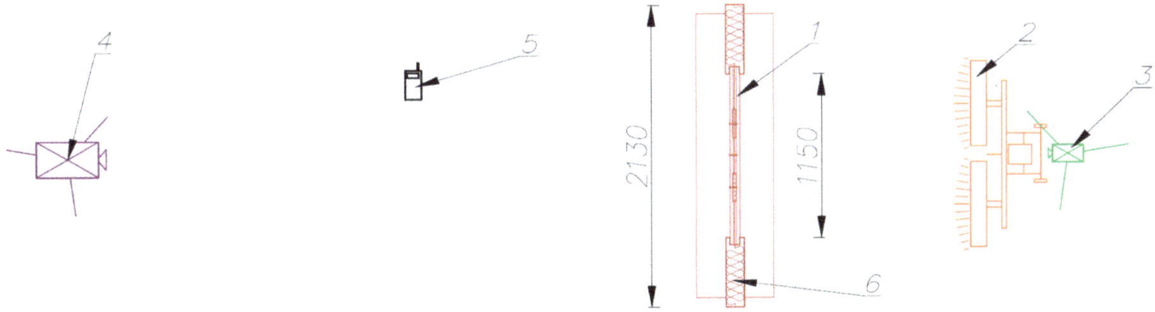

Figure 2. Schematic of test stand (given units in mm): (**a**) side view and (**b**) top view. Designations in the figure: 1—tested partition model, 2—infrared radiator, 3—thermal imaging camera Optris PI400, 4—thermal imaging camera Flir P65, 5—air temperature and relative humidity sensor, 6—Styrofoam band.

As part of the experiment, two building partition models, one made of OSB and the other of GB, were investigated. The experiment consisted in thermally exciting the partition with a 30-minute long continuous heat pulse. The heating was effected at two different infrared radiator distances from the excited surface, i.e., 1500 and 500 mm. The heating of the investigated surface was followed by its cooling during which the temperature field distribution was cyclically measured every 20 s on both sides of the tested element (the transmission mode and the reflection mode). The thermal imaging measurement was conducted for up to eight hours, counting from switching the radiator off. A simple analysis based on the obtained thermograms was carried out. For this purpose, the absolute contrast was calculated from the formula below.

$$C_a(t) = T_p(t) - T_{pj}(t) \quad [^\circ C] \tag{1}$$

where:

Tp(t)—the temperature on a surface point of the cross-section with a defect [°C],
Tpj(t)—the temperature on a surface point of the homogeneous cross section without a defect [°C].

Measuring points in the thermograms were situated in the center of the area with a defect and at the center of the tested element in the defect-free cross-section. The thermogram analysis software "in-point reading" function was used to read temperature values. The size of the in-point function area was 3 × 3 pixels.

Various building materials were used in the experiment. Besides the kind of thermal excitation and the tested element geometry, the thermal parameters of the materials used also influenced the distribution of the temperature field. The thermal parameters of the materials used in the experiment are specified (on the basis of various literature sources) in Table 1.

Table 1. Thermal parameters of building materials used in the experiment (values based on various literature sources).

Type of Material	Bulk Density	Specific Heat	Heat Capacity	Thermal Conductivity	Thermal Diffusivity
[–]	ϱ_{vol} [kg·m^{-3}]	c_w [J·kg^{-1}·K^{-1}]	C_{vol} [J·m^{-3}·K^{-1}]	λ [W·m^{-1}·K^{-1}]	a [m^2·h^{-1}]
Styrofoam	30	1460	0.04×10^6	0.033	0.75×10^{-6}
Granite	2600	920	2.39×10^6	2.80	1.17×10^{-6}
Steel	7900	500	3.95×10^6	17.0	4.30×10^{-6}
GB	1000	1000	1.00×10^6	0.23	0.23×10^{-6}
OSB	650	1700	1.11×10^6	0.13	0.12×10^{-6}

The thermal parameters of the tested materials affected the test results the most. The homogeneous materials and the material inclusions were appropriately matched in order to compare the influence of the differences between the thermal parameters on the temperature field distribution recorded during the experiment.

3. Results

3.1. Thermograms

The obtained temperature field distribution results are presented as thermograms recorded at characteristic instants. Figure 3 shows the results for the measurement performed in the reflection mode with the heating from a distance of 0.5 m for the homogeneous OSB material (row I) and the GB material (row II), and in the transmission mode at the infrared radiator-surface distance of 1.5 m for the GB partition model (row III) and the OSB partition model (row IV) as well as for the measurement performed in the reflection mode with heating from a distance of 1.5 m for the GB partition (row V) and the OSB partition (row VI). The thermogram recording time, counting from the beginning of the element cooling down, is specified above each of the columns (A–D). The numbers from 1 to 18 indicate the temperature reading positions at the center of the cross section with a defect, while "R" designates the temperature reading position at the defect-free cross section.

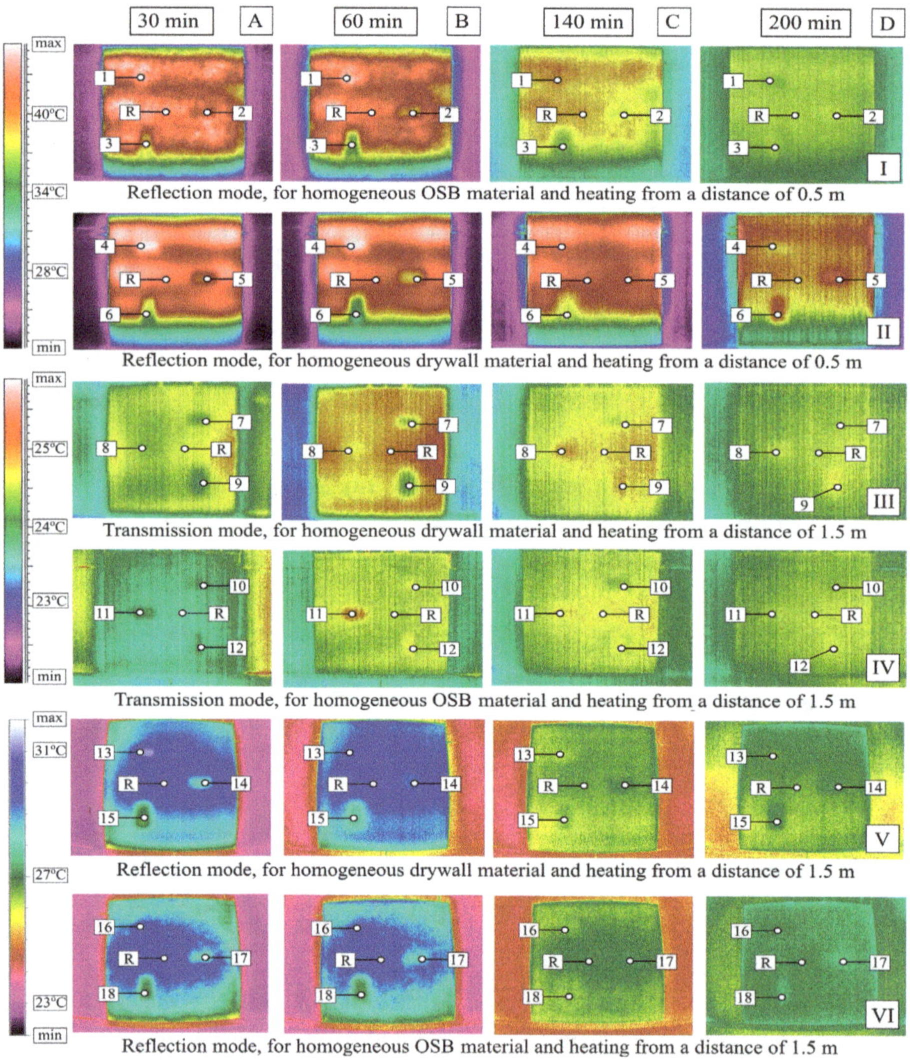

Figure 3. Thermograms at selected instants (time from the beginning of the tested element cooling down is specified above the figure). Numbers from 1 to 18 indicate the temperature reading position at the center of the cross section with a defect, while "R" designates the temperature reading position at the defect-free cross section.

3.2. Absolute Contrasts

On the basis of the obtained temperature field distribution on the investigated surface over time, the absolute contrasts were calculated from Formula (1). The absolute contrast values were calculated in the thermogram points designated with numbers from 1 to 18—the cross sections with defects—whereas "R" designates the temperature value in the defect-free cross section (see Figure 3). Figures 4–9 correspond to the next rows, according to Figure 3 (rows I–VI), for the tested partition model surfaces. The characteristic point numbers 1–18 and the cross section designations A–D are the same in Figures 3 and 4. The thick trend line was plotted as a 6th degree polynomial.

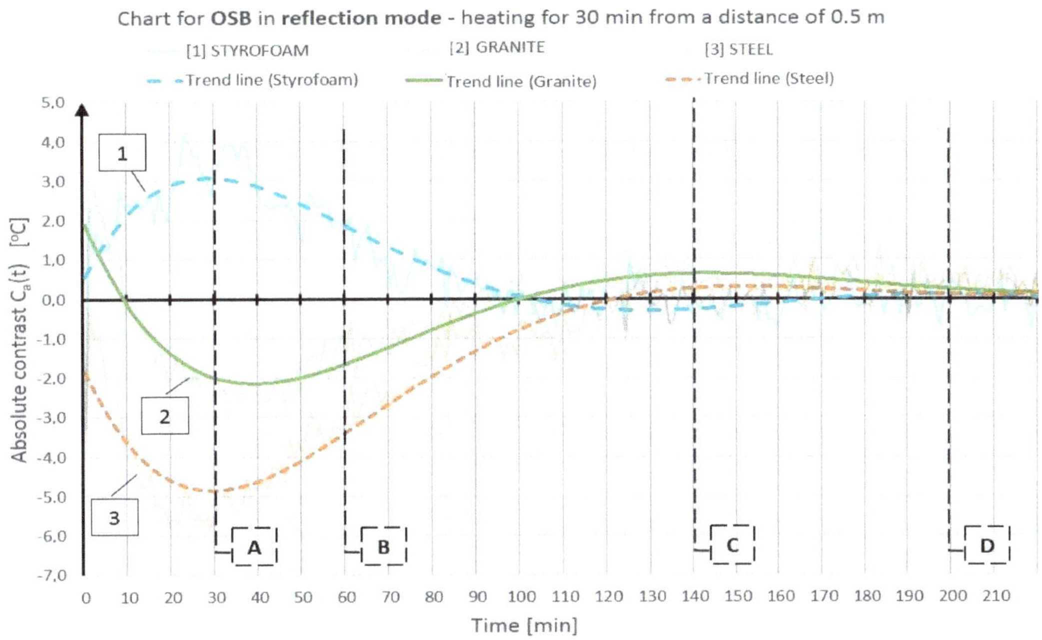

Figure 4. Diagram of the absolute contrast for measurement in a reflection mode for the OSB partition model with heating from a distance of 0.5 m (based on thermograms in Figure 3, row I).

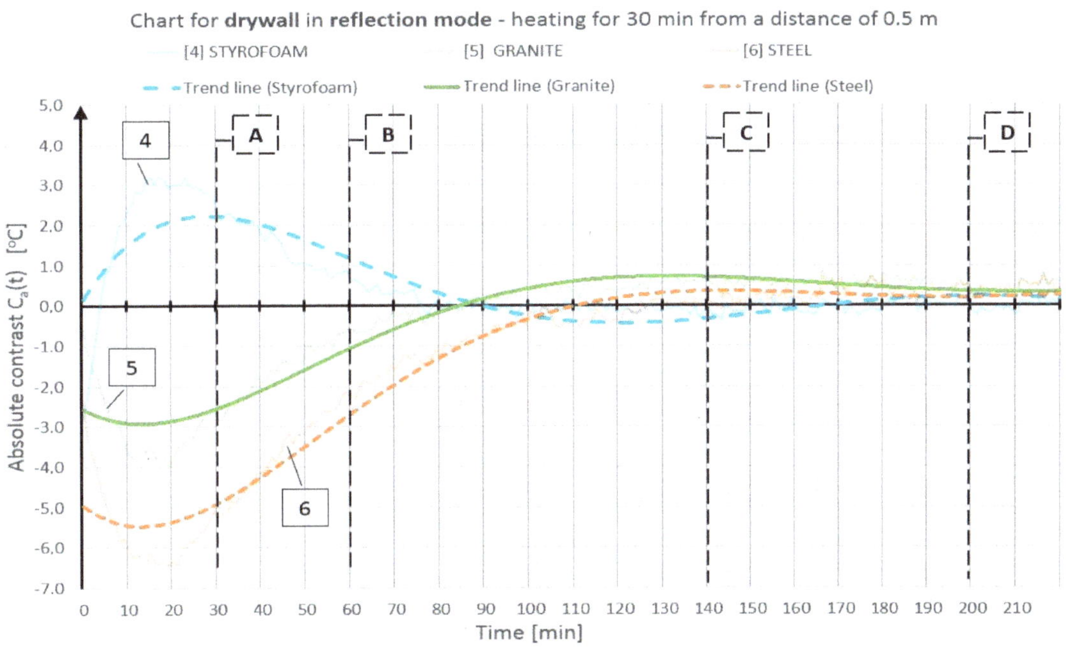

Figure 5. Diagram of absolute contrast for measuring the reflection mode for the GB partition model with heating from a distance of 0.5 m (based on thermograms in Figure 3, row II).

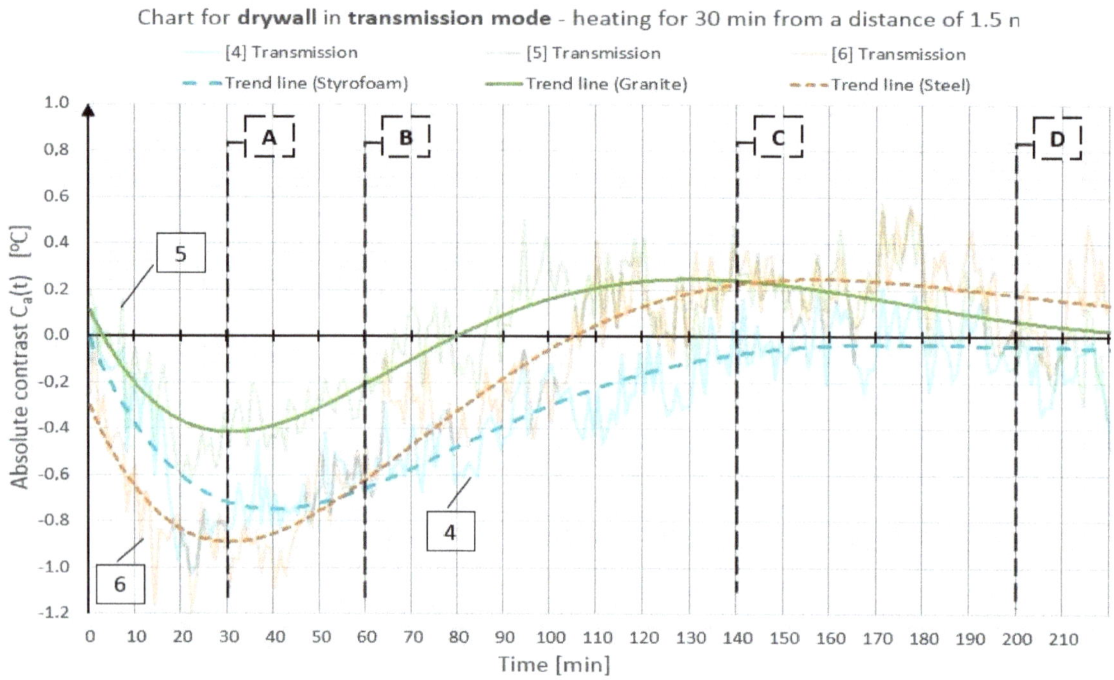

Figure 6. Diagram of absolute contrast for measuring in the transmission mode for the GB partition model with heating from a distance of 1.5 m (based on thermograms in Figure 3, row III).

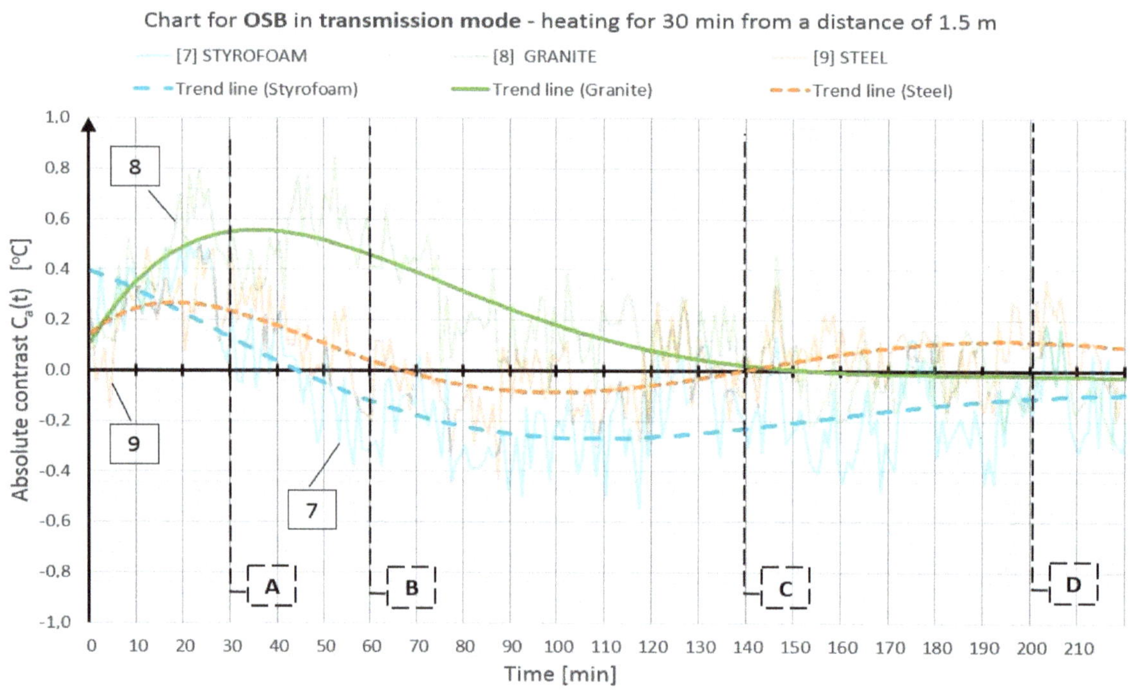

Figure 7. Diagram of absolute contrast for measuring in the transmission mode for the OSB partition model with heating from a distance of 1.5 m (based on thermograms in Figure 3, row IV).

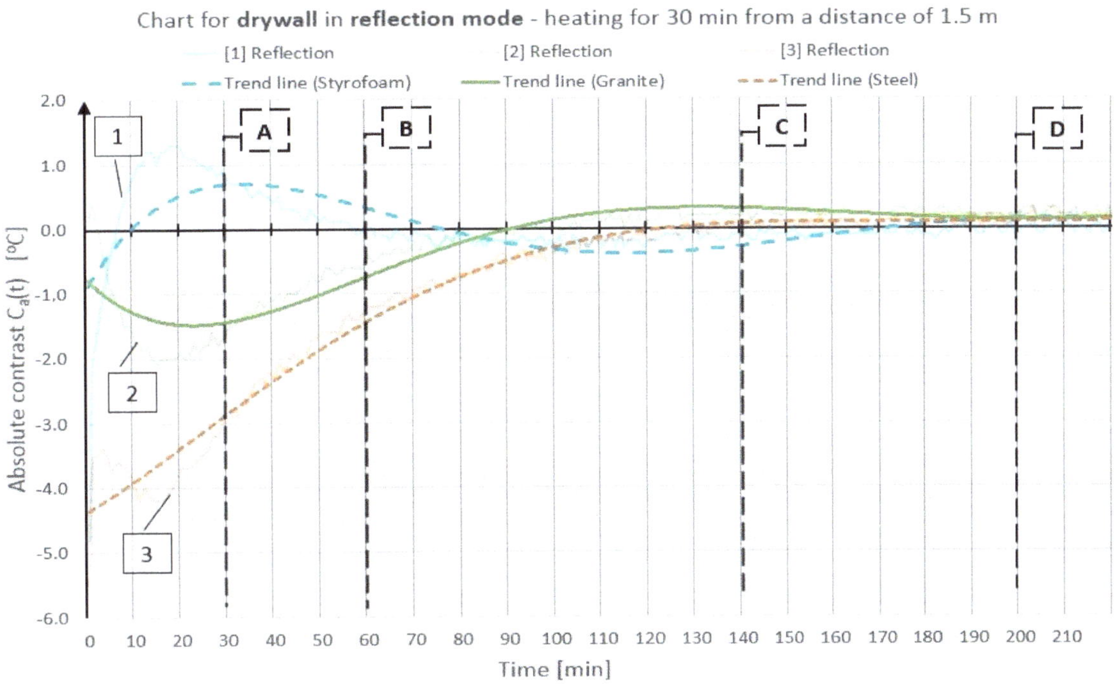

Figure 8. Diagram of absolute contrast for measuring in the reflection mode for a GB partition model with heating from a distance of 1.5 m (based on thermograms in Figure 3, row V).

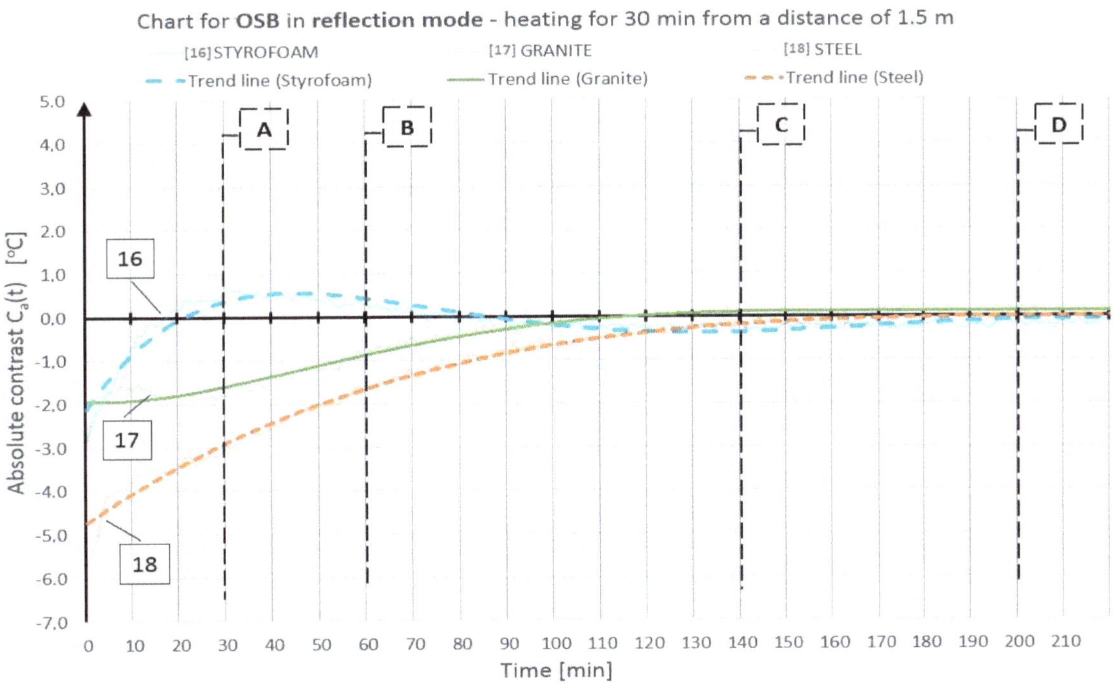

Figure 9. Diagram of absolute contrast for measuring in a reflection mode for the OSB partition model with heating from a distance of 1.5 m (based on thermograms in Figure 3, row VI).

In order to illustrate better and compare the differences in absolute contrast values for the particular types of material inclusions, the obtained data are presented in individual diagrams for styrofoam (Figure 10), granite (Figure 11), and steel (Figure 12).

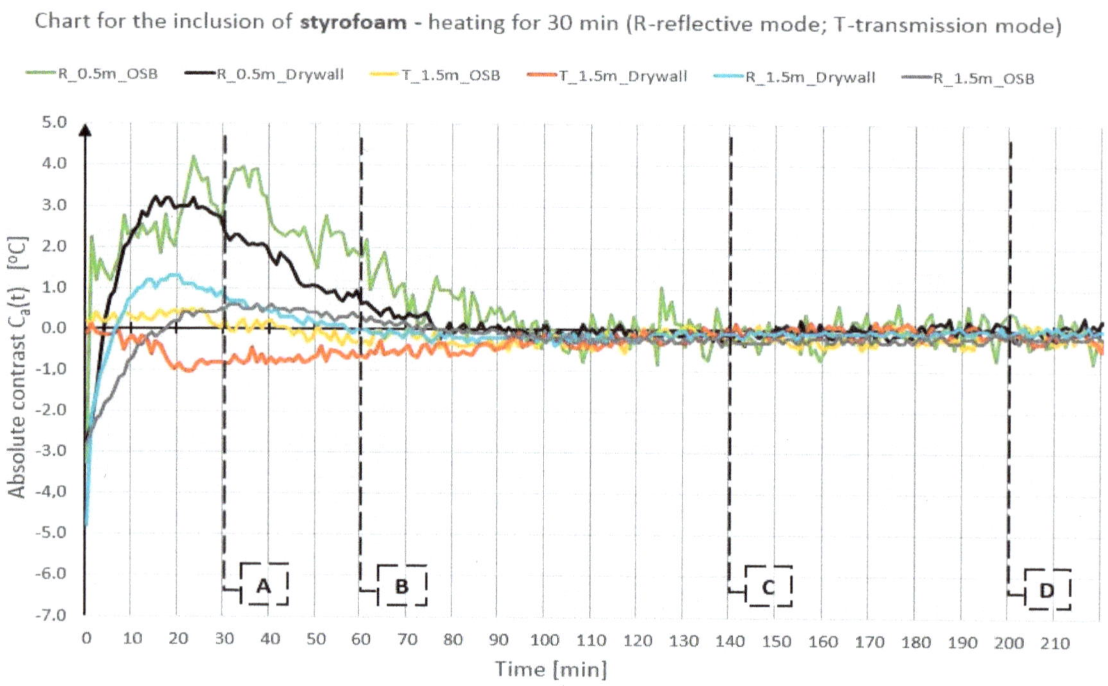

Figure 10. Comparison of absolute contrasts in a reflection mode and in a transmission mode for material inclusion made of Styrofoam.

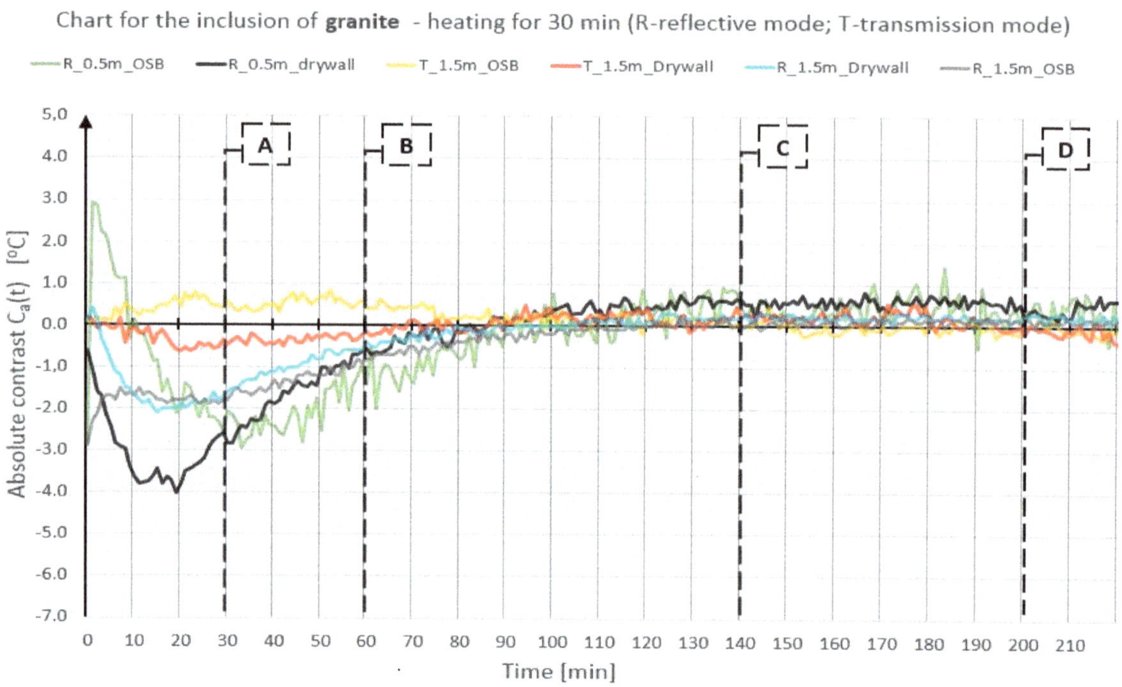

Figure 11. Comparison of absolute contrasts in a reflection mode and in a transmission mode for material inclusion made of granite.

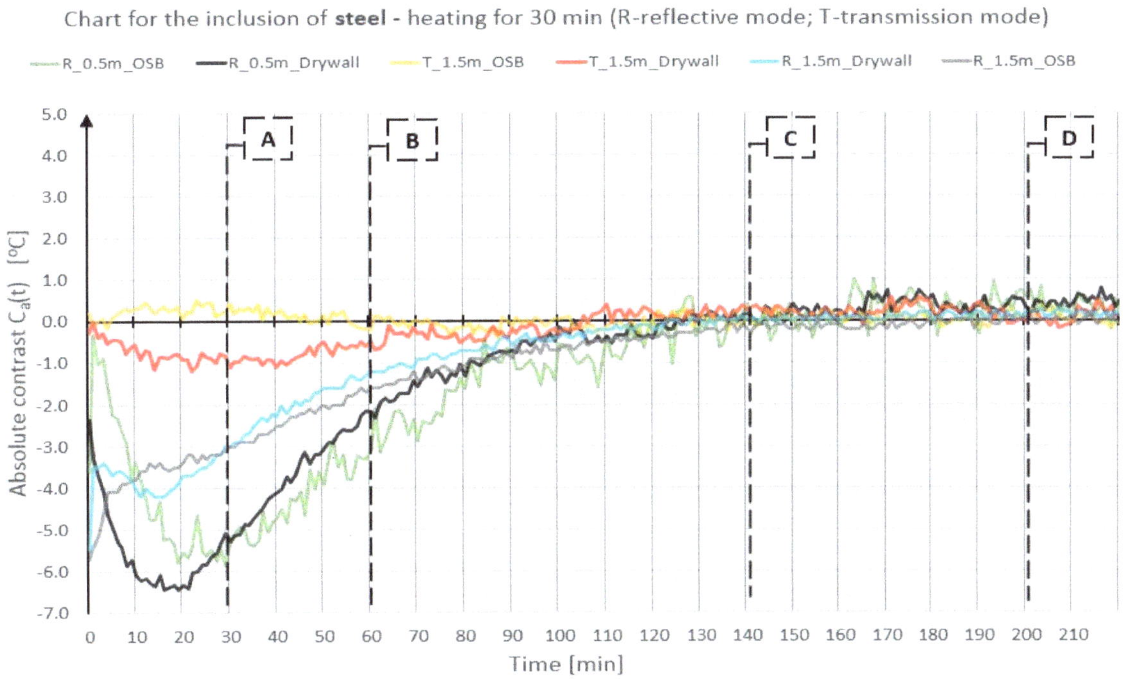

Figure 12. Comparison of absolute contrasts in a reflection mode and in a transmission mode for material inclusion made of steel.

4. Discussion

4.1. Absolute Contrasts and Absolute Inverse Contrasts

The inclusions incorporated into the OSB and GB partition models, which is invisible from the outside, can be located using the active thermography method. The different thermal properties of the inclusions in comparison with those of the basic material in which these inclusions occur cause disturbances in the temperature field distribution visible in the thermograms. The temperature is uniform in those places where there are no inclusions. The only non-uniformities, which can occur there, are due to the non-uniform heating of the surface (when the surface is heated for up to 60 min from a small distance—Figure 3, rows I and II). In the cross sections with inclusions, one can observe warmer or cooler areas in comparison with the homogeneous cross sections. In the case of measurement in the reflection mode, for both the partition materials (OSB and GB), the Styrofoam inclusion was visible as an area with a higher temperature, whereas the granite inclusion and the steel inclusion were observed as areas with a temperature lower than that of the areas without an inclusion. This dependence was observed during heating from both the distance of 0.5 m and 1.5 m. In the case of heating from the distance of 0.5 m for 30 min, the investigated surfaces was heated up to a temperature of about 90 °C, whereas, for heating from the distance of 1.5 m, the temperature on the surface did not exceed 66 °C. In both cases, the inclusions were clearly visible and locatable, but at surface temperatures, when higher than 90 °C, the investigative method is no longer non-destructive since the excessively high temperature damages the structure of the tested materials. Moreover, in order to reach such a high temperature of the surface, the heat source must be placed close to this surface, which results in the non-uniform heating of the latter, consequently, and reduces the visibility of the inner inclusions. When the surface is excited more intensively, the obtained maximum absolute thermal contrasts range from about 3 °C to over 6 °C (Figures 4 and 5). At less intensive heating of the surface, the maximum temperature contrasts range from about 1 °C to over 4 °C (Figures 8 and 9). As one can see in Figure 3, regardless of the degree to which the surface is heated up, at the 30th min from the beginning of the cooling down phase, all the material inclusions are clearly visible and locatable. Hence, one can conclude that, for this geometrical configuration of the building partition and the material inclusions, it suffices to heat up the surface to a temperature of about 60 °C (using the 30-min long continuous heating by an infrared radiator) in order to locate defects situated under a 22-mm thick layer of the basic material. With regard to the thermal parameters of the materials (Table 1), the two basic materials (OSB and GB) have similar heat capacity, but GB is characterized by a higher thermal diffusivity at nearly twice the value. This means that heat will propagate faster in the partition model made of GB. Styrofoam has a very low heat capacity in comparison with that of the partition material, whereby this cross section heats up very quickly and is visible as an area with a temperature higher than that of the defect-free area. By contrast, granite and steel have a heat capacity much higher than that of GB and OSB, which means that the cross section containing such inclusions heats up more slowly than the homogeneous area around the inclusions. The temperature distribution described above is visible in the reflection mode for both strong and weak heating, up to approximately the 90th min from the beginning of cooling down. After this time, inverse contrasts were observed, i.e., the areas with inclusions, which, at the beginning are visible as cooler areas, and, in the second phase of cooling down (after approximately the 90th min), begin to be visible as warmer areas and vice versa. This phenomenon is visible due to the very long recording of the cooling down of the tested elements. It occurs as a result of the difference in heat capacity between the inclusion and the basic material: steel and granite characterized by the high heat capacity cool down much slower than the partitions made of GB and OSB, respectively. In the case of Styrofoam, due to its lower heat capacity, the cross section with a defect cools down quicker than the homogeneous cross section without a defect. In the experiment, the inverse contrast phenomenon can be best observed for the steel inclusion and the granite inclusion when the GB partition model is heated from a distance of 0.5 m (Figure 3 row II). The change of the absolute contrasts over time is shown in Figures 4–9. The sharpest contrasts (the most clearly visible areas with an inclusion) would occur from approximately the 15th to approximately the 40th min since switching the heat source off. For the excitation from a distance of 0.5 m (for both the OSB partition (Figure 4) and the GB partition (Figure 5)) the contrast reached about +4 °C while the cross sections with the steel defect and the granite defect induced a contrast amounting to about −6 °C and about −4 °C, respectively (the minus sign indicates that the defect is visible as a cooler area, whereas the plus sign represents an area warmer than the surrounding area without a defect). In the case of heating from a distance of 1.5 m, the absolute contrast was lower by about 3 °C for the Styrofoam and by about 2 °C for the steel and the granite. Contrasts were also observed during the recording of thermograms in the transmission mode (Figure 3, rows III and IV). The values of the contrasts were considerably lower and amount to ±1 °C (Figures 6 and 7). Even at such low contrasts, however, the defects could be located up to the 60th min from the beginning of the cooling of the partition model.

Absolute contrasts occurring as the element cools down are marked in the diagrams (Figures 13–15). In the initial cooling down phase, a normal absolute contrast (positive for material inclusions with low heat capacity and negative for material inclusions with high heat capacity in comparison with the basic partition material) occurs. In the next cooling down phase, inversion of the contrasts occurs. This is referred to as an inverse absolute contrast. The inverse absolute contrasts are much smaller than the normal absolute contrasts, but they still allow the location of defects within the tested partition, even a few hours after the thermal stimulation of the surface has ended. In the adopted geometric partition model, an interesting dependence was observed. In the case of materials with high heat capacity (granite and steel), the contrasts have the same sign in both the reflection mode and the transmission mode, whereas, for defects with low heat capacity (Styrofoam), the contrast is positive in the reflection mode and negative in the transmission mode. Regardless of heat capacity, in the transmission mode, the defects were always perceived as areas warmer than the homogeneous areas without defects.

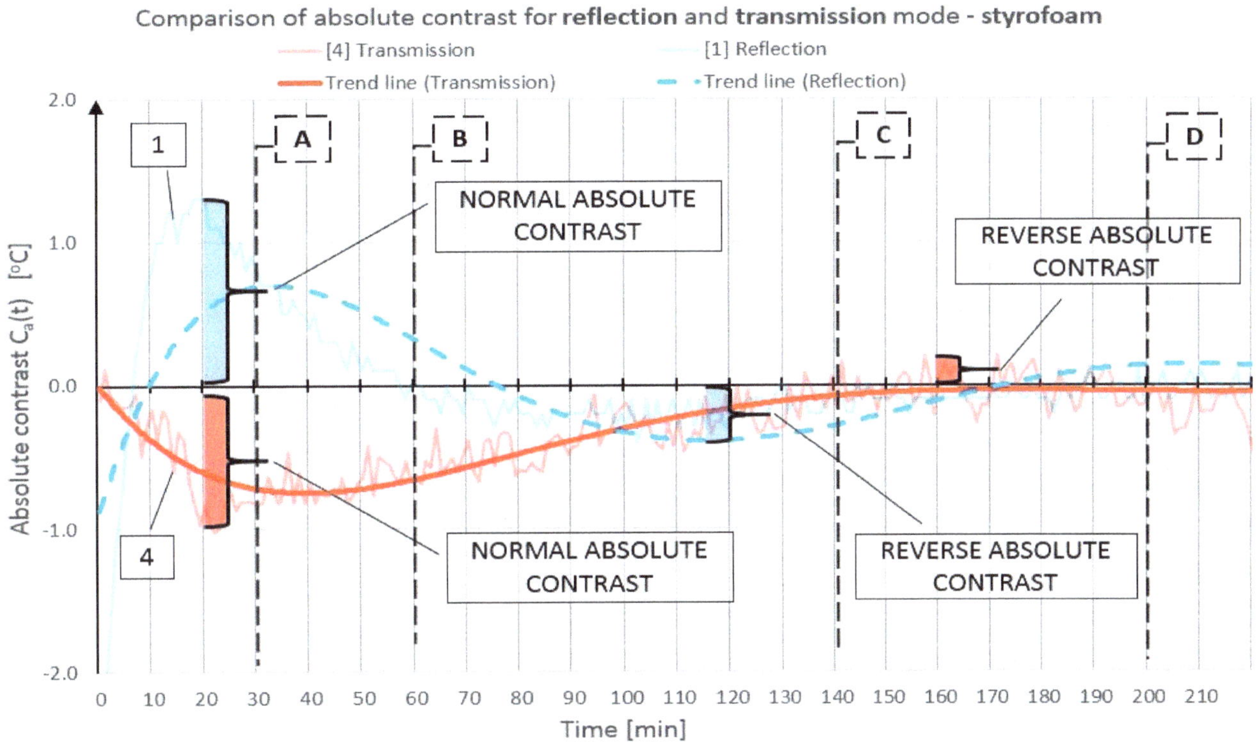

Figure 13. Diagram of absolute contrast in a reflection mode and in a transmission mode for Styrofoam inclusion, with marked places where normal absolute contrast and inverse absolute contrast occur (heating from a distance of 1.5 m).

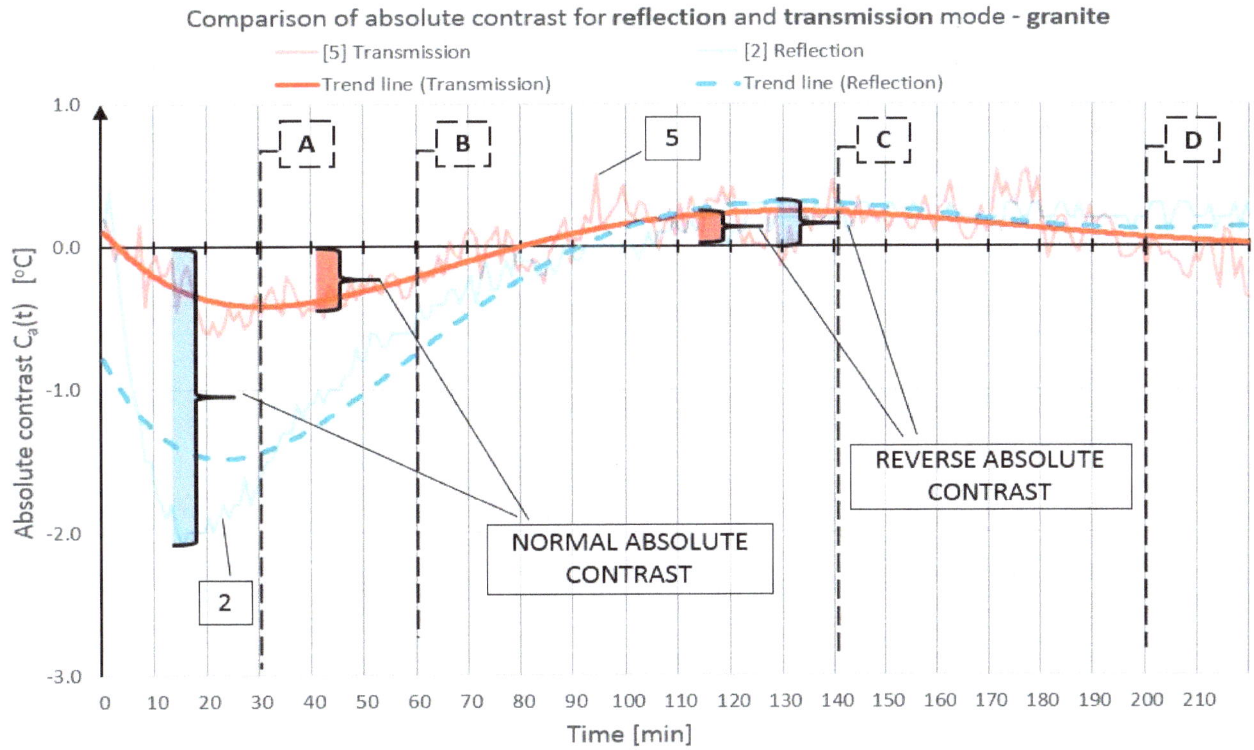

Figure 14. Diagram of absolute contrast in a reflection mode and in a transmission mode for granite inclusion, with marked places where normal absolute contrast and inverse absolute contrast occur (heating from a distance of 1.5 m).

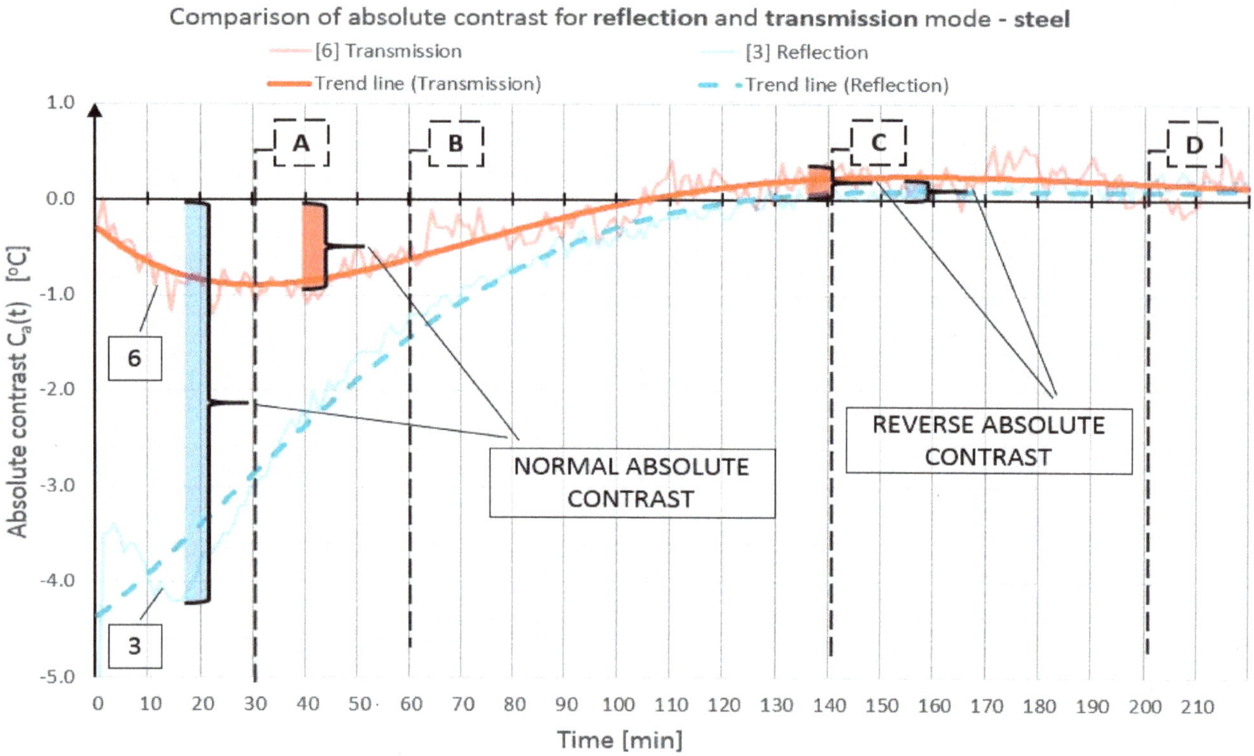

Figure 15. Diagram of absolute contrast in a reflection mode and in a transmission mode for steel inclusion, with marked places where normal absolute contrast and inverse absolute contrast occur (heating from a distance of 1.5 m).

4.2. Defect Location Depth

In addition, other parameters are used in defect detection. In References [16,17], it was noted that parameters "m" and R^2 could be used for this purpose.

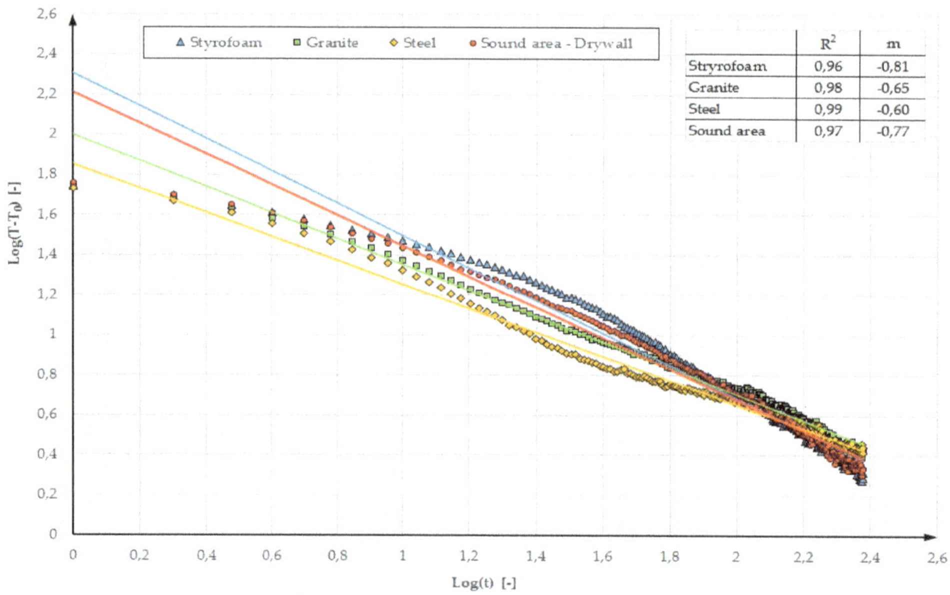

Figure 16. Log–log graph of the cooling phase and R^2 and m values (stimulated period of 30 min) for the OSB partition model with heating from a distance of 0.5 m.

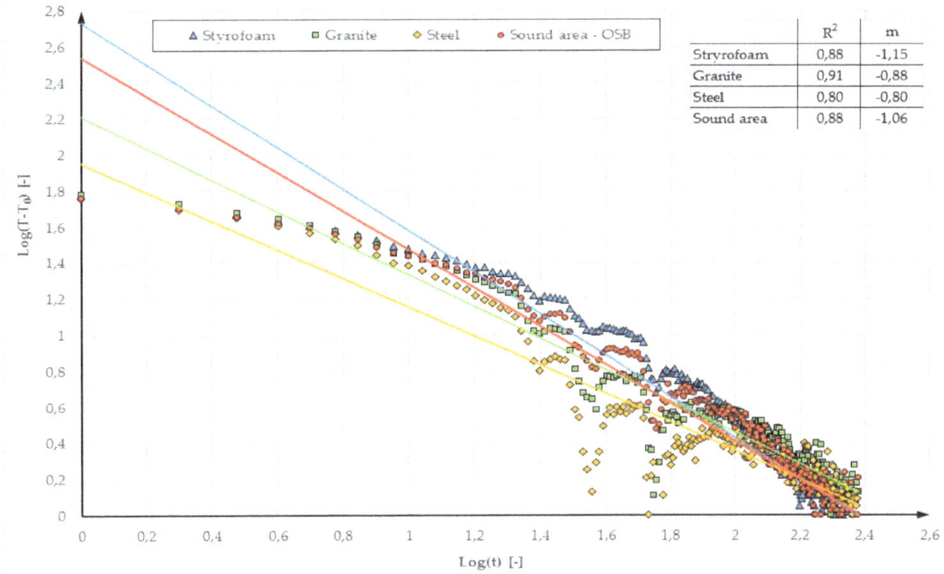

Figure 17. Log–log graph of the cooling phase and R^2 and m values (stimulated period 30 min) for the GB partition model with heating from a distance of 0.5 m.

In the above publications, parameter "m" is defined as the inclination and R^2 as a cooling curve determination coefficient. The cooling curve plot is a Log-Log graph. Time axis X represents Log(t) while axis Y represents the contrast logarithm Log(T − T_0), where T is the temperature varying over time and T_0 is the temperature measured before the stimulation. For these assumptions, Figures 16 and 17 showed the determined values of parameters "m" and R^2 for the Log-Log graph. In Reference [16], when testing a thin plate (16.2 mm) containing defects with a diameter of a few mm, located at a depth of a few mm (the first cycle of testing using the lock-in method was analysed), the determined R^2 parameter values were in the range of 0.92 to 0.98 (for the selected defects) and the deflections ranged from −0.43 to −0.76. One can notice that both when small elements with a low thermal capacity and massive partitions with a high thermal capacity are tested, similar temperature-time dependences are obtained in the characteristic cross sections with and without defects. A better fit of the cooling line was obtained for the partition model made of gypsum board (R^2 ranging from 0.96 to 0.99). A better cooling line fit means a greater possibility of locating defects on the thermograms. This dependence is visible in Figure 3 where defects in the GB partition are more distinct and their edges are "sharper" in comparison with the partition model made of OSB.

The echo defect shape (EDS) method can be used to determine the depth at which a defect is located in the tested material. This method is described in References [49,50] in more detail where the depth of defect location is calculated using Equation (2).

$$d = \sqrt{-\frac{\lambda \cdot t}{\rho \cdot c_w} \cdot ln(C_{rel}(t))} \ \ [\mathrm{m}] \tag{2}$$

where:

λ—the thermal conductivity of the tested material (above the defect) $[\mathrm{W \cdot m^{-1} \cdot K^{-1}}]$,
t—the time elapsed since the end of heating [s],
ρ—the bulk density of the tested material (above the defect) $[\mathrm{kg \cdot m^{-3}}]$,
c_w—the specific heat of the tested material (above the defect) $[\mathrm{J \cdot kg^{-1} \cdot K^{-1}}]$,
$C_{rel}(t)$—the relative contrast in relation to the reference area (the cross section with the defect) [-].

The relative contrast is a ratio of the absolute contrast to the temperature in a surface point of the homogenous cross section without a defect. When calculating "d", it is highly important to select a proper relative contrast, which must be high enough to reduce the measurement noise, but still be below the maximum values [50]. In Reference [49], it is recommended to adopt $C_{rel}(t)$ equal to 0.025 if the thermal noise is in the range of 0.01–0.015. In Reference [50], this value is optimized and selecting the value of 0.07 is recommended. It sometimes happens that the relative contrast does not reach the value of 0.07 during the test. For testing elements with a large thermal mass by means of a long thermal pulse, the authors optimized $C_{rel}(t)$ value selection to the value that occurs 3 min before the first peak of this contrast. The relative contrast and defect location depth versus time dependence (in the GB partition model heated from a distance of 0.5 m) is shown in Figures 18–20.

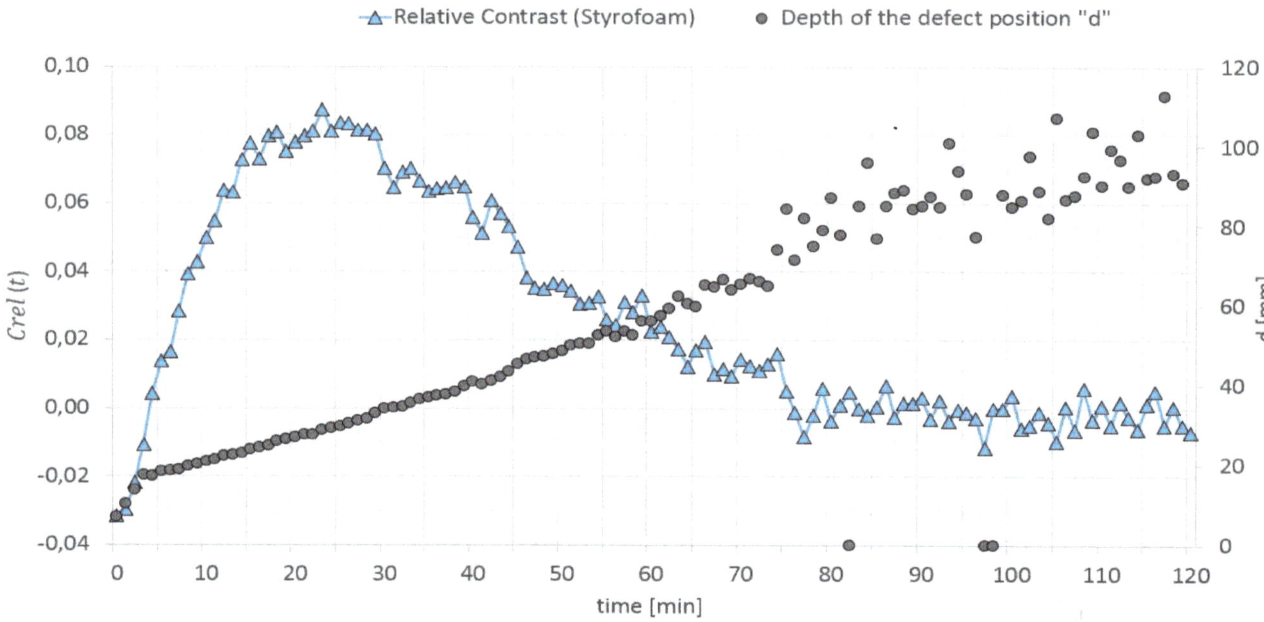

Figure 18. Relative contrast and defect location depth versus time (Styrofoam).

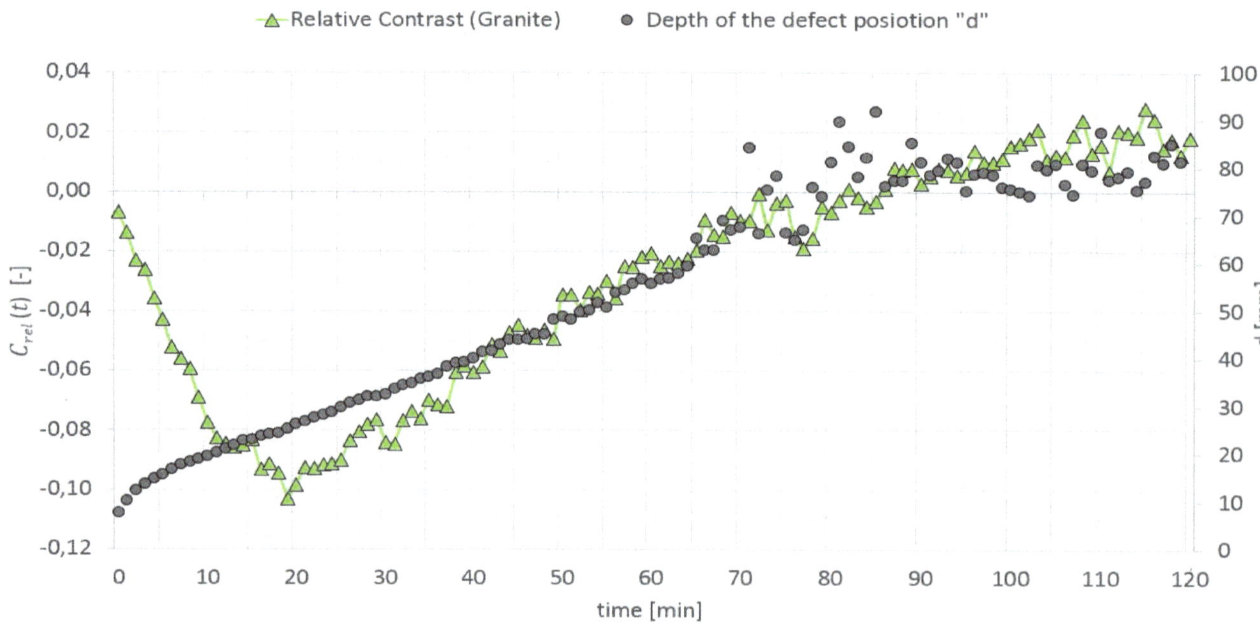

Figure 19. Relative contrast and defect location depth versus time (Granite).

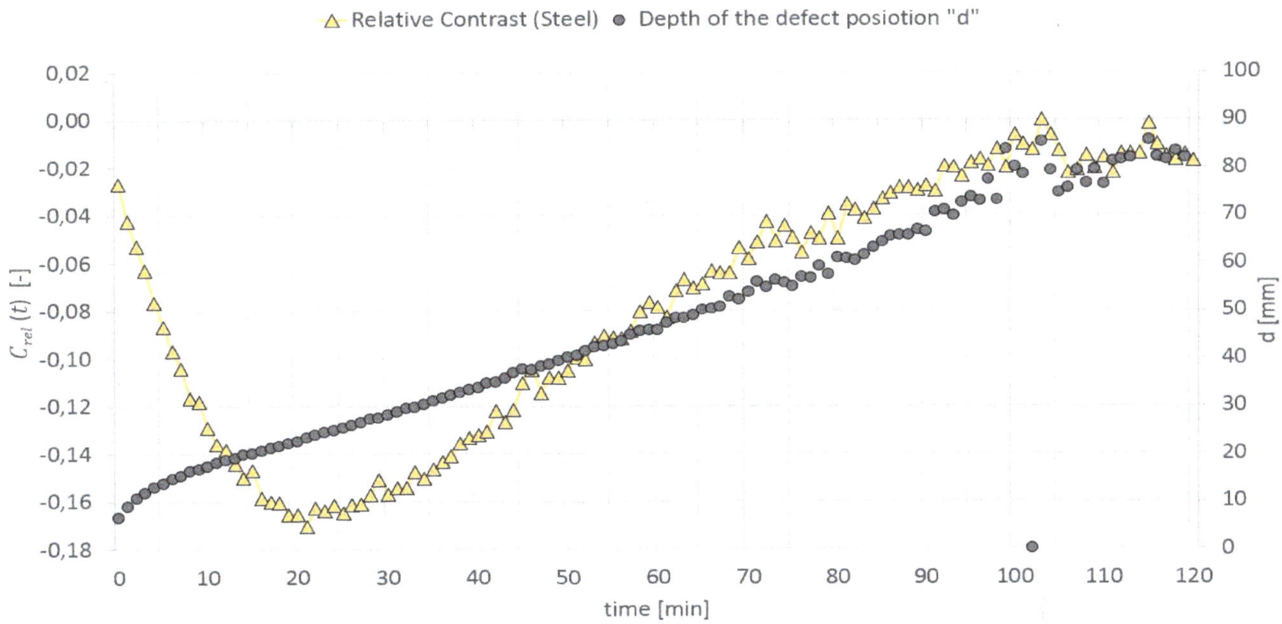

Figure 20. Relative contrast and defect location depth versus time (Steel).

For the above assumptions, the defect location depth was calculated for two partition models thermally stimulated from different distances. The GB and OSB values contained in Table 1 were used in the calculations. The results are shown in Table 2. The actual depth at which the defects are located amounts to 22 mm. For the presented investigative methodology, the calculated values do not exceed the relative error of 10% (except for the Styrofoam defect in the OSB model, heated from the distance of 1.5 m. In this case, the relative error amounts to 23%).

Table 2. Defect location depths calculated using the EDS method.

Defect	OSB, Heating from Distance 0.5 m	OSB, Heating from Distance 1.5 m	GK, Heating from Distance 0.5 m	GK, Heating from Distance 1.5 m
–	d [mm]	d [mm]	d [mm]	d [mm]
Styrofoam	21.67	27.14	22.51	22.93
Granite	24.15	20.63	22.89	21.92
Steel	19.97	22.36	21.32	19.51

4.3. Dimensions of the Defect

The measurement by means of the Flir P65 camera with a 24° × 18° lens was carried out from a distance of 4.0 m, which means that the size of a single pixel in the thermogram is 5.2 mm. For the Optris PI400 thermal imaging camera with a 60° × 45° lens and measurement taken from a distance of 2.0 m, the pixel size is 4.6 mm. Using the thermograms, one can easily assess qualitatively the location of the material inclusions. A disturbance in the temperature field distribution indicates the place where the inclusions occur. On the basis of the thermograms, one can also carry out a simple quantitative assessment and estimate the dimensions of the defects inside the tested partition. The procedure for estimating the width and height of a defect is schematically shown in Figure 16. First, one should select the thermogram in which the defect is most clearly visible. The best thermogram is the one in which the thermal contrast between the area with the defect and the defect-free area as well as the potential edges of the defect is the sharpest (most distinct). Then, virtual lines representing the predicted edge of the defect are assumed. In the program for analyzing thermograms, one can read the x, y coordinates of the particular lines and, on this basis, calculate the potential dimensions (in pixels) of the perceived defect. Lastly, knowing the size of a single pixel, one can estimate the size of the material inclusion.

Figure 21 shows how the defect made of granite is estimated using thermograms obtained from the Optris thermal imaging camera. The estimated dimensions are 211.6 mm × 124.2 mm, while the actual dimensions of the defect are 200 mm × 100 mm. It can be assumed that the relative estimation error does not exceed 6% for the width and 25% for the height. If the above procedure is used, the obtained dimensions will always exceed the actual dimensions of the defect since the assumed area delimiting the place where defects occur is an area of the temperature field disturbance. This disturbance always decays from the defect's edge outwards instead of inwards. Quantitative procedures for estimating defect dimensions by using pulsed thermography can be found, for example, in References [52,53].

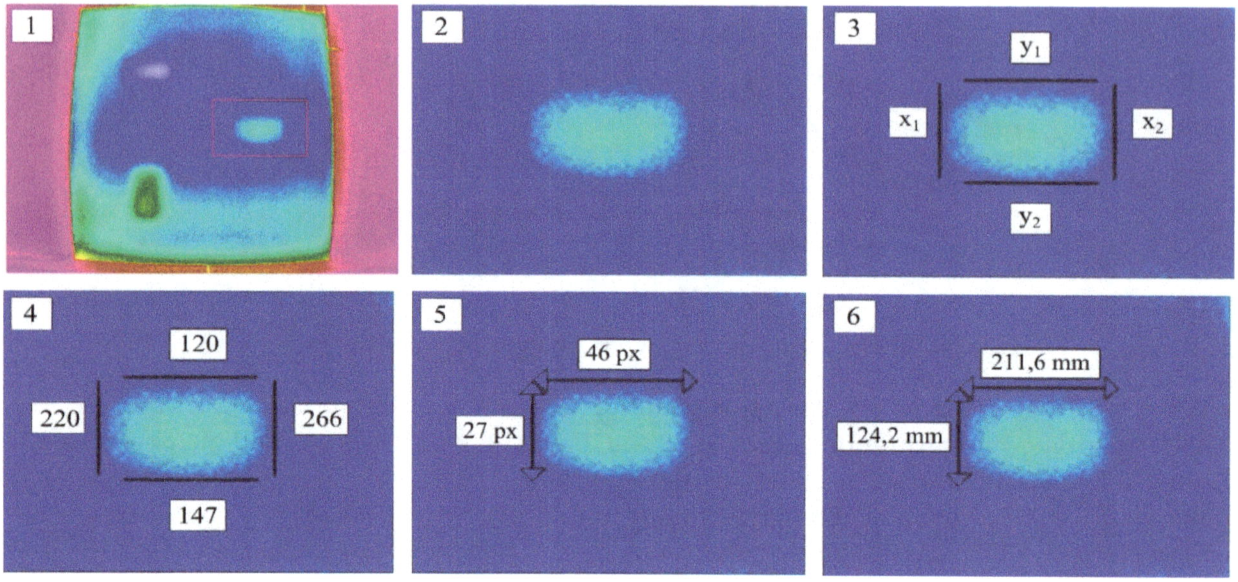

Figure 21. Procedure for estimating the defect size on the basis of the known size of a single pixel (4.6 mm).

5. Conclusions

From the experiment carried out on the building partition with modeled inclusions of various materials as well as from the analysis of the obtained thermograms and the calculated absolute contrasts, the following conclusions can be drawn (for the experimental geometry of the partition and the inclusions).

1. Using the non-destructive active thermography technique, one can locate material inclusions in building partitions (elements with a large thermal capacity), situated at a depth of at least 22 mm below the surface of cladding made of wood or gypsum;
2. In order to locate defects in a building partition, it is sufficient to heat its surface to about 60 °C whereby the test does not lose its non-destructive character;
3. The highest temperature contrasts between the cross sections with defects and the cross sections without defects (the inclusions are most clearly visible then) arise at the time interval from the 15th to the 40th min since the beginning of cooling down;
4. During testing in the reflection mode, inclusions with a considerably higher heat capacity than that of the basic materials are visible as areas whose temperature is lower than that of the areas without inclusions;
5. During testing in the reflection mode, inclusions with a considerably lower heat capacity than that of the basic materials are visible as areas whose temperature is higher than that of the areas without inclusions;

6. During testing in the transmission mode, the areas with defects are visible as areas whose temperature is higher than that of the areas without inclusions, regardless of the heat capacity of the material inclusions;
7. Approximately 90 min after the heating source is switched off, the second phase of cooling down takes place. In this phase, an inverse absolute contrast appears—the sign of the contrast changes and the inclusions initially visible as a warmer area are perceived as cooler areas, and vice versa (this phenomenon occurs on the same side of the partition);
8. Using the Echo Defect Shape method, one can successfully (with an error below 10%) determine the depth at which a defect is located in the building partition;
9. Knowing the specifications of the thermal imaging camera lens (the size [in mm] of 1 pixel in the thermogram), one can estimate the size of the defects present inside the partition by indicating the place in the thermogram where the defect's edges are likely situated.

The experiment shows that active thermography has great potential for materials testing in construction. It has been demonstrated that defect detection can be successfully conducted over large areas of massive members (large thermal capacity), using an appropriately powerful heat source and a long heat pulse. By recording the distribution of the temperature field during the cooling down of the tested partition over a long time, inverse contrasts (on the same front side of the partition) were observed. As part of further research, inverse heat conduction problems should be more broadly solved and the depth at which inclusions occur, as well as their thermal parameters, should be estimated on the basis of the obtained test results (thermograms). The further research will deal with applying inverse contrasts to improve the defect detection method (quantitative analysis). It is anticipated that the proposed investigative technique will be increasingly used for the non-destructive testing of materials, especially large-sized building units.

Author Contributions: Conceptualization, P.N. and H.N. Methodology, H.N. Validation, P.N. Formal Analysis, P.N. and H.N. Investigation, P.N. Resources, P.N. Data Curation, P.N. Writing-Original Draft Preparation, P.N. and H.N. Writing-Review & Editing, H.N. Visualization, P.N. and H.N. Supervision, H.N. Project Administration, H.N.

Acknowledgments: The authors thank Prof. Jacek Kasperski—the Head of the Department of Construction and Flow Machines at the Wrocław University of Science and Technology—for lending the thermographic camera used in the experiment.

References

1. Maierhofer, C.; Reinhardt, H.W.; Dobmann, G. *Non-Destructive Evaluation of Reinforced Concrete Structures—Non-Destructive Testing Methods*; Woodhead Publishing: Cambridge, UK, 2010; Volume 2, ISBN 978-18-4569-950-5.
2. Hoła, J.; Bień, J.; Sadowski, Ł.; Schabowicz, K. Non-destructive and semi-destructive diagnostics of concrete structures in assessment of their durability. *Bull. Pol. Acad. Sci.* **2015**, *63*, 87–96. [CrossRef]
3. Lüthi, T. *Non-Destructive Evaluation Methods. EMPA—Swiss Federal Laboratories for Materials Science and Technology*; Self-Published: Dübendorf, Switzerland, 2013.
4. Balageas, D.; Maldague, X.; Burleigh, D.; Vavilow, V.P.; Oswald-Tranta, B.; Roche, J.M.; Pradere, C.; Carlomagno, G.M. Thermal (IR) and Other NDT Techniques for Improved Material Inspection. *J. Nondestruct. Eval.* **2016**, *35*, 18. [CrossRef]
5. Nowak, H. *Application of Infrared Thermography in Construction*; Wrocław University of Science and Technology Publishing House: Wrocław, Poland, 2012; ISBN 978-83-7493-676-7. (In Polish)
6. Barreira, E.; Freitas, V.P. Evaluation of building materials using infrared thermography. *Constr. Build. Mater.* **2007**, *21*, 218–224. [CrossRef]
7. Barreira, E.; Almeida, R.M.S.F.; Delgado, J.M.P.Q. Infrared thermography for assessing moisture related phenomena in building components. *Constr. Build. Mater.* **2016**, *110*, 251–269. [CrossRef]

8. Wróbel, A.; Kisilewicz, T. Detection of thermal bridges—Aims, possibilities and conditions. In Proceedings of the QIRT 2008 Proceedings—9-th Quantitative InfraRed Thermography—International Conference, Krakow, Poland, 2–5 July 2008; pp. 227–232, ISBN 978-83-908655-1-5.
9. Kisilewicz, T.; Wróbel, A. Quantitative infrared wall inspection. In Proceedings of the 10-th Edition of the Quantitative InfraRed Thermography—International Conference, Québec-City, QC, Canada, 27–30 June 2010; pp. 589–594, ISBN 978-2-9809199-1-6.
10. Khodayar, F.; Sojasi, S.; Maldague, X. Infrared thermography and NDT: 2050 horizon. *Quant. Infrared Thermogr. J.* **2016**, *13*, 210–231. [CrossRef]
11. Vavilov, V. Thermal NDT: Historical milestones, state-of-the-art and trends. *Quant. Infrared Thermogr. J.* **2014**, *11*, 66–83. [CrossRef]
12. Wu, D.; Karpen, W.; Busse, G. Lockin thermography for multiplex photothermal nondestructive evaluation. In Proceedings of the QIRT Conference, Paris, France, 7–9 July 1992.
13. Ibarra-Castanedo, C.; Piau, J.M.; Guilbert, S.; Avdelidis, N.P.; Genest, M.; Bendada, A.; Maldague, X. Comparative Study of Active Thermography Techniques for the Nondestructive Evaluation of Honeycomb Structures. *Res. Nondest. Eval.* **2009**, *20*, 1–31. [CrossRef]
14. Maierhofer, C.; Myrach, P.; Steinfurth, H.; Reischel, M.; Röllig, M. Development of standards for flash thermography and lock-in thermography. In Proceedings of the 14th International Conference on Quantitative InfraRed Thermography (QIRT14), Bordeaux, France, 7–11 July 2014.
15. Dudzik, S. *Determining the Depth of Material Defects by Means of Dynamic Active Thermography and Artificial Neural Networks*; Częstochowa University of Technology Publishing House: Częstochowa, Poland, 2013; ISBN 978-83-7193-572-5. (In Polish)
16. Palumbo, D.; Galietti, U. Damage Investigation in Composite Materials by Means of New Thermal Data Processing Procedures. *Strain* **2016**, *52*, 276–285. [CrossRef]
17. Palumbo, D.; Cavallo, P.; Galietti, U. An investigation of the stepped thermography technique for defects evaluation in GFRP materials. *NDT E Int.* **2019**, *102*, 254–263. [CrossRef]
18. Maldague, X. *Theory and Practice of Infrared Technology of Non-Destructive Testing*; John Wiley & Sons: New York, NY, USA, 2001; ISBN 978-0-471-18190-3.
19. *EN 17119:2018 Non-Destructive Testing—Thermographic Testing—Active Thermography*; European Standard: Brussels, Belgium, 2018.
20. Milovanović, B.; Banjad Pečur, I. Review of Active IR Thermography for Detection and Characterization of Defects in Reinforced Concrete. *J. Imaging* **2016**, *2*, 11. [CrossRef]
21. Arndt, R.W. Square pulse thermography in frequency domain as adaptation of pulsed phase thermography for qualitative and quantitative applications in cultural heritage and civil engineering. *Infrared Phys. Technol.* **2010**, *53*, 246–253. [CrossRef]
22. Ibarra-Castanedo, C.; Sfarra, S.; Klein, M.; Maldague, X. Solar loading thermography: Time-lapsed thermographic survey and advanced thermographic signal processing for the inspection of civil engineering and cultural heritage structures. *Infrared Phys. Technol.* **2017**, *82*, 56–74. [CrossRef]
23. Noszczyk, P.; Nowak, H. Active thermography as a state-of-the-art method of testing reinforced concrete units (in Polish). *J. Civ. Eng. Environ. Arch.* **2016**, *33*, 279–286. [CrossRef]
24. Maierhofer, C.; Brink, A.; Röllig, M.; Wiggenhauser, H. Quantitative impulse-thermography as non-destructive testing method in civil engineering—Experimental results and numerical simulations. *Constr. Build. Mater.* **2005**, *19*, 731–737. [CrossRef]
25. Maierhofer, C.; Arndt, R.; Röllig, M.; Rieck, C.; Walther, A.; Scheel, H.; Hillemeier, B. Application of impulse-thermography for non-destructive assessment of concrete structures. *Cem. Conrete Compos.* **2006**, *28*, 393–401. [CrossRef]
26. Weiser, M.; Röllig, M.; Arndt, R.; Erdmann, B. Development and test of a numerical model for pulse thermography in civil engineering. *Heat Mass Transf.* **2010**, *46*, 1419–1428. [CrossRef]
27. Maierhofer, C.; Brink, A.; Röllig, M.; Wiggenhauser, H. Transient thermography for structural investigation of concrete and composites in the near surface region. *Infrared Phys. Technol.* **2002**, *43*, 271–278. [CrossRef]
28. Maierhofer, C.; Wiggenhauser, H.; Brink, A.; Röllig, M. Quantitative numerical analysis of transient IR-experiments on buildings. *Infrared Phys. Technol.* **2004**, *46*, 173–180. [CrossRef]
29. Maierhofer, C.; Brink, A.; Röllig, M.; Wiggenhauser, H. Detection of shallow voids in concrete structures with impulse thermography and radar. *NDT E Int.* **2003**, *36*, 257–263. [CrossRef]

30. Khan, F.; Bolhassani, M.; Kontsos, A.; Hamid, A.; Bartoli, I. Modeling and experimental implementation of infrared thermography on concrete masonry structures. *Infrared Phys. Technol.* **2015**, *69*, 228–237. [CrossRef]
31. Donatelli, A.; Aversa, P.; Luprano, V.A.M. Set-up of an experimental procedure for the measurement of thermal transmittances via infrared thermography on lab-made prototype walls. *Infrared Phys. Technol.* **2016**, *79*, 135–143. [CrossRef]
32. Rumbayan, R.; Washer, G.A. Modeling of Environmental Effects on Thermal Detection of Subsurface Damage in Concrete. *Res. Nondest. Eval.* **2014**, *25*, 235–252. [CrossRef]
33. Sfarra, S.; Marcucci, E.; Ambrosini, D.; Paoletti, D. Infrared exploration of the architectural heritage: From passive infrared thermography to hybrid infrared thermography (HIRT) approach. *Mater. Constr.* **2016**, *66*, 94. [CrossRef]
34. Aggelis, D.G.; Kordatos, E.Z.; Strantza, M.; Soulioti, D.V.; Matikas, T.E. NDT approach for characterization of subsurface cracks in concrete. *Constr. Build. Mater.* **2011**, *25*, 3089–3097. [CrossRef]
35. Lai, W.W.-L.; Lee, K.-K.; Poon, C.-S. Validation of size estimation of debonds in external wall's composite finishes via passive Infrared thermography and a gradient algorithm. *Constr. Build. Mater.* **2015**, *7*, 113–124. [CrossRef]
36. Kurita, K.; Oyado, M.; Tanaka, H.; Tottori, S. Active infrared thermographic inspection technique for elevated concrete structures using remote heating system. *Infrared Phys. Technol.* **2009**, *52*, 208–213. [CrossRef]
37. Halabe, U.B.; Vasudevan, A.; Klinkhachorn, P.; GangaRao, H.V.S. Detection of subsurface defects in fiber reinforced polymer composite bridge decks using digital infrared thermography. *Nondestruct. Test. Eval.* **2007**, *22*, 155–175. [CrossRef]
38. Keo, S.A.; Brachelet, F.; Breaban, F.; Defer, D. Steel detection in reinforced concrete wall by microwave infrared thermography. *NDT E Int.* **2014**, *62*, 172–177. [CrossRef]
39. Cotič, P.; Kolarič, D.; Bosiljkov, V.B.; Bosiljkov, V.; Jagličić, Z. Determination of the applicability and limits of void and delamination detection in concrete structures using infrared thermography. *NDT E Int.* **2015**, *74*, 87–93. [CrossRef]
40. Scott, M.; Kruger, D. Infrared Thermography as a Diagnostic Tool for Subsurface Assessments of Concrete Structures. In Proceedings of the ITC User Conference, Stockholm, Sweden, 29–30 October 2014.
41. Scott, M.; Kruger, D. Effects of Solar Loading on the Limits of Predictability of Internal Delamination Defects in Concrete Using Infrared Thermography. In Proceedings of the Conference InfraMation, Orlando, FL, USA, 4–7 November 2013.
42. Brachelet, F.; Keo, S.; Defer, D.; Breaban, F. Detection of reinforcement bars in concrete slabs by infrared thermography and microwaves excitation. In Proceedings of the Conference QIRT, Bordeaux, France, 7–11 July 2014.
43. Brink, A.; Maierhofer, C.; Röllig, M.; Wiggenhauser, H. Application of quantitative impulse thermography for structural evaluation in civil engineering—Comparison of experimental results and numerical simulations. In Proceedings of the 6th Quantitative InfraRed Thermography Conference, Dubrovnik, Croatia, 24–27 September 2002.
44. Arndt, R.; Maierhofer, C.; Röllig, M.; Weritz, F.; Wiggenhauser, H. Structural Investigation of Concrete and Masonry Structures behind Plaster by means of Pulse Phase Thermography. In Proceedings of the 7th International Conference on Quantitative Infrared Thermography (QIRT), Brussels, Belgium, 5–8 July 2004.
45. Vavilov, V.P.; Pan, Y.; Moskovchenko, A.I.; Čapka, A. Modeling, detecting and evaluating water ingress in aviation honeycomb panels. *Quant. Infrared Thermogr. J.* **2017**, *14*, 206–217. [CrossRef]
46. Lai, W.L.; Poon, C.-S. Boundary and size estimation of debonds in external wall finishes of high-rise buildings using Infrared thermography. In Proceedings of the QIRT2012—11th international conference on quantitative infrared thermography, Naples, Italy, 11–14 June 2012.
47. Vavilov, V.P.; Burleigh, D.D. Review of pulsed thermal NDT: Physical principles, theory and data processing. *NDT E Int.* **2015**, *73*, 28–52. [CrossRef]
48. Szymanik, B.; Frankowski, P.K.; Chady, T.; John Chelliah, C.R.A. Detection and Inspection of Steel Bars in Reinforced Concrete Structures Using Active Infrared Thermography with Microwave Excitation and Eddy Current Sensors. *Sensors* **2016**, *16*, 234. [CrossRef]
49. Lugin, S.; Netzelmann, U. A defect shape reconstruction algorithm for pulsed thermography. *NDT E Int.* **2007**, *40*, 220–228. [CrossRef]

50. Richter, R.; Maierchofer, C.; Kreutzbruck, M. Numerical method of active thermography for the reconstruction of back wall geometry. *NDT E Int.* **2013**, *54*, 189–197. [CrossRef]
51. Vavilov, V. Infrared Techniques for Materials Analysis and Nondestructive Testing. In *Infrared Methodology and Technology*; Maldague, X.P.V., Ed.; Gordon and Breach Publishers: London, UK, 1994; pp. 230–309.
52. Balageas, D.L.; Roche, J.M.; Leroy, H. Comparative Assessment of Thermal NDT Data Processing Techniques for Carbon Fiber Reinforced Polymers. *Mater. Eval.* **2017**, *75*, 1019–1031.
53. Giorleo, G.; Meola, C. Comparison between pulsed and modulated thermography in glass-epoxy laminates. *NDT E Int.* **2002**, *35*, 287–292. [CrossRef]

Permissions

All chapters in this book were first published in MDPI; hereby published with permission under the Creative Commons Attribution License or equivalent. Every chapter published in this book has been scrutinized by our experts. Their significance has been extensively debated. The topics covered herein carry significant findings which will fuel the growth of the discipline. They may even be implemented as practical applications or may be referred to as a beginning point for another development.

The contributors of this book come from diverse backgrounds, making this book a truly international effort. This book will bring forth new frontiers with its revolutionizing research information and detailed analysis of the nascent developments around the world.

We would like to thank all the contributing authors for lending their expertise to make the book truly unique. They have played a crucial role in the development of this book. Without their invaluable contributions this book wouldn't have been possible. They have made vital efforts to compile up to date information on the varied aspects of this subject to make this book a valuable addition to the collection of many professionals and students.

This book was conceptualized with the vision of imparting up-to-date information and advanced data in this field. To ensure the same, a matchless editorial board was set up. Every individual on the board went through rigorous rounds of assessment to prove their worth. After which they invested a large part of their time researching and compiling the most relevant data for our readers.

The editorial board has been involved in producing this book since its inception. They have spent rigorous hours researching and exploring the diverse topics which have resulted in the successful publishing of this book. They have passed on their knowledge of decades through this book. To expedite this challenging task, the publisher supported the team at every step. A small team of assistant editors was also appointed to further simplify the editing procedure and attain best results for the readers.

Apart from the editorial board, the designing team has also invested a significant amount of their time in understanding the subject and creating the most relevant covers. They scrutinized every image to scout for the most suitable representation of the subject and create an appropriate cover for the book.

The publishing team has been an ardent support to the editorial, designing and production team. Their endless efforts to recruit the best for this project, has resulted in the accomplishment of this book. They are a veteran in the field of academics and their pool of knowledge is as vast as their experience in printing. Their expertise and guidance has proved useful at every step. Their uncompromising quality standards have made this book an exceptional effort. Their encouragement from time to time has been an inspiration for everyone.

The publisher and the editorial board hope that this book will prove to be a valuable piece of knowledge for researchers, students, practitioners and scholars across the globe.

List of Contributors

Daniel Ferrández
Departamento de Ingeniería de Organización, Administración de Empresas y Estadística, Universidad Politécnica de Madrid, 28660 Madrid, Spain

Carlos Morón and Engerst Yedra
Departamento de Tecnología de la Edificación, Universidad Politécnica de Madrid, 28040 Madrid, Spain

Pablo Saiz
Departamento de Economía Financiera, Contabilidad e Idioma Moderno, Universidad Rey Juan Carlos, 28032 Madrid, Spain

Evangelina Atanes-Sánchez
Departamento de Ingeniería Mecánica, Química y Diseño Industrial, Universidad Politécnica de Madrid, 28012 Madrid, Spain

Yaser Abdellatef
Civil Engineering Department, University of Ottawa, 161 Louis Pasteur, Ottawa, ON K1N 6N5, Canada
Mechanical Power Department, Faculty of Engineering, Cairo University, Giza 12613, Egypt

Mohammad Amil Khan, Asif Khan and Mehdi Iftekharul Alam
Civil Engineering Department, University of Manitoba, 15 Gillson St., Winnipeg, MB R3T 5V6, Canada

Miroslava Kavgic
Civil Engineering Department, University of Ottawa, 161 Louis Pasteur, Ottawa, ON K1N 6N5, Canada
Civil Engineering Department, University of Manitoba, 15 Gillson St., Winnipeg, MB R3T 5V6, Canada

Patryk Ziolkowski and Maciej Niedostatkiewicz
Faculty of Civil and Environmental Engineering, Gdansk University of Technology, Gabriela Narutowicza 11/12, 80-233 Gdansk, Poland

Janusz Juraszek
Faculty of Materials, Civil and Environmental Engineering, University of Bielsko-Biala; 43-309 Bielsko-Biala, Poland

Richard Slávik
Department of Wood Science and Technology, Faculty of Forestry and Wood Technology, 613 00 Brno, Czech Republic

Miroslav Čekon
Department of Physics, Faculty of Civil Engineering, Slovak University of Technology, 810 05 Bratislava, Slovakia
AdMaS Centre, Faculty of Civil Engineering, Brno University of Technology, 602 00 Brno, Czech Republic

Jan Štefaňák
AdMaS Centre, Faculty of Civil Engineering, Brno University of Technology, 602 00 Brno, Czech Republic

Krzysztof Schabowicz, Tomasz Gorzelańczyk and Mateusz Szymków
Faculty of Civil Engineering, Wrocław University of Science and Technology, Wybrzeże Wyspiańskiego 27, 50-370 Wrocław, Poland

Barbora Doušová, David Koloušek and Miloslav Lhotka
University of Chemistry and Technology Prague, Faculty of Chemical Technology, Technická 5, 166 28 Praha 6, Czech Republic

Martin Keppert
Faculty of Civil Engineering, Czech Technical University in Prague, Thákurova 7, 166 29 Praha 6, Czech Republic

Martina Urbanová, Libor Kobera and Jiří Brus
Institute of Macromolecular Chemistry AS CR, Heyrovského nám. 2, 162 06 Praha 6, Czech Republic

Marek Maj and Andrzej Ubysz
Faculty of Civil Engineering,Wroclaw University of Science and Technology, Wroclaw 50-370, Poland

Hala Hammadeh
Faculty of Engineering, Middle East University, Amman 11831, Jordan

Farzat Askifi
Department of Structure Engineering, Faculty of Civil Engineering, Damascus University, Damascus, Syria

Jarosław Michałek, Michał Pachnicz and Maciej Sobótka
Faculty of Civil Engineering, Wrocław University of Science and Technology, 50-370 Wrocław, Poland

Qiuwei Yang, Chaojun Wang, Na Li and Wei Wang
School of Civil Engineering, Shaoxing University, Shaoxing, Zhejiang 312000, China

Yong Liu
State Key Lab Water Resources & Hydropower Engineering School, Wuhan University, Wuhan 430072, China

Monika Zielińska
Department of Technical Fundamentals of Architectural Design, Faculty of Architecture, Gdansk University of Technology, Narutowicza 11/12, 80-233 Gdansk, Poland

Magdalena Rucka
Department of Mechanics of Materials and Structures, Faculty of Civil and Environmental Engineering, Gdansk University of Technology, Narutowicza 11/12, 80-233 Gdansk, Poland

Dalibor Kocáb, Petr Misák and Petr Cikrle
Faculty of Civil Engineering, Brno University of Technology, Veveří 331/95, 602 00 Brno, Czech Republic

Paweł Noszczyk and Henryk Nowak
Department of Building Physics and Computer Design Methods, Faculty of Civil Engineering, Wroclaw University of Science and Technology, 50-370 Wrocław, Poland

Index

www.ingramcontent.com/pod-product-compliance
Lightning Source LLC
LaVergne TN
LVHW061512220926
848485LV00017B/878